Studies in Systems, Decision and Control

Volume 245

The series "Studies in Systems, Decision and Control" (SSDC) covers both new developments and advances, as well as the state of the art, in the various areas of broadly perceived systems, decision making and control–quickly, up to date and with a high quality. The intent is to cover the theory, applications, and perspectives on the state of the art and future developments relevant to systems, decision making, control, complex processes and related areas, as embedded in the fields of engineering, computer science, physics, economics, social and life sciences, as well as the paradigms and methodologies behind them. The series contains monographs, textbooks, lecture notes and edited volumes in systems, decision making and control spanning the areas of Cyber-Physical Systems, Autonomous Systems, Sensor Networks, Control Systems, Energy Systems, Automotive Systems, Biological Systems, Vehicular Networking and Connected Vehicles, Aerospace Systems, Automation, Manufacturing, Smart Grids, Nonlinear Systems, Power Systems, Robotics, Social Systems, Economic Systems and other. Of particular value to both the contributors and the readership are the short publication timeframe and the world-wide distribution and exposure which enable both a wide and rapid dissemination of research output.

** Indexing: The books of this series are submitted to ISI, SCOPUS, DBLP, Ulrichs, MathSciNet, Current Mathematical Publications, Mathematical Reviews, Zentralblatt Math: MetaPress and Springerlink.

More information about this series at http://www.springer.com/series/13304

Paweł D. Domański

Control Performance Assessment: Theoretical Analyses and Industrial Practice

Paweł D. Domański
Institute of Control and Computation Engineering
Warsaw University of Technology
Warsaw, Poland

ISSN 2198-4182 ISSN 2198-4190 (electronic)
Studies in Systems, Decision and Control
ISBN 978-3-030-23595-6 ISBN 978-3-030-23593-2 (eBook)
https://doi.org/10.1007/978-3-030-23593-2

This Springer imprint is published by the registered company Springer Nature Switzerland AG
The registered company address is: Gewerbestrasse 11, 6330 Cham, Switzerland

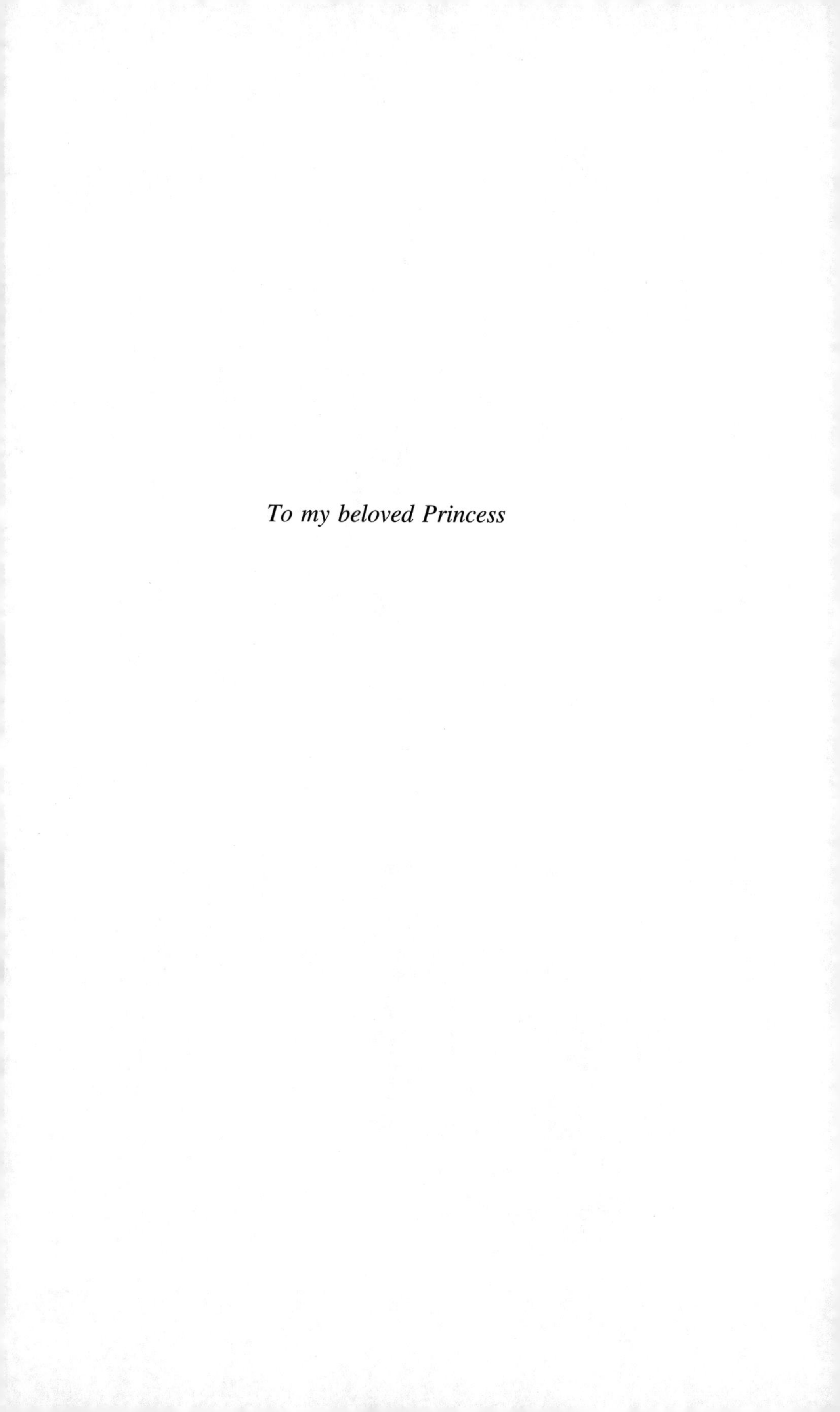

To my beloved Princess

Preface

And we're changing our ways,
Taking different roads.
—Ian Curtis

Control Performance Assessment aims to measure the quality of a control system. Once we have information, whether a system works properly or not, an engineer may undertake appropriate steps to improve the situation if needed. The assessment task requires methodologies and indexes, often called KPIs (Key Performance Indicators) that enable to measure how good the system is. They allow to benchmark different systems and prioritize the maintenance actions. Furthermore, some of the measures may show a reason for the inappropriate operation and give indications of what should be done.

The need for control performance measures exists as long as the control system's history. The task is commonly performed in industry, even without the specific awareness. That's the thing that just must be done. Control Performance Assessment originates from industry and it appeared in the scientific research in *60s* with benchmarking. The methods in the beginning were relatively simple and achievable. They mostly used the step response as such data was available. The notions of the overshoot and the settling time are still valid and are used as they represent a commonly understood reference platform. The availability of the sampled data that could be acquired from the discrete-time-computer-based control systems has allowed to execute more complicated calculations and to develop new methods. Engineers have gained the possibility to use integral measures, as for instance such as popular mean square error or absolute error. Also statistical factors, like mean, standard deviation, variance skewness, or kurtosis, were able to be calculated and used. Engineers have obtained new tools to measure control loop quality.

Although these methods are very informative and popular they share one common feature. In fact, they are not very constructive. An engineer getting mean square error or variance value has no clue how good or how bad the system is. Once the measure is calculated several times, the relative knowledge of what is better or worse reveals. We do not know how far away from the ideal performance we are. Introduction of the minimum variance benchmarking approach in 1989 constituted a major step in that direction. Opening of the path has initiated dozens of papers following the underlying benchmarking idea. The research has covered various feedback and feedforward control structures. Additionally, other benchmarks, like LQG, MPC, PID, have been proposed. Furthermore, what is the most valuable, this approach has entered industry. Specialized software packages have been developed and the scientific ideas have commercialized. New methods extended the scope of the available methods allowing further penetration of the market.

Expanding popularity and the CPA installation base started to reveal the limitations and new challenges. The need to go beyond well-known area of the linear and Gaussian world appeared together with exponentially increasing amount of the available datasets. The Big Data era has come. During last years several new, alternative approaches appeared, drawing inspiration from other science areas and contexts. Non-Gaussian and robust statistics, entropy, fractal, and persistence measures started to be investigated and utilized in the industrial applications.

This book is dedicated to different kinds of readers. The intention that text should be readable for the industrial reader limited the amount of the theoretical aspects, though the aim to show new scientific research areas has been intended as well. It is hoped that a good compromise, if possible, has been achieved.

This book has two main narration goals. The gathering in one place of the representatives of the available CPA approaches and their comparison using well-known benchmarks constitutes one dimension of the book. Synthesis of the observation and development of the comprehensive assessment procedure proposal forms the second objective of the work. Industrial infinite diversity of numerous *unknown pleasures* is well known. Thus, the task to develop single methodology covering all possibilities is simply mission impossible. The hope for a step in good direction remains.

Finally, I would like to thank all the people who contributed in some way to this work. There is always a challenge to enlist everybody, and even a bigger one to omit somebody. I will not take that risk. Except mentioning one short evening talk, June 2009, with Prof. Roman Galar, when I have heard that the *black swans* exist. I am also aware that this *blink* would not work without thousands of hours spent day and night at dozens of industrial sites for more than 20 years. Last but not least, it is just obvious that nothing would happen without my Family.

Thank you!

Warsaw, Poland
April 2019

Paweł D. Domański

Contents

Part II Validation with Simulation Examples

Acronyms

CPA	Control Performance Assessment
PO	Process Optimization
APC	Advanced Process Control
I&C	Instrumentation and Control
PLC	Programmable Logic Controller
KPI	Key Performance Indicator
DCS	Distributed Control System
SCADA	Supervisory Control And Data Acquisition
MES	Manufacturing Execution System
ERP	Enterprise Resource Planning
SAMA	Scientific Apparatus Makers Association
HMI	Human Machine Interface
FAT	Factory Acceptance Test
SAT	Site Acceptance Test
ISE	Integral of Square Error
MSE	Mean Square Error
IAE	Integral of Absolute Error
TSV	Total Squared Variation
ITAE	Integral Time Absolute Error
ISTC	Integral of Square Time derivative of the Control input
CPI	Control Performance Index
EWMA	Exponentially Weighted Moving Average
VarBand	Variance Band
AMP	Amplitude Index
RDR	Reference to Disturbance Ratio
ARMAV	Auto-Regressive Moving-Average Vector
ITAE	Integral Time Absolute Error
ARMA	Auto-Regressive Moving Average
AR	Auto-Regressive
PID	Proportional-Integral-Derivative

MPC	Model Predictive Control
DMC	Dynamic Matrix Control
GPC	Generalized Predictive Control
MAC	Model Algorithmic Control
IMC	Internal Model Control
MRAC	Model Reference Adaptive Control
DoC	Degree of Controllability
SVD	Singular Value Decomposition
HVAC	Heat, Ventilation and Air Conditioning
DFA	Detrended Fluctuation Analysis
$g(t)$	Impulse response
$h(t)$	Step response
$G(j\omega)$	Frequency response
$\delta(t)$	Dirac's impulse function
$\vartheta(t)$	Step function
ε_∞	Steady state error
DR	Decay Ratio
PV	Peak Value
t_p	Peak time
t_r	Rise time
t_s	Settling time
κ	Overshoot
RHP	Right Half s-Plane
r_{us}	Relative undershoot
AI	Area Index
OI	Output index
RI	R-index
$y_0(t)$	Setpoint
$y(t)$	Process output
$u(t)$	Controller output
$\Delta u(t)$	Controller output change
$\varepsilon(t)$	Control error
OPC	OLE for Process Control
PDF	Probabilistic Distribution Function
MLE	Maximum Likelihood Estimator
x_0	Mean value
x_m	Median value
σ	Standard deviation
σ^{rob}	M-estimator of standard deviation
$\psi_H(x,b)$	Huber ψ function
$\psi_{log}(x)$	Logistic smooth ψ function
$Var(x),\ \sigma^2$	Variance
α	Stability factor (characteristic exponent) of α-stable PDF
β	Skewness factor of α-stable PDF

δ	Position factor of α-stable PDF
γ^L	Scale (dispersion) factor of α-stable PDF
γ^C	Scale (dispersion) factor of Cauchy PDF
b	Scale (dispersion) factor of Laplace PDF
NLI	NonLinearity Index
NGI	Non-Gaussianity Index
RVI	Relative Variance Index
LQG	Linear Quadratic Gaussian
ARMA	Auto-Regressive Moving Average
ARMAX	Auto-Regressive Moving Average with auXiliary input
ARIMA	Auto-Regressive Integrated Moving Average
ARIMAX	Auto-Regressive Integrated Moving Average with auXiliary input
CARIMA	Controlled Auto-Regressive Integrated Moving Average
ARFIMA	Auto-Regressive Fractional Integrated Moving Average
$\upsilon(k)$	White noise signal
$g(z^{-1})$	Impulse response of the closed loop system
σ^2_{MV}	Minimum variance lower constraint of the assumed control quality
$\eta_0,\ \eta_1,\ \eta_2,\ \eta_3$	Minimum variance CPA indicators
$\rho_{\upsilon y}(k)$	Cross correlation coefficients
$\hat{\rho}_{\upsilon y}(k)$	Estimates for the cross correlation coefficients
J_o^{PID}	Optimal PID benchmark performance index
J_o^{IMC}	Optimal IMC benchmark performance index
J_o^{GMV}	Optimal GMV benchmark performance index
J_o^{LQG}	Optimal LQG benchmark performance index
J_o^{MPC}	Optimal MPC benchmark performance index
FCOR	Filtering and CORrelation analysis
PCA	Principal Component Analysis
DoC	Degree of Controllability
MSPCA	Multiscale Principal Component Analysis
$\rho_{xy}(k)$	Cross correlation
GMV	Generalized Minimum Variance
EHPI	Extended Horizon Performance Index
HMM	Hidden Markov Models
DFT	Discrete Fourier Transform
RDR	Reference to Disturbance Ratio
DFA	Detrended Fluctuation Analysis
FD	Fractal (Fractional) Dimension
H, H^0	Hurst (single) exponent
H^1, H^2, H^3	Multiple Hurst exponents in short, middle and long range scales
LRD	Long-Range Dependence
fGn	fractional Gaussian noise
MF-DFA	Multi-Fractal Detrended Fluctuation Analysis
MIE	Minimum Information Entropy

H^{DE}	Differential entropy
H^{RE}	Rational entropy
TCO	Total Cost of Operation
QCC	Quality Control Charts
UCL	Upper Control Limit
LCL	Lower Control Limit
CSTR	Continuously Stirred Tank Reactor
TITO	Two Input Two Output
IoT	Internet of Things
IoS	Internet of Services
IoD	Internet of (very big) Data
IIoT	Industrial Internet of Things
TQM	Total Quality Management
TPM	Total Productivity Maintenance
ROI	Return Of Investment
P&ID	Piping and Instrumentation Diagram

Part I
Methods and Measures

The subject of the Control Performance Assessment exists in practice from the beginning of control engineering. Users always want to know how well a controller works and even further, they want to know how much better it could be. Scientists have noticed that interest quite early. Automatically, once they started designing and evaluating new control algorithms and philosophies they required tools to calculate how much better their achievement is.

The idea behind this presentation is to put together all existing important ideas and achievements in the area of the Control Performance Assessment. This approach entails the structure of the book. The text consists of three main parts. First, the methods and approaches that might be found in both scientific research and industrial practice are introduced and presented. The second part presents the evaluation of presented methods in a simulated environment, based on the frequently used control benchmarks reflecting industrial examples well. Finally the validation of considered methods with industrial examples and using real process data is shown.

Following the above scheme this part presents the summary of the CPA methodologies. It starts with the introduction of the subject (Chap. 1). Next two chapters discuss the measures defined in the time domain. Chapter 2 introduces the methods using the most popular approach, i.e. step response. Chapter 3 presents the general approaches in the time domain with the main focus on integral indexes.

The data-based methods are expanded with Chap. 4 deliberating statistical approaches, both classical Gaussian and alternative non-Gaussian. Chapter 5 discusses the model-based methods, being mostly dedicated to the Minimum Variance approach. It is followed by the analysis in the frequency domain sketched in Chap. 6.

The presentation of the alternative methods, such as persistence, fractal and entropy is included in Chap. 7, while business-like methods, generally called the Key Performance Indicators - KPIs, are presented in Chap. 8. The presentation of methods and measures concludes with Chap. 9, dedicated to the specific activity closely associated with the CPA task, i.e. the estimation of benefits that may be achieved through improved control quality.

Chapter 1
Does Control Quality Matter?

L'essentiel est invisible pour les yeux.

– Antoine de Saint-Exupéry

Don't tell people how to do things, tell them what to do and let them surprise you with their results.

– George S. Patton Jr.

1.1 Process Improvement Versus Control System

Nowadays the industry is witnessing winds of change. The era of Industry 4.0 transformation approaches. The issues of product throughput maximization, increased environmental protection and efficient energy management that have been pushing systems towards their technological constraints are not enough. We want go even beyond that. A modern plant has to fulfill varying stringent regulations to operate at the edge of technological limitations. It is accompanied by production horizontal and vertical integration, simulations, autonomous operation and flexibility. Most of these paradigms require the backbone of control engineering technologies and cannot exist without a properly designed and maintained control system. Due to such industrial demands, high control system performance must be closely coupled with the task of Control Performance Assessment (CPA).

Control issues need to be addressed during the whole period of the process operational life-cycle, however they are most clearly seen during the plant construction and modernization phases. Construction of a new plant is a long-term, complex and very expensive process. The main focus is put on the capital costs mostly connected to the equipment hardware. Control system and infrastructure need to be acquired and

P. D. Domański, *Control Performance Assessment: Theoretical Analyses and Industrial Practice*, Studies in Systems, Decision and Control 245,
https://doi.org/10.1007/978-3-030-23593-2_1

installed. Unfortunately, the control system design, tuning and stabilization phase end the entire sequence, and thus, are often not subject to necessary amount of attention. As time runs short, it pushes decision makers to complete the process within the schedule. The schedules are cramped, required resources are being minimized and all hands on board are just used to start up the plant. Once the installation is running, further initiatives are limited due to the warranty period.

Often, only minor issues are solved during daily plant operation, although the acquired experience frequently reveals the problems originating from improper control system design or tuning. It is mostly due to the shortage of the on-site control engineering personnel. The understanding that control quality plays crucial role in the production performance of process and manufacturing industries provokes projects solely dedicated to the base control system rehabilitation or optimization. Initiatives for control system rehabilitation, Process Optimization (PO) or application of the Advanced Process Control (APC) are a common practice in industry in the last 30 years [8]. They provide an occasion to reconsider modifications inside of the control philosophy. The good practice of the control system improvement uses *bottom-up* direction starting from the control Instrumentation and Control (I&C) modernization followed by the base regulatory control update, APC and PO.

Surveys show that a large number of control systems (66–80%) cannot achieve the desired (achievable) performance [68]. Any kind of an improvement initiative for the already operating plant is needed, but it requires financial justification. Firstly, because it alters regular operation and stops production (lost production profit). Secondly, it is due to the management business initiative. Nowadays the decisions are in hands of financial managers, not engineers. It is not enough to say that the improvement makes the control better and safer giving more comfort to the operators. The project always has to bring significant financially tangible results. We need to have good measures (KPI—Key Performance Indicator) to be able to deliver such a project. KPIs originate from technical variables, like pressure, temperature, flow, concentration, mass, etc. These variables often reflect the technical effect of the control system improvement. Finally the process measures need to be transferred into economic numbers in monetary units.

One has to remember that real industrial processes are in general non-stationary, *evolving* organisms. Single shot project control upgrade, modification or tuning solves the problems only in a short-term perspective in most of the cases. Actually, well maintained control system requires permanent attention and close supervision to sustain high performance numbers. There are many industrial cases showing that base control regulation fine tuning brings major installation improvements and significant financial benefits [34, 73, 102], while APC/PO application brings further gains [41, 89].

In practice this subject should be covered by on-site expert personnel through perpetual assessment, validation, modifications and fine-tuning of the deficient control loops. Unfortunately, the number of loops and complexity of the systems increases, while simultaneously there is noticeable shortage of experienced personnel. Control engineers start to be a limiting factor in covering all of tuning and maintenance needs.

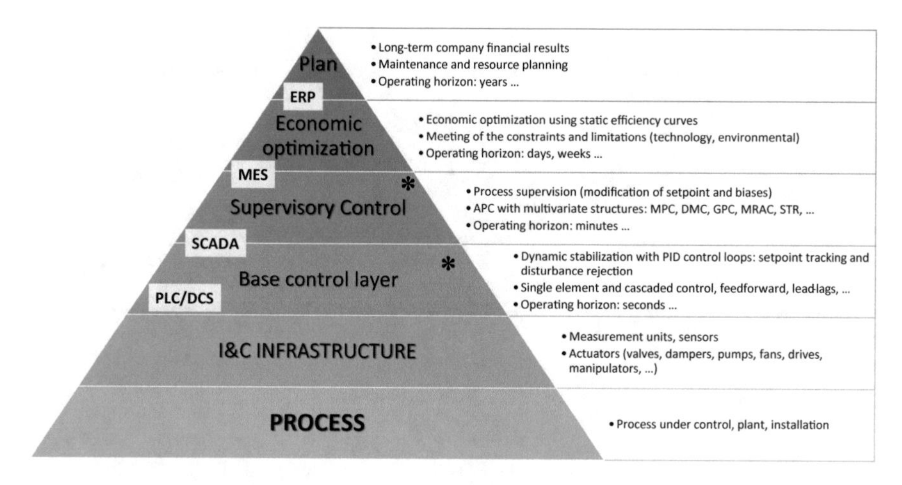

Fig. 1.1 Hierarchical functional levels of the control system (* denotes levels, where CPA is used)

This problem is especially observed in the advanced solutions using PO and APC [139]. Nonetheless, it is noticeable as well in the area of base control.

Above issues form very strong arguments supporting significance of and increasing need for control performance monitoring and diagnostic tools. Assessment solutions and procedures should constitute an inevitable element of the plant I&C infrastructure (see Fig. 1.1). The control quality validation should be considered very seriously during each phase of the process life-cycle: design and construction, normal operation, maintenance, upgrades and reconstruction. Simultaneously, management should embed CPA into the daily operating procedures, maintenance plans and plant software infrastructure within hierarchical structure of PLC/DCS/SCADA/MES/ERP systems [10] (see Fig. 1.2). Control system quality significantly affects process performance and should not be neglected nor forgotten.

The scope of the process improvement by means of the control should addresses all the functional levels of the system, i.e. hardware infrastructure and control logical philosophy. The hardware elements include the regulation system itself (PLC/DCS) together with plant instrumentation, i.e. actuators and measurement sensors. Control philosophy includes all the logical aspects of the control system including structures, algorithms and tuning setups often described using SAMA (Scientific Apparatus Makers Association) diagrams.

The CPA task may be positioned in two dimensions of the system: process control life-cycle and infrastructure hierarchy. It is often used in a relatively simple way, like for instance using only some elements according to the daily operation or incidental needs. However, there is one situation, where the CPA task should be performed in its full scope. It is control system feasibility study. Control system life-cycle process consists of several phases separated with decision milestones. Each milestone is associated with assessment validating current status, the performance baseline and improvement potential. We may distinguish five main stages.

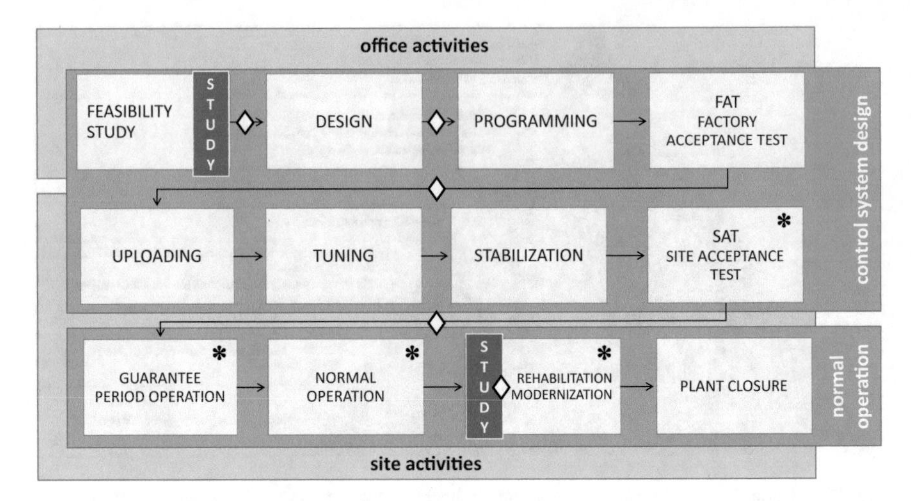

Fig. 1.2 Control system implementation process life-cycle (* denotes phases, when the CPA is used, ◇ the decision milestone)

(1) Initial feasibility study delivers a picture of the installation operating conditions from the perspective of instrumentation and control system. It consists of the evaluation of both control system dynamical performance indexes and baseline economic KPIs.
(2) Solution design allows to formulate initial assumptions into the functional solution description, followed by the technical design. SAMA drawings deliver the language and graphic presentation format. Additionally, the operating screens for the Human machine Interface (HMI) interface are designed.
(3) Programming enables to code the technical control solution design into the PLC/DCS language, like ladder languages or DCS embedded functionals, often similar to the AutoCAD drawings. This programming also includes operating screens and all required interfaces to the auxiliary subsystems.
(4) Factory Acceptance Test (FAT) is performed at the development level and enables simulation validation of the proposed solution. It is necessary milestone for further on-site implementation.
(5) Programmed solution is then downloaded into the site control system. It has to be noted that the downloading procedure is performed during the installation outage, as a safety precaution.
(6) Regulatory system site implementation of the proper control philosophy is the iterative process consisting of the tuning and stabilization period (at least few weeks) enabling results settlement.
(7) On-site implementation is followed by the Site Acceptance Test (SAT), which includes solution commissioning, comparing the obtained results with the planned numbers.

(8) System operation starts with the warranty period. During the warranty the control improvement vendor is responsible for the results, while afterwards it remains in the hands of the plant personnel.
(9) The operation often shows the shortcuts of the applied solution or there might appear new possibilities unknown during system implementation. Control system rehabilitation or modernization initiative can be run in such situations.
(10) Once the installation has exhausted all the possibilities of the profitable operation, the site closes.

Control system feasibility study may be initiated due to many reasons. The implementation of the assessment software is one of such situations. Some of the large industrial sites, that have to manage thousands of the control loops with the very limited human resources decide to invest and implement dedicated software packages that automatically monitor and assess control loops. There is a significant number of commercially available software packages, algorithms, case studies and benchmarks within all of the above areas [47, 69, 160]. They are developed and offered by large control vendors or by domain-based companies. The accessibility of solutions and number of successful implementations is relatively low, despite global agreement about importance of the control system quality and sustainability. The market is penetrated only to a minimal degree. There are still areas that require closer attention, perspicacious analysis and development of appropriate methods addressing uncovered situations.

Dedicated control performance study is a second such a situation, where the control system assessment is mostly done in full scope. Study initiative is mostly performed in the situation, where there is a will or a need to improve the process [34, 56, 61]. Such a performance or feasibility study delivers a momentary picture of the installation operating conditions from the perspective of instrumentation and control system. It often consists of the evaluation for both control system dynamical performance indexes and baseline economic KPIs. It has to be reflected in the monetary units, as it proves the rationality of the improvement.

This short introduction tries to show the rationale for the Control Performance Assessment and its positioning within the control system infrastructure and life-cycle processes. The need seems to be unquestionable, however the introduction would not be complete without the profound presentation of the historical background and the detailed industrial coverage.

1.2 The Assessment and Benchmarking History

We observe ongoing research on evaluation of various approaches and measures supporting the task of process performance assessment caused the regulation. The reason is simple. Control systems often perform inefficiently due to several internal and external reasons [37, 72]. We may enumerate the most important causes, like insufficient daily supervision and maintenance, process fluctuations, instrumentation

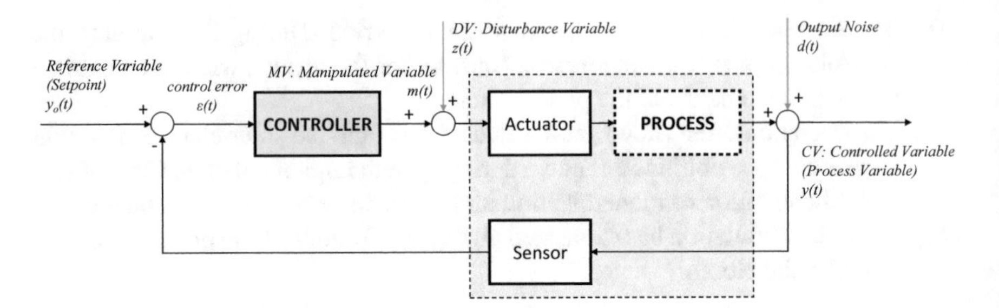

Fig. 1.3 SISO control loop diagram

malfunction, inappropriate control logic, poor tuning, variable operating conditions, shortage of experienced personnel, disturbances, noises, human interactions etc. [10, 149].

Human maintenance is never sufficient and does not cover all needs. Researchers and engineers continuously work on development of automatic and autonomous supporting tools that would evaluate clear measures reflecting situation and translating technical numbers into economic ones. This process leads towards the design of commercially available software packages.

The CPA research started from the analysis of the simple and standard formulation of the SISO control loop (see Fig. 1.3). First report in this area was presented in 1967 by Åström [4] for pulp and paper plant. Although the paper focuses on the issues of the controller design and tuning, the obtained algorithms are compared with single performance measure: process variable standard deviation in the considered case. Benchmarking as such, was introduced for business processes by Rank Xerox [23] in late 70s.

The use of control error integral-based measures [96], as for instance Mean Square Error (MSE), Integral of the Absolute Error (IAE) and the others forms the second group of the early introduced indexes.

Research continued in the 70's [28], interestingly also for the paper machine. The authors proposed to measure control effectiveness by comparing the observed output variation with an estimate of the theoretical minimum variation obtained from ARMAV time-series models for a paper machine process. In the 80's Hang et al. has been analyzing the performance of the delayed Smith predictor control using PI benchmarking and different performance measures, like step response, Integral of Time Absolute Error (ITAE) and MSE. Seborg et al. [129] has summarized the research and proposed a systematic method for assessing closed-loop performance by using time domain measures, including overshoot, Decay Ratio (DR) and setting time.

Pryor has proposed [120] the use of the power spectrum in the analysis of process data. Tolfo [159] has been assessing advanced control strategies with the use of the standard deviation of the variable distribution. Moreover the author introduced a method to estimate possible benefits due to better control. The method is named as the hard credit evaluation by reduced σ method. Ohtsu [113] in 1984 has applied statistical methods to evaluate performance of the ship autopilot.

It is interesting to notice that the majority of these works had been conducted for the pulp and paper, chemical or petrochemical industry. It is due to the fact that these industries have been significantly penetrated with the APC and the emerging awareness, that control quality matters.

The research has accelerated in 1989. It has been focused up to this moment on the simple measures, like for instance integral indexes or normal distribution variance properties. Such approaches are simple and informative, however they share one common drawback. They are simply not constructive. They do not inform about the distance from the best achievable or desired performance. Thus we only have relative knowledge, whether the control has improved or not.

Harris [57] proposed a method based on the minimum variance notion, which has allowed to measure the distance from the best available control. Simultaneously it has significantly increased the interest and applicability of the CPA. The theory of the minimum variance control has been enabled with works of Kalman [81], Åström [5] and Box and Jenkins [17]. The method has used normal operation data to calculate minimum variance benchmark. The method has applied simple process model with the delay (known or estimated) [133]. It requires to calculate the coefficients of the impulse response from noise-to-output transfer function with regressive models, as for instance of ARMA-type.

The research followed the minimum variance path covering various aspects of the control, as for instance: univariate feedback [26, 38] and feedforward control [27], unstable and nonminimumphase systems [162], multivariate MIMO and MISO cases [39, 59, 65, 132], varying setpoint [116], cascaded control [87], aspects of setpoint tracking versus disturbance rejection [157], linear time-variant systems [94]. The research often followed this path with an introduction of several improvements to the method as for instance Variability Matrix [40], an iterative solution for PID benchmark [88], LQG benchmarks for stationary stochastic disturbances [13, 14], optimization based approaches solving a series of convex programs using sums of squares programming [131], LQG benchmarks for various of continuous space, PID and state space controllers [49, 50, 79], the method based on multi-model mixing time-variant minimum variance [95] and many others. The index has found and interesting application in the design of the PID control [85] as well.

Markov parameters of the closed loop transfer function has been used to evaluate Harris index and to assess the closed loop performance of the regulatory and tracking switching control systems in [108]. Many further modifications had been added to the index as for instance generalized minimum variance control law [48] or hybrid approach using different estimators [98].

As the model information is crucial to the minimum variance performance assessment the research was conducted in supporting aspects of model-structure detection [7], disturbance/load-change detection [53] and moving-window-based performance assessment [114].

The minimum variance approach has rapidly gained large and common acceptance, despite the fact that the knowledge about process delay can be considered as method deficiency. It has been approved in industry as well. Several different control software packages have been released [69, 72] supporting plant owners with automatically generated measures, identified control bottlenecks and suggesting solutions.

Finally, the aspect of the CPA task considered to be a part of the process has to be addressed. The most of the research has addressed the above considerations in the batch-like application, when the measures are evaluated every some selected time period. On-line evaluation and interpretation of the control quality analysis has been addressed in fewer works, as for instance [30, 111]. An interesting solution with on-line operation considers temporally correlated slow features analysis through the statistical properties and according monitoring indexes based on contribution plots had been proposed in [45]. Next, the combination of co-integration analysis and slow feature analysis, had been adopted for concurrent monitoring of operation condition and process dynamics for non-stationary dynamic processes subject to time variant conditions [198] had been analyzed. It should be noted that the CPA task is quite often considered jointly with the task of the controller tuning or retuning [134].

CPA task only exists inside of the comprehensive control infrastructure. The control quality is integrated with sensors and actuators. The aspect of the impact of the actuators has been extensively covered within the research with dozen of publications dedicated to the nonlinearities resulting from the actuators characteristics. On the contrary, the impact of the measurements on the control performance has been seldom addressed in the research. The authors of the [151] have prepared the review of various measurement strategies and their impact on control quality. The effect of the sampling jitter associated with the measurements [189] has been covered in the work as well. The further research has followed four different directions:

- evaluation of the other methods, apart from minimum variance or classical methods causing the research for distinct and unorthodox areas and contexts,
- exploitation of the developed approaches towards various control algorithms and loop structure aspects [179, 197],
- exploration of the CPA potential in industry [91],
- extension of the loop quality assessment towards large scale perspective of the root-cause diagnosis or data-based causality analysis.

These areas are brought closer in the following paragraphs.

1.2.1 Further CPA Research

There have appeared measures and approaches using methods from the alternative research contexts such as non-Gaussian statistics, approaches using functions in different domains, fractal and persistence measures, entropy and many others, apart from the classical approaches, i.e. the methods using step response measures, base Gaussian statistics, process variable integrals or minimum variance approach.

The first works have addressed approaches relatively close to the classical ones. Tyler and Morari [163] have proposed the likelihood method applying constraints on the closed-loop transfer function impulse response coefficients. Huang and Shah [64] have proposed performance assessment with user defined robust benchmark under a unified H_2 framework. A multi-resolution spectral independent component analysis has been proposed to detect and isolate the sources of multiple oscillations in [177]. Nonlinear principal component analysis enables to recognize nonlinear correlations in process data [193] and as such has been applied to detect valve stiction as the reason of the poorly operating control loop. Control performance assessment for a class of nonlinear systems modeled by autoregressive second-order Volterra series with a general linear additive disturbance [100] has been also investigated using the nonlinear generalized minimum variance controller.

Exploitation of the frequency-based approaches has started with the work of Kendra [82]. The closed-loop performance variation measure has been defined as its maximum singular value frequency by frequency. It has been shown that this measure is essential and informative in characterizing closed-loop performance variation [171]. Two-dimensional Fourier transformation method has been compared with ARMA-model based approaches in [101]. Zhang et al. [196] have applied discrete Fourier transform to detect multiple oscillations using properties of the Raleigh distribution. Close loop frequency approach has been extended towards the nonlinear model based methodology using the Hammerstein models [145] to assess loop nonlinearities and valve stiction.

There can be found in the literature an extension of the standard measures with the more advanced approaches. The introduction of the Internal Model Control (IMC) control as the benchmark has allowed the formulation of the IMC-IAE-based index proposed to assess the setpoint tracking performance of PID control loops [191]. PID control with a focus on unstable open loop systems has been addressed in [11]. The shape related performance measures based on the concept of monotonicity has been discussed in [67].

The tensor space modeling representation for switched systems has been developed [74] to model high coupling interaction of control and protection systems. The data driven tensor space algorithm based on higher order SVD has also been developed to assess the performance of switched control systems. By using orthogonal projection in tensor space extended from the matrix space, prediction error approach has been employed to obtain the optimal prediction error variance formulating the control performance benchmark for quality assessment.

Veronesi and Visioli [168] have applied final value theorem to asses and fine tune the PID controllers, however the proposed measure uses integral absolute value index (IAE). Recalde and Yue in [122] have applied state-dependent models derived with Kalman filter to evaluate closed-loop variance/covariance measures for nonlinear systems.

The dynamics of a well-controlled control loop should correspond to its maximum entropy. The entropy of control related variables decreases as loop performance quality degrades. Any loss of dynamic complexity in the control system gives rise to an increase of the predictability of the control error time series [46]. Minimum entropy approach has been proposed in [104] and further developed for SISO structures [183, 194], cascaded loops [195].

Fractional models have been used for the PID quality analysis by Škarda et al. [138]. Further research in this direction has led toward the persistence and fractional measures. Srinivasan et al. [144] has proposed in 2012 the application of the Hurst exponent evaluated using Detrended Fluctuation Analysis (DFA). The approach has been validated for SISO control loop with PI and PID controller. The same area has been exploited by Spinner [141, 142] followed by the analysis of multivariate systems [25]. The authors have used DFA analysis to the process output. The reason why detrending has been applied was caused by large variability of the process output as it often has to track variations in setpoint. Different approach has been proposed by Pillay in 2014 [117]. He has proposed to apply DFA analysis and Hurst exponent evaluation to the MSE (Mean Square Error) of the control error signal. His examples have also been based on the SISO loops controlled by PID regulator.

Soft computing approaches to the control performance assessment task are relatively rare. We may find only limited number of works investigating artificial neural networks [97, 207], fuzzy-neural [18] or fuzzy clustering [24] approaches. They do not require any model and can be used on-line using normal operation data.

Hybrid multi-methods approaches can improve the results over independent single method as in many other situations, like time series prediction. Similar approaches can also be found for the control performance assessment task, as for instance the multi-index method using combination of historical prediction error covariance indexes [135].

An interesting aspect of the Control performance tasks aspects in the situation of the large amount of data (Big Data) has been lately addressed in [44, 176]. The aspect of the autonomous systems operation has also been reflected in the research. There has emerged an interesting and challenging research area associated with the controllability of such systems and switching between autonomous and manually driven operating modes [109].

1.2.2 CPA Analysis for Various Control Strategies

The continuation of the CPA research towards the exploitation of the various control strategies, algorithms and structures is considered as the natural area of the further research. The researches analyze the most popular configurations of the PID loops as in the initial works cited earlier, like single-element, cascaded [126] and three-point control with the feedforward element [63, 148] (see Fig. 1.4). Authors have for instance been using cross-correlation analysis for this task. Various aspects of the disturbance decoupling has been taken into consideration as well.

Output stochastic distribution systems that have a goal to control output probabilistic distribution function rather than the output are discussed in [205]. In [174] there has been proposed a method for batch processes controlled by iterative learning control based on two-dimensional linear quadratic Gaussian benchmark.

PID algorithm constitutes the majority of the industrial process control structures, nonetheless other algorithms and strategies started to penetrate industry as

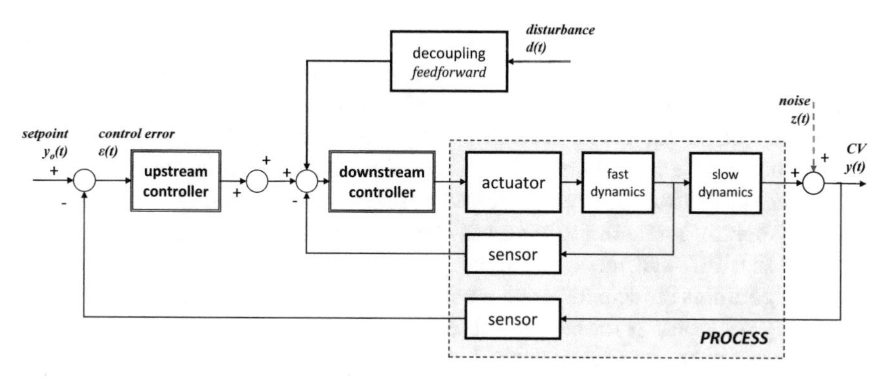

Fig. 1.4 Three element control cascaded loop with disturbance decoupling feedforward

well. Multivariate scenarios have been considered as well [66], naturally driving the interest towards Advanced Process Control. APC includes the variety of different methods and algorithms, as predictive control, adaptive structures or any kind of the soft computing approaches, like fuzzy logic, artificial neural networks, evolutionary computation or any non-standard heuristics. Once we consider process industry, predictive control mostly hidden under the name of Model Predictive Control (MPC) is considered as the most important one.

Many different structures of the predictive control have attracted research interest in both analytical or nonlinear optimization configurations and has been conducted for several years. First works have applied knowledge-based system for the DMC predictive controller [127]. Further works followed the path similar to the PID control with benchmarking approaches [78, 140]. Zhao et al. [199, 200] has proposed the LQG benchmarking to estimate achievable variability reduction through control system improvement and Chen has proposed statistical approach to detect model mismatch [20].

Predictive DMC structures have been compared and assessed in different application configurations as a single controller or the supervisory level over PID regulatory control [84, 119]. Further works continued in various directions. Model based approaches [6, 19, 152] are accompanied with minimum variance methods [175, 179, 180] that also require some process knowledge. Statistical approach through correlation analysis of optimal and working controller were proposed in [6], while prediction error benchmarking was used in [184]. Industrial validation of the multivariate MPC performance assessment at para-xylene production and poly-propylene splitter column processes in [43], kerosene and naphtha hydrotreating units is presented in [76] and delayed coking plant [15].

Different aspect of economic and non-dynamic controller performance has been addressed in [1, 90]. Comprehensive review of various approaches is presented in [35]. Finally non-Gaussian statistical [32] and fractal [31] methodologies have been investigated for the GPC predictive control algorithm. Application in the nonlinear

DMC predictive control has been compared with other measures in [33]. Comprehensive review of the assessment and diagnostic methods may be found in [77].

Some researchers also focused their attention to the adaptive control investigating quality assessment for Model Reference Adaptive Control (MRAC) [42]. Aerospace application to the adaptive control with neural network model has been addressed in [52]. Fuzzy control has been addressed in several papers from the early stages of the fuzzy control research [16] and has been continued in different configurations of fuzzy PD, PID and other structures [12, 80, 103, 165]. Apart from the strict control algorithms the scientists and practitioners explored other auxiliary areas and supporting solutions, as for instance observers and Kalman filers [161].

Simultaneously the infrastructure issues impacting control properties, like for instance actuator nonlinearities impeding loop [146, 185, 188] and multi-loop oscillations [186, 196] had been considered in different configurations. Valve forms the majority of actuators used in process industry. The specific but a very important effect of valve stiction has been reflected in many works [21, 60, 70, 71, 164, 187] any other nonlinearities of various classes also brought the attention of scientists.

There is also a large group of the research activities dedicated to the detection of the loop oscillations. They are mostly data-driven using time or frequency domain. The list of these approaches is relatively long, however the main important ones use evaluation of zero crossings of the IAE index [155], filtered autocorrelation function of the error signal [158], correlation analysis of the power spectrum [153], spectral principal component analysis [156] and spectral independent component analysis [177], spectral envelope method [75], discrete cosine transformation [92, 172], modified empirical mode decomposition [143, 147], GA-based factorization [36], intrinsic time-scale decomposition [51, 178] or hidden Markov model [181].

The impact of the external noise has been analyzed as well [106] showing that the rise time and settling time of the setpoint response, as well as the peak deviation and the shape of the response following the load step disturbance can be accurately predicted. The remote implementation of the CPA has been developed and presented in [182]. An interesting aspect of the CPA has been investigated by the Vatanski et al. [166], when the connection between technical measures of economic KPIs has been analyzed.

1.2.3 Summary of the Research Status

Nowadays, the CPA research covers almost all aspects of control met in the control engineering research and industrial practice. Different categories and domain groups of methods have been evaluated [112, 115, 136] during such a long history and wide research scope. Concluding, one can enumerate the following different CPA areas of interest:

1. indexes based on setpoint step response: overshoot, undershoot, rise, peak and settling time [55], decay ratio, offset (steady state error), peak value [142],

2. indexes based on the disturbance step response, such as Idle Index [55], Area Index, Output Index [170], R-index [123],
3. integral time measures: Mean Square Error (MSE) [206], Integral Absolute Error (IAE) [137, 167], Integral Time Absolute Value (ITAE) [202], Integral of Square Time derivative of the Control input (ISTC) [203], Total Squared Variation (TSV) [190], Amplitude Index (AMP) [142],
4. correlation analysis measures, as for instance oscillation detection index [60], relative damping index [62],
5. statistical factors utilizing different probabilistic distribution function (standard deviation, variance, skewness, kurtosis, scale or shape, ...) [22, 61], variance band index [93] or the factors of other probabilistic distributions [204],
6. minimum variance and normalized Harris index [57], Control Performance Index (CPI) [48] and other variance benchmarking methods [58],
7. all types of the model-based measures [105, 121], derived from close loop identification, like aggressive/oscillatory and sluggishness indexes [124],
8. fusion CPA measures using sensor combination [83] or the exponentially weighted moving averages (EWMA) evaluated for other indexes [125, 130],
9. various benchmarking methods [54],
10. frequency methods starting from classical Bode, Nyquist and Nichols charts with phase and gain margins [136, 169] followed by deeper investigations, like with the use of the Fourier transform [128], sensitivity function [154] or the reference to disturbance ratio (RDR) index [2], singular spectrum analysis [192],
11. alternative indexes using wavelets [110], orthonormal Laguerre networks [99] and other functions [72], neural networks [207], support vector machines [118], Hurst exponent [117], persistence measures [29], entropy [194], ...
12. graphic visualization and patter recognition methods [62, 107, 150],
13. case-specific business KPIs like for instance number of alarms or human interventions, time in manual mode [86] and many others often expressed in currency units [10].

1.3 Control Assessment in Industrial Life

Industrial use of the control assessment can be met in different stages of the control system life-cycle process. The main areas of CPA application are the process improvement feasibility analysis often includes control performance assessment initiative and the control system maintenance process. The first one is mostly a single shot action. It is initiated in conjunction with the wider initiative associated with any process or installation improvement. Control system and its rehabilitation have to be considered as the control system is often included in any process perfecting project. On the other hand, the control system maintenance is a daily process. The CPA and the assessment of the control system quality is crucial from the perspective of the normal process operation and daily benefits. Thus, a control engineer requires tools (measures, methods, software) that may support him in this task.

There is one more task that is connected with the CPA incorporating these methods and applications. Industrial processes are often non-stationary and time-varying systems with many correlations impacted by disturbances. They cause many challenges for the control system implementation, design and tuning. Thus, there arises a need for methodologies to compare control rehabilitation cost against expected economic benefits. Such decisions are mostly based on the financial basis. Estimation techniques allowing calculation of the benefits resulting from the control system improvement have been proposed by [9, 173]. The cost element of the decision is simple as it may be easily derived from past projects or explicitly obtained from control system vendor. The benefit part is evaluated specifically in each case. The algorithm is based on the mitigation of process variability, leading towards quantitative results [3]. Frequently, one may assume upper or lower limitation for the variable. Reduction of its variability through better control enables to shift it closer to the constraint and thus to generate benefit. As the variable is explicitly linked with the performance, benefits may be calculated. The method assumes that the shape of the variable histogram is Gaussian and standard deviation is used as a variability measure. Apart from that, other methods were proposed, like probabilistic optimization approach [201].

References

1. Agarwal, N., Huang, B., Tamayo, E.C.: Assessing model prediction control (MPC) performance. 2. Bayesian approach for constraint tuning. Indus. Eng. Chem. Res. **46**(24), 8112–8119 (2007)
2. Alagoz, B.B., Tan, N., Deniz, F.N., Keles, C.: Implicit disturbance rejection performance analysis of closed loop control systems according to communication channel limitations. IET Control Theor. Appl. **9**(17), 2522–2531 (2015)
3. Ali, M.K.: Assessing economic benefits of advanced control. In: 5th Asian Control Conference, Process Control in the Chemical Industries, Chemical Engineering Department, pp. 146–159. King Saud University, Riyadh, Kingdom of Saudi Arabia (2002)
4. Åström, K.J.: Computer control of a paper machine—an application of linear stochastic control theory. IBM J. **11**, 389–405 (1967)
5. Åström, K.J.: Introduction to Stochastic Control Theory. Mathematics in Science and Engineering, vol. 70. Academic Press (1970)
6. Badwe, A.S., Gudi, R.D., Patwardhan, R.S., Shah, S.L., Patwardhan, S.C.: Detection of model-plant mismatch in MPC applications. J. Process Control **19**(8), 1305–1313 (2009)
7. Basseville, M.: On-board component fault detection and isolation using the statistical local approach. Automatica **34**(11), 1391–1415 (1998)
8. Bauer, M., Craig, I.K.: Economic assessment of advanced process control—a survey and framework. J. Process Control **18**(1), 2–18 (2008)
9. Bauer, M., Craig, I.K., Tolsma, E., de Beer, H.: A profit index for assessing the benefits of process control. Indus. Eng. Chem. Res. **46**(17), 5614–5623 (2007)
10. Bauer, M., Horch, A., Xie, L., Jelali, M., Thornhill, N.: The current state of control loop performance monitoring—a survey of application in industry. J. Process Control **38**, 1–10 (2016)
11. Begum, G.K., Rao, S.A., Radhakrishnan, T.K.: Performance assessment of control loops involving unstable systems for set point tracking and disturbance rejection. J. Taiwan Inst. Chem. Eng. **85**, 1–17 (2018)

12. Bhandare, D.S., Kulkarni, N.R.: Performances evaluation and comparison of PID controller and fuzzy logic controller for process liquid level control. In: 2015 15th International Conference on Control, Automation and Systems (ICCAS), pp. 1347–1352 (2015)
13. Bialic, G.: Methods of control performance assessment for sampled data systems working under stationary stoachastics disturbances. Ph.D. thesis, Dissertation of Technical University of Opole, Poland (2006)
14. Bialic, G., Błachuta, M.J.: Performance assessment of control loops with PID controllers based on correlation and spectral analysis. In: Proceedings of the 12th IEEE International Conference on Methods and Models in Automation and Robotics, MMAR 2006 (2006)
15. Botelho, V.R., Trierweiler, J.O., Farenzena, M., Longhi, L.G.S., Zanin, A.C., Teixeira, H.C.G., Duraiski, R.G.: Model assessment of MPCs with control ranges: an industrial application in a delayed coking unit. Control Eng. Pract. **84**, 261–273 (2019)
16. Boverie, S., Demaya, B., Ketata, R., Titli, A.: Performance evaluation of fuzzy controllers. IFAC Proc. Vol. **25**(6), 69–74 (1992). IFAC Symposium on Intelligent Components and Instruments for Control Applications (SICICA'92), Malaga, Spain, 20–22 May 1992
17. Box, G.E.P., Jenkins, G.M.: Time Series Analysis: Forecasting and Control. Holden-Day, San Francisco, CA (1970)
18. Cano-Izquierdo, J.M., Ibarrola, J., Kroeger, M.A.: Control loop performance assessment with a dynamic neuro-fuzzy model (dFasArt). IEEE Trans. Autom. Sci. Eng. **9**(2), 377–389 (2012)
19. Carelli, A.C., da Souza Jr, M.B.: GPC controller performance monitoring and diagnosis applied to a diesel hydrotreating reactor. IFAC Proc. Vol. **42**(11), 976–981 (2009)
20. Chen, J.: Statistical methods for process monitoring and control. Master's thesis, McMaster University, Hamilton, Ontario, Canada (2014)
21. Choudhury, M.A.A.S., Shah, S.L., Thornhill, N.F.: Detection and quantification of control valve stiction. IFAC Proc. **37**(9), 865–870 (2004a). 7th IFAC Symposium on Dynamics and Control of Process Systems 2004 (DYCOPS -7), Cambridge, USA, 5–7 July 2004
22. Choudhury, M.A.A.S, Shah, S.L, Thornhill, N.F.: Diagnosis of poor control-loop performance using higher-order statistics. Automatica **40**(10), 1719–1728 (2004b)
23. Cross, R., Iqbal, A.: The rank xerox experience: benchmarking ten years on. In: IFIP Advances in Information and Communication Technology, Benchmarking—Theory and Practice, pp. 3–10 (1995)
24. Daneshwar, M.A., Noh, N.M.: Detection of stiction in flow control loops based on fuzzy clustering. Control Eng. Pract. **39**, 23–34 (2015)
25. Das, L., Srinivasan, B., Rengaswamy, R.: Multivariate control loop performance assessment with Hurst exponent and mahalanobis distance. IEEE Trans. Control Syst. Technol. **24**(3), 1067–1074 (2016)
26. Desborough, L., Harris, T.J.: Performance assessment measures for univariate feedback control. Can. J. Chem. Eng. **70**(6), 1186–1197 (1992)
27. Desborough, L., Harris, T.J.: Performance assessment measures for univariate feedforward/feedback control. Can. J. Chem. Eng. **71**(4), 605–616 (1993)
28. DeVries, W., Wu, S.: Evaluation of process control effectiveness and diagnosis of variation in paper basis weight via multivariate time-series analysis. IEEE Trans. Autom. Control **23**, 702–708 (1978)
29. Domański, P.D.: Non-Gaussian and persistence measures for control loop quality assessment. Chaos Interdiscip. J. Nonlinear Sci. **26**(4), 043105 (2016)
30. Domański, P.D.: On-line control loop assessment with non-Gaussian statistical and fractal measures. In: Proceedings of 2017 American Control Conference, Seattle, USA, pp. 555–560 (2017)
31. Domański, P.D., Ławryńczuk, M.: Assessment of predictive control performance using fractal measures. Nonlinear Dyn. **89**, 773–790 (2017a)
32. Domański, P.D., Ławryńczuk, M.: Assessment of the GPC control quality using non-Gaussian statistical measures. Int. J. Appl. Math. Comput. Sci. **27**(2), 291–307 (2017b)
33. Domański, P.D., Ławryńczuk, M.: Control quality assessment of nonlinear model predictive control using fractal and entropy measures. In: Preprints of the First International Nonlinear Dynamics Conference NODYCON 2019, Rome, Italy (2019)

34. Domański, P.D., Golonka, S., Jankowski, R., Kalbarczyk, P., Moszowski, B.: Control rehabilitation impact on production efficiency of ammonia synthesis installation. Ind. Eng. Chem. Res. **55**(39), 10366–10376 (2016)
35. Duarte-Barros, R.L., Park, S.W.: Assessment of model predictive control performance criteria. J. Chem. Chem. Eng. **9**, 127–135 (2015)
36. El-Ferik, S., Shareef, M.N., Ettaleb, L.: Detection and diagnosis of plant-wide oscillations using GA based factorization. J. Process Control **22**(1), 321–329 (2012)
37. Ender, D.: Process control performance: not as good as you think. Control Eng. **40**(10), 180–190 (1993)
38. Eriksson, P.G., Isaksson, A.J.: Some aspects of control loop performance monitoring. In: 1994 Proceedings of IEEE International Conference on Control and Applications, vol. 2, pp. 1029–1034 (1994)
39. Ettaleb, L.: Control loop performance assessment and oscillation detection. Ph.D. thesis, University of British Columbia, Canada (1999)
40. Farenzena, M.: Novel methodologies for assessment and diagnostics in control loop management. Ph.D. thesis, Dissertation of Universidade Federal do Rio Grande do Sul, Porto Alegre, Brazil (2008)
41. Gabor, J., Pakulski, D., Domański, P.D., Świrski, K.: Closed loop NOx control and optimization using neural networks. In: IFAC Symposium on Power Plants and Power Systems Control, Brussels, Belgium, pp. 188–196 (2000)
42. Galvez, J.M.: Performance assessment of model reference adaptive control—a benchmark based comparison among techniques. In: ABCM Symposium Series in Mechatronics, vol. 5, pp. 186–195 (2012)
43. Gao, J., Patwardhan, R., Akamatsu, K., Hashimoto, Y., Emoto, G., Shah, S.L., Huang, B.: Performance evaluation of two industrial MPC controllers. Control Eng. Pract. **11**(12), 1371–1387 (2003). 2002 IFAC World Congress
44. Gao, X., Yang, F., Shang, C., Huang, D.: A review of control loop monitoring and diagnosis: prospects of controller maintenance in big data era. Chin. J. Chem. Eng. **24**(8), 952–962 (2016)
45. Gao, X., Yang, F., Shang, C., Huang, D.: A novel data-driven method for simultaneous performance assessment and retuning of PID controllers. Ind. Eng. Chem. Res. **56**(8), 2127–2139 (2017)
46. Ghraizi, R.A., Martínez, E.C., de Prada, C.: Control loop performance monitoring using the permutation entropy of error residuals. IFAC Proc. Vol. **42**(11), 494–499 (2009). 7th IFAC Symposium on Advanced Control of Chemical Processes
47. Gomez, D., Moya, E.J., Baeyens, E.: Control performance assessment: a general survey. In: de Leon, F., de Carvalho, A.P., Rodriguez-Gonzalez, S., de Paz Santana, J.F., Rodriguez, J.M.C. (eds.) Distributed Computing and Artificial Intelligence: 7th International Symposium, pp. 621–628. Springer, Berlin (2010)
48. Grimble, M.J.: Controller performance benchmarking and tuning using generalised minimum variance control. Automatica **38**(12), 2111–2119 (2002a)
49. Grimble, M.J.: Restricted structure controller tuning and performance assessment. IEE Proc. Control Theor. Appl. **149**(1), 8–16 (2002b)
50. Grimble, M.J.: Restricted structure control loop performance assessment for PID controllers and state-space systems. Asian J. Control **5**(1), 39–57 (2003)
51. Guo, Z., Xie, L., Ye, T., Horch, A.: Online detection of time-variant oscillations based on improved ITD. Control Eng. Pract. **32**, 64–72 (2014)
52. Gupta, P., Guenther, K., Hodgkinson, J., Jacklin, S., Richard, M., Schumann, J., Soares, F.: Performance monitoring and assessment of neuro-adaptive controllers for aerospace applications using a Bayesian approach. In: Proceedings of the AIAA Guidance, Navigation, and Control Conference and Exhibit, San Francisco, CA, USA, AIAA 2005, p. 6451 (2005)
53. Gustafsson, F.: Adaptive filtering and change detection. Wiley, New York (2000)
54. Hadjiiski, M., Georgiev, Z.: Benchmarking of process control performance. In: Problems of Engineering, Cybernetics and Robotics, vol. 55, pp. 103–110. Bulgarian Academy of Sciences, Sofia, Bulgaria (2005)

55. Hägglund, T.: Automatic detection of sluggish control loops. Control Eng. Pract. **7**(12), 1505–1511 (1999)
56. Hang, C.C., Tan, C.H., Chan, W.P.: A performance study of control systems with dead time. IEEE Trans. Ind. Electron. Control Instrum. IECI **27**(3), 234–241 (1980)
57. Harris, T.J.: Assessment of closed loop performance. Can. J. Chem. Eng. **67**, 856–861 (1989)
58. Harris, T.J., Seppala, C.T.: Recent developments in controller performance monitoring and assessment techniques. In: Proceedings of the Sixth International Conference on Chemical Process Control, pp. 199–207 (2001)
59. Harris, T.J., Seppala, C.T., Desborough, L.D.: A review of performance monitoring and assessment techniques for univariate and multivariate control systems. J. Process Control **9**(1), 1–17 (1999)
60. Horch, A.: A simple method for detection of stiction in control valves. Control Eng. Pract. **7**(10), 1221–1231 (1999)
61. Horch, A., Heiber, F.: On evaluating control performance on large data sets. IFAC Proc. Vol. **37**(9), 535–540 (2004)
62. Howard, R., Cooper, D.: A novel pattern-based approach for diagnostic controller performance monitoring. Control Eng. Pract. **18**(3), 279–288 (2010)
63. Huang, B.: Multivariate statistical methods for control loop performance assessment. Ph.D. thesis, University of Alberta, Department of Chemical and Material Engineering, Canada (1997)
64. Huang, B., Shah, S.L.: Limits of control loop performance: practical measures of control loop performance assessment. In: AIChE Annual Meeting (1996)
65. Huang, B., Shah, S.L., Kwok, E.K.: On-line control performance monitoring of MIMO processes. In: Proceedings of the 1995 American Control Conference, vol. 2, pp. 1250–1254 (1995)
66. Huang, Q., Zhang, Q.: Research on multivariable control performance assessment techniques. In: 2011 International Symposium on Advanced Control of Industrial Processes (ADCONIP), pp. 508–511. IEEE (2011)
67. Huba, M.: Performance measures and the robust and optimal control design. IFAC-PapersOnLine **51**(4):960–965 (2018). 3rd IFAC Conference on Advances in Proportional-Integral-Derivative Control PID 2018
68. Hugo, A.: Process control performance monitoring and assessment. Control Arts Inc. http://www.controlarts.com/ (2001)
69. Jelali, M.: An overview of control performance assessment technology and industrial applications. Control Eng. Pract. **14**(5), 441–466 (2006)
70. Jelali, M.: Estimation of valve stiction in control loops using separable least-squares and global search algorithms. J. Process Control **18**(7), 632–642 (2008)
71. Jelali, M.: Estimation of valve stiction using separable least-squares and global search algorithms. In: Jelali, M., Huang, B. (eds.) Detection and Diagnosis of Stiction in Control Loops: State of the Art and Advanced Methods, pp. 205–228. Springer, London (2010)
72. Jelali, M.: Control Performance Management in Industrial Automation: Assessment, Diagnosis and Improvement of Control Loop Performance. Springer-Verlag, London (2013)
73. Jelali, M., Thormann, M., Wolff, A., Müller, T., Loredo, L.R., Sanfilippo, F., Zangari, G., Foerster, P.: Enhancement of product quality and production system reliability by continuous performance assessment of automation systems (AUTOCHECK). Technical Report Final report to Contract No RFS-CR03045, EUR 23205, Office for Official Publications of the European Communities, Luxemburg (2008)
74. Jiang, D.Y., Hu, L.S., Shi, P.: Performance assessment of switched control systems based on tensor space approach. Int. J. Adap. Control Sig. Process. **30**(4), 634–663 (2016)
75. Jiang, H., Choudhury, M.A.A.S., Shah, S.L.: Detection and diagnosis of plant-wide oscillations from industrial data using the spectral envelope method. J. Process Control **17**(2), 143–155 (2007)
76. Jiang, H., Shah, S.L., Huang, B., Wilson, B., Patwardhan, R., Szeto, F.: Performance assessment and model validation of two industrial MPC controllers. IFAC Proc. Vol. **41**(2), 8387–8394 (2008)

77. Jimoh, M.T.: A vision for MPC performance maintenance. Ph.D. thesis, Dissertation of the College of Science and Engineeiring, University of Glasgow (2013)
78. Julien, R.H., Foley, M.W., Cluett, W.R.: Performance assessment using a model predictive control benchmark. J. Process Control **14**(4), 441–456 (2004)
79. Kadali, R., Huang, B.: Controller performance analysis with LQG benchmark obtained under closed loop conditions. ISA Trans. **41**(4), 521–537 (2002)
80. Kala, H., Deepakraj, D., Gopalakrishnan, P., Vengadesan, P., Iyyar, M.K.: Implicit disturbance rejection performance analysis of closed loop control systems according to communication channel limitations. Int. J. Innov. Res. Electric. Electron. Instrum. Control Eng. **2**(3), 1311–1314 (2014)
81. Kalman, R.E.: Contribution to the theory of optimal control. Bol. Soc. Math. Mex. **5**, 102–119 (1960)
82. Kendra, S.J., Cinar, A.: Controller performance assessment by frequency domain techniques. J. Process Control **7**(3), 181–194 (1997)
83. Khamseh, S.A., Sedigh, A.K., Moshiri, B., Fatehi, A.: Control performance assessment based on sensor fusion techniques. Control Eng. Pract. **49**, 14–28 (2016)
84. Khan, M., Tahiyat, M., Imtiaz, S., Choudhury, M.A.A.S., Khan, F.: Experimental evaluation of control performance of MPC as a regulatory controller. ISA Trans. **70**, 512–520 (2017)
85. Kinoshita, T., Ohnishi, Y., Yamamoto, T., Shah, S.L.: Design of a performance-driven control system based on the control assessment. In: 44th Annual Conference of the IEEE Industrial Electronics Society, IECON 2018, pp. 5383–5388 (2018)
86. Knierim-Dietz, N., Hanel, L., Lehner, J.: Definition and verification of the control loop performance for different power plant types. Institute of Combustion and Power Plant Technology, University of Stutgart, Technical report (2012)
87. Ko, B.S., Edgar, T.F.: Performance assessment of cascade control loops. AIChE J. **46**(2), 281–291 (2000)
88. Ko, B.S., Edgar, T.F.: PID control performance assessment: the single-loop case. AIChE J. **50**(6), 1211–1218 (2004)
89. Laing, D., Uduehi, D., Ordys, A.: Financial benefits of advanced control. Benchmarking and optimization of a crude oil production platform. In: Proceedings of American Control Conference, vol. 6, pp. 4330–4331 (2001)
90. Lee, K.H., Huang, B., Tamayo, E.C.: Sensitivity analysis for selective constraint and variability tuning in performance assessment of industrial MPC. Control Eng. Pract. **16**(10), 1195–1215 (2008a)
91. Lee, K.H., Xu, F., Huang, B., Tamayo, E.C.: Controller performance analysis technology for industry: implementation and case studies. IFAC Proc. Vol. **41**(2), 14912–14919 (2008b). 17th IFAC World Congress
92. Li, X., Wang, J., Huang, B., Lu, S.: The DCT-based oscillation detection method for a single time series. J. Process Control **20**(5), 609–617 (2010)
93. Li, Y., O'Neill, Z.: Evaluating control performance on building HVAC controllers. In: International Building Performance Simulation Association, pp. 962–967, Hyderabad, India (2015)
94. Li, Z., Evans, R.J.: Minimum-variance control of linear time-varying systems. Automatica **33**(8), 1531–1537 (1997)
95. Liu, M.C.P., Wang, X., Wang, Z.L.: Performance assessment of control loop with multiple time-variant disturbances based on multi-model mixing time-variant minimum variance control. In: Proceeding of the 11th World Congress on Intelligent Control and Automation, pp. 4755–4759 (2014)
96. Lopez, A., Murrill, P., Smith, C.: Controller tuning relationships based on integral performance criteria. Instrum. Technol. **14**(12), 57–62 (1967)
97. Loquasto, F., Seborg, D.E.: Monitoring model predictive control systems using pattern classification and neural networks. Ind. Eng. Chem. Res. **42**(20), 4689–4701 (2003)
98. Lynch, C.B., Dumont, G.A.: Control loop performance monitoring. In: Proceedings of IEEE International Conference on Control and Applications, vol. 2, pp. 835–840 (1993)

99. Lynch, C.B., Dumont, G.A.: Control loop performance monitoring. IEEE Trans. Control Syst. Technol. **4**(2), 185–192 (1996)
100. Maboodi, M., Khaki-Sedigh, A., Camacho, E.F.: Control performance assessment for a class of nonlinear systems using second-order volterra series models based on nonlinear generalised minimum variance control. Int. J. Control **88**(8), 1565–1575 (2015)
101. Mäkelä, M., Manninen, V., Heiliö, M., Myller, T.: Performance assessment of automatic quality control in mill operations. In: 2006 Proceedings of Control Systems (2006)
102. Marlin, T.E., Perkins, J.D., Barton, G.W., Brisk, M.L.: Benefits from process control: results of a joint industry-university study. J. Process Control **1**(2), 68–83 (1991)
103. Mason, D.G., Edwards, N.D., Linkens, D.A., Reilly, C.S.: Performance assessment of a fuzzy controller for atracurium-induced neuromuscular block. Br. J. Anaesth. **7**(3), 396–400 (1996)
104. Meng, Q.W., Fang, F., Liu, J.Z.: Minimum-information-entropy-based control performance assessment. Entropy **15**(3), 943–959 (2013a)
105. Meng, Q.W., Gu, J.Q., Zhong, Z.F., Ch, S., Niu, Y.G.: Control performance assessment and improvement with a new performance index. In: 2013 25th Chinese Control and Decision Conference (CCDC), pp. 4081–4084 (2013b)
106. Micic, A.D., Matausek, M.R.: Closed-loop PID controller design and performance assessment in the presence of measurement noise. Chem. Eng. Res. Design **104**(Supplement C), 513–518 (2015)
107. Mitchell, W., Shook, D., Shah, S.L.: A picture worth a thousand control loops: an innovative way of visualizing controller performance data. Invited Plenary Presentation, Control Systems (2004)
108. Moridi, A., Armaghan, S., Sedigh, A.K., Choobkar, S.: Design of switching control systems using control performance assessment index. In: 2011 Proceedings of the World Congress on Engineering, vol. II, pp. 186–195 (2011)
109. Naujoks, F., Wiedemann, K., Schőmig, N., Jarosch, O., Gold, C.: Expert-based controllability assessment of control transitions from automated to manual driving. MethodsX **5**, 579–592 (2018)
110. Nesic, Z., Dumont, G., Davies, M., Brewster, D.: CD control diagnostics using a wavelet toolbox. In: Proceedings CD Symposium, IMEKO, vol. XB, pp. 120–125 (1997)
111. Nissinen, A., Nuyan, S., Virtanen, P.: On-line performance monitoring in process analysis. In: 2006 Proceedings of Control Systems (2006)
112. O'Connor, N., O'Dwyer, A.: Control loop performance assessment: a classification of methods. In: Proceedings of the Irish Signals and Systems Conference, Queens University Belfast, pp. 530–535 (2002)
113. Ohtsu, K., Kitagawa, G.: Statistical analysis of the AR type ship's autopilot system. J. Dyn. Syst. Meas. Control **106**(3), 193–202 (1984)
114. Olaleye, F., Huang, B., Tamayo, E.: Industrial applications of a feedback controller performance assessment of time-variant processes. Ind. Eng. Chem. Res. **43**(2), 597–607 (2004)
115. O'Neill, Z., Li, Y., Williams, K.: HVAC control loop performance assessment: a critical review (1587-rp). Sci. Technol. Built Environ. **23**(4), 619–636 (2017)
116. Perrier, M., Roche, A.A.: Towards mill-wide evaluation of control loop performance. In: Proceedings of the Control Systems, pp. 205–209 (1992)
117. Pillay, N., Govender, P.: A data driven approach to performance assessment of PID controllers for setpoint tracking. Proc. Eng. **69**, 1130–1137 (2014)
118. Pillay, N., Govender, P.: Multi-class SVMs for automatic performance classification of closed loop controllers. J. Control Eng. Appl. Inf. **19**(3), 3–12 (2017)
119. Pour, N.D., Huang, B., Shah, S.: Performance assessment of advanced supervisory-regulatory control systems with subspace LQG benchmark. Automatica **46**(8), 1363–1368 (2010)
120. Pryor, C.: Autocovariance and power spectrum analysis derive new information from process data. Control Eng. **2**, 103–106 (1982)
121. Qin, S.J.: Control performance monitoring—a review and assessment. Comput. Chem. Eng. **23**(2), 173–186 (1998)

122. Recalde, L.F., Yue, H.: Control performance monitoring of state-dependent nonlinear processes. IFAC-PapersOnLine **50**(1), 11313–11318 (2017). 20th IFAC World Congress
123. Salsbury, T.I.: A practical method for assessing the performance of control loops subject to random load changes. J. Process Control **15**(4), 393–405 (2005)
124. Salsbury, T.I.: Continuous-time model identification for closed loop control performance assessment. Control Eng. Pract. **15**(1), 109–121 (2007)
125. Salsbury, T.I., Alcala, C.F.: Two new normalized EWMA-based indices for control loop performance assessment. In: 2015 American Control Conference (ACC), pp. 962–967 (2015)
126. Scali, C., Marchetti, E., Esposito, A.: Effect of cascade tuning on control loop performance assessment. In: IFAC Conference on Advances in PID Control PID'12 (2012)
127. Schäfer, J., Cinar, A.: Multivariable MPC system performance assessment, monitoring, and diagnosis. J. Process Control **14**(2), 113–129 (2004)
128. Schlegel, M., Skarda, R., Cech, M.: Running discrete Fourier transform and its applications in control loop performance assessment. In: 2013 International Conference on Process Control (PC), pp. 113–118 (2013)
129. Seborg, D., Edgar, T.F., Mellichamp, D.: Process Dynamics & Control. Wiley, New York (2006)
130. Seem, J.E., House, J.M.: Integrated control and fault detection of air-handling units. HVAC & R Res. **15**(1), 25–55 (2009)
131. Sendjaja, A.Y., Kariwala, V.: Achievable PID performance using sums of squares programming. J. Process Control **19**(6), 1061–1065 (2009)
132. Seppala, C.T.: Dynamic analysis of variance methods for monitoring control system performance. Ph.D. thesis, Queen's University Kingston, Ontario, Canada (1999)
133. Shah, S.L., Patwardhan, R., Huang, B.: Multivariate controller performance analysis: methods, applications and challenges. In: Proceedings of the Sixth International Conference on Chemical Process Control, pp. 199–207 (2001)
134. Shang, C., Huang, B., Yang, F., Huang, D.: Slow feature analysis for monitoring and diagnosis of control performance. J. Process Control **39**, 21–34 (2016)
135. Shang, L., Tian, X., Cai, L.: A multi-index control performance assessment method based on historical prediction error covariance. IFAC-PapersOnLine **50**(1), 13892–13897 (2017). 20th IFAC World Congress
136. Shardt, Y., Zhao, Y., Qi, F., Lee, K., Yu, X., Huang, B., Shah, S.: Determining the state of a process control system: current trends and future challenges. Can. J. Chem. Eng. **90**(2), 217–245 (2012)
137. Shinskey, F.G.: How good are our controllers in absolute performance and robustness? Meas. Control **23**(4), 114–121 (1990)
138. Skarda, R., Cech, M., Schlegel, M.: Simultaneous control loop performance assessment and process identification based on fractional models. IFAC—PapersOnLine **48**(8), 859–864 (2015). 9th IFAC Symposium on Advanced Control of Chemical Processes ADCHEM 2015
139. Smuts, J.F., Hussey, A.: Requirements for successfully implementing and sustaining advanced control applications. In: Proceedings of the 54th ISA POWID Symposium, pp. 89–105 (2011)
140. Sotomayor, O.A.Z., Odloak, D.: Performance assessment of model predictive control systems. IFAC Proc. Vol. **39**(2), 875–880 (2006). 6th IFAC Symposium on Advanced Control of Chemical Processes
141. Spinner, T.: Performance assessment of multivariate control systems. Ph.D. thesis, Texas Tech University (2014)
142. Spinner, T., Srinivasan, B., Rengaswamy, R.: Data-based automated diagnosis and iterative retuning of proportional-integral (PI) controllers. Control Eng. Pract. **29**, 23–41 (2014)
143. Srinivasan, B., Rengaswamy, R.: Automatic oscillation detection and characterization in closed-loop systems. Control Eng. Pract. **20**(8), 733–746 (2012)
144. Srinivasan, B., Spinner, T., Rengaswamy, R.: Control loop performance assessment using detrended fluctuation analysis (DFA). Automatica **48**(7), 1359–1363 (2012)

145. Srinivasan, B., Spinner, T., Rengaswamy, R.: A reliability measure for model based stiction detection approaches. IFAC Proc. Vol. **45**(15), 750–755 (2012b). 8th IFAC Symposium on Advanced Control of Chemical Processes
146. Srinivasan, B., Nallasivam, U., Rengaswamy, R.: An integrated approach for oscillation diagnosis in linear closed loop systems. Chem. Eng. Res. Design **93**, 483–495 (2015)
147. Srinivasan, R., Rengaswamy, R., Miller, R.: A modified empirical mode decomposition (EMD) process for oscillation characterization in control loops. Control Eng. Pract. **15**(9), 1135–1148 (2007)
148. Stanfelj, N., Marlin, T.E., MacGregor, J.F.: Monitoring and diagnosing process control performance: the single-loop case. Ind. Eng. Chem. Res. **32**(2), 301–314 (1993)
149. Starr, K.D., Petersen, H., Bauer, M.: Control loop performance monitoring—experience over two decades. IFAC-PapersOnLine **49**(7), 526–532 (2016). 11th IFAC Symposium on Dynamics and Control of Process SystemsIncluding Biosystems DYCOPS-CAB 2016 (2016)
150. Stockmann, M., Habera, R., Schmitz, U.: Pattern recognition for valve stiction detection with principal component analysis. IFAC Proc. Vol. **42**(8), 1438–1443 (2009). 7th IFAC Symposium on Fault Detection, Supervision and Safety of Technical Processes
151. Su, A.J., Yu, C.C., Ogunnaike, B.A.: On the interaction between measurement strategy and control performance in semiconductor manufacturing. J. Process Control 18(3):266–276 (2008). Festschrift honouring Professor Dale Seborg
152. Sun, Z., Qin, S.J., Singhal, A., Megan, L.: Performance monitoring of model-predictive controllers via model residual assessment. J. Process Control **23**(4), 473–482 (2013)
153. Tangirala, A.K., Shah, S.L., Thornhill, N.F.: PSCMAP: a new tool for plant-wide oscillation detection. J. Process Control **15**(8), 931–941 (2005)
154. Tepljakov, A., Petlenkov, E., Belikov, J.: A flexible MATLAB tool for optimal fractional-order PID controller design subject to specifications. In: 2012 31st Chinese Control Conference (CCC) (2012)
155. Thornhill, N.F., Hägglund, T.: Detection and diagnosis of oscillation in control loops. Control Eng. Pract. **5**(10), 1343–1354 (1997)
156. Thornhill, N.F., Shah, S.L., Huang, B., Vishnubhotla, A.: Spectral principal component analysis of dynamic process data. Control Eng. Pract. **10**(8), 833–846 (2002)
157. Thornhill, N.F., Huang, B., Shah, S.L.: Controller performance assessment in set point tracking and regulatory control. Int. J. Adap. Control Sig. Process. **17**(7–9), 709–727 (2003a)
158. Thornhill, N.F., Huang, B., Zhang, H.: Detection of multiple oscillations in control loops. J. Process Control **13**(1), 91–100 (2003b)
159. Tolfo, F.: A methodology to assess the economic returns of advanced control projects. In: 1983 American Control Conference, pp. 1141–1146. IEEE (1983)
160. Torres, B.S., de Carvalho, F.B., de Oliveira Fonseca, M., Filho, C.S.: Performance assessment of control loops - case studies. In: IFAC ADCHEM Conference (2006)
161. Tulsyan, A., Huang, B., Gopaluni, R.B., Forbes, J.F.: Performance assessment, diagnosis, and optimal selection of non-linear state filters. J. Process Control **24**(2), 460–478 (2014). ADCHEM 2012 Special Issue
162. Tyler, M.L., Morari, M.: Performance assessment for unstable and nonminimum-phase systems. IFAC Proc. Vol. **28**(12), 187–192 (1995)
163. Tyler, M.L., Morari, M.: Performance monitoring of control systems using likelihood methods. Automatica **32**(8), 1145–1162 (1996)
164. Ulaganathan, N., Rengaswamy, R.: Blind identification of stiction in nonlinear process control loops. In: 2008 American Control Conference, pp. 3380–3384 (2008)
165. Kumar, V., Rana, K.P., Gupta, V.: Real-time performance evaluation of a fuzzy PI + fuzzy PD controller for liquid-level process. Int. J. Innov. Res. Electric. Electron. Instrum. Control Eng. **13**(2), 89–96 (2008)
166. Vatanski, N., Jämsä-Jounela, S.L., Rantala, A., Harju, T.: Control loop performance measures in the evaluation of process economics. In: 16th Triennial IFAC World Congress (2005)
167. Veronesi, M., Visioli, A.: Performance assessment and retuning of PID controllers for load disturbance rejection. IFAC Proc. Vol. **45**(3), 530–535 (2012). 2nd IFAC Conference on Advances in PID Control

168. Veronesi, M., Visioli, A.: Process parameters estimation, performance assessment and controller retuning based on the final value theorem: Some extensions. IFAC-PapersOnLine **50**(1), 9198–9203 (2017). 20th IFAC World Congress
169. Vishnubhotla, A.: Frequency and time-domain techniques for control loop performance assessment. Ph.D. thesis, University of Alberta, Department of Chemical and Material Engineering, Canada (1997)
170. Visioli, A.: Method for proportional-integral controller tuning assessment. Ind. Eng. Chem. Res. **45**(8), 2741–2747 (2006)
171. Wan, S., Huang, B.: Robust performance assessment of feedback control systems. Automatica **38**(1), 33–46 (2002)
172. Wang, J., Huang, B., Lu, S.: Improved DCT-based method for online detection of oscillations in univariate time series. Control Eng. Pract. **21**(5), 622–630 (2013)
173. Wei, D., Craig, I.: Development of performance functions for economic performance assessment of process control systems. In: 9th IEEE AFRICON, pp. 1–6 (2009)
174. Wei, S., Cheng, J., Wang, Y.: Data-driven two-dimensional LQG benchmark based performance assessment for batch processes under ILC. IFAC-PapersOnLine **48**(8), 291–296 (2015). 9th IFAC Symposium on Advanced Control of Chemical Processes ADCHEM 2015
175. Wei, W., Zhuo, H.: Research of performance assessment and monitoring for multivariate model predictive control system. In: 2009 4th International Conference on Computer Science Education, pp. 509–514 (2009)
176. Wu, Z., Ran, Z., Xu, Q., Wang, W.: Dynamic performance monitoring of current control system for fused magnesium furnace driven by big data. In: 2018 IEEE International Conference on Consumer Electronics-Taiwan, pp. 1–2 (2018)
177. Xia, C., Howell, J., Thornhill, N.F.: Detecting and isolating multiple plant-wide oscillations via spectral independent component analysis. Automatica **41**(12), 2067–2075 (2005)
178. Xie, L., Lang, X., Horch, A., Yang, Y.: Online oscillation detection in the presence of signal intermittency. Control Eng. Pract. **55**, 91–100 (2016)
179. Xu, F., Huang, B., Tamayo, E.C.: Assessment of economic performance of model predictive control through variance/constraint tuning. IFAC Proc. Vol. **39**(2), 899–904 (2006). 6th IFAC Symposium on Advanced Control of Chemical Processes
180. Xu, F., Huang, B., Akande, S.: Performance assessment of model pedictive control for variability and constraint tuning. Ind. Eng. Chem. Res. **46**(4), 1208–1219 (2007)
181. Yan, Z., Chen, J., Zhang, Z.: Using hidden markov model to identify oscillation temporal pattern for control loops. Chem. Eng. Res. Design **119**, 117–129 (2017)
182. Yang, S.H., Dai, C., Knott, R.P.: Remote maintenance of control system performance over the Internet. Control Eng. Pract. **15**(5), 533–544 (2007)
183. You, H., Zhou, J., Zhu, H., Li, D.: Performance assessment based on minimum entropy of feedback control loops. In: 2017 6th Data Driven Control and Learning Systems (DDCLS), pp. 593–598 (2017)
184. Yu, J., Qin, S.J.: Statistical MIMO controller performance monitoring. Part II: Performance diagnosis. J. Process Control **18**(3–4), 297–319 (2008)
185. Yu, W., Wilson, D.I., Young, B.R.: Eliminating valve stiction nonlinearities for control performance assessment. IFAC Proc. Vol. **42**(11), 506–511 (2009). 7th IFAC Symposium on Advanced Control of Chemical Processes
186. Yu, W., Wilson, D., Young, B.: Control performance assessment for block-oriented nonlinear systems. In: IEEE ICCA 2010, pp. 1151–1156 (2010a)
187. Yu, W., Wilson, D.I., Young, B.R.: Nonlinear control performance assessment in the presence of valve stiction. J. Process Control **20**(6), 754–761 (2010b)
188. Yu, W., Wilson, D.I., Young, B.R.: A comparison of nonlinear control performance assessment techniques for nonlinear processes. Can. J. Chem. Eng. **90**(6), 1442–1449 (2012a)
189. Yu, W., Wilson, D.I., Young, B.R.: Control performance assessment in the presence of sampling jitter. Chem. Eng. Res. Design Trans. Inst. Chem. Eng. Part A **90**(1), 129–137 (2012b)
190. Yu, Z., Wang, J.: Performance assessment of static lead-lag feedforward controllers for disturbance rejection in PID control loops. ISA Trans. **64**, 67–76 (2016)

191. Yu, Z., Wang, J., Huang, B., Bi, Z.: Performance assessment of PID control loops subject to setpoint changes. J. Process Control **21**(8), 1164–1171 (2011)
192. Yuan, H.: Process analysis and performance assessment for sheet forming processes. Ph.D. thesis, Queen's University, Kingston, Ontario, Canada (2015)
193. Zabiri, H., Ramasamy, M.: NLPCA as a diagnostic tool for control valve stiction. J. Process Control **19**(8), 1368–1376 (2009). Special Section on Hybrid Systems: Modeling, Simulation and Optimization
194. Zhang, J., Jiang, M., Chen, J.: Minimum entropy-based performance assessment of feedback control loops subjected to non-Gaussian disturbances. J. Process Control **24**(11), 1660–1670 (2015a)
195. Zhang, J., Zhang, L., Chen, J., Xu, J., Li, K.: Performance assessment of cascade control loops with non-Gaussian disturbances using entropy information. Chem. Eng. Res. Design **104**, 68–80 (2015b)
196. Zhang, K., Huang, B., Ji, G.: Multiple oscillations detection in control loops by using the DFT and Raleigh distribution. IFAC-PapersOnLine **48**(21), 529–534 (2015c). 9th IFAC Symposium on Fault Detection, Supervision and Safety for Technical Processes SAFEPROCESS 2015
197. Zhang, Y., Henson, M.A.: A performance measure for constrained model predictive controllers. In: Proceedings of the 1999 European Control Conference, pp. 918–923 (1999)
198. Zhao, C., Huang, B.: Control performance monitoring with temporal features and dissimilarity analysis for nonstationary dynamic processes. IFAC-PapersOnLine **51**(18), 357–362 (2018). 10th IFAC Symposium on Advanced Control of Chemical Processes ADCHEM 2018
199. Zhao, C., Su, H., Gu, Y., Chu, J.: A pragmatic approach for assessing the economic performance of model predictive control systems and its industrial application. Chin. J. Chem. Eng. **17**(2), 241–250 (2009a)
200. Zhao, C., Zhao, Y., Su, H., Huang, B.: Economic performance assessment of advanced process control with LQG benchmarking. J. Process Control **19**(4), 557–569 (2009b)
201. Zhao, C., Xu, Q., Zhang, D., An, A.: Economic performance assessment of process control: a probability optimization approach. In: International Symposium on Advanced Control of Industrial Processes, pp. 585–590 (2011)
202. Zhao, Y., Xie, W., Tu, X.: Performance-based parameter tuning method of model-driven PID control systems. ISA Trans. **51**(3), 393–399 (2012)
203. Zheng, B.: Analysis and auto-tuning of supply air temperature PI control in hot water heating systems. Ph.D. thesis, Dissertation of University of Nebraska (2007)
204. Zhong, L.: Defect distribution model validation and effective process control. In: Proceedings of the SPIE, vol. 5041, pp. 5041–5048 (2003)
205. Zhou, J.L., Wang, X., Zhang, J.F., Wang, H., Yang, G.H.: A new measure of uncertainty and the control loop performance assessment for output stochastic distribution systems. IEEE Trans. Autom. Control **60**(9), 2524–2529 (2015)
206. Zhou, Q., Liu, M.: An on-line self-tuning algorithm of pi controller for the heating and cooling coil in buildings. In: Proceedings of the 11th Symposium on Improving Building Systems in Hot and Humid Climates, Fort Worth, TX (1998)
207. Zhou, Y., Wan, F.: A neural network approach to control performance assessment. Int. J. Intell. Comput. Cybern. **1**(4), 617–633 (2008)

Chapter 2
Step Response Indexes

I think the first things that are relevant are that things should work well;
they should function.

– Robin Day

Control system assessment analyzing the shape and parameters of the step response form the oldest, most classical and the most widely used method. It has many reasons, but the main one is the common understanding and straightforward interpretation of the results. Additionally, these measures enable to conduct root-cause analysis and show directions for controller retuning and further loop improvement. One has to remember about the method shortcut, apart from all above advantages. We need to have the step response. Its existence is not so obvious.

Further presentation of the step response based approaches is divided into two parts. At first the methodologies and algorithms for step response estimation are sketched (Sect. 2.1). It is followed by Sect. 2.2 describing loop step response measures, Sect. 2.3 presenting loop disturbance step methods and finally concluded with the measures based on the loop impulse response based indexes (Sect. 2.4).

2.1 Step Response Identification

Identification of the step response is in focus of the control engineers for years. We may consider two groups of the approaches: direct and indirect. Direct method assumes that we are able to perform active site experiment, i.e. we excite the loop input (setpoint) with the step signal $\vartheta(t)$ sometimes called the Heaviside function $\mathscr{H}$

P. D. Domański, *Control Performance Assessment: Theoretical Analyses and Industrial Practice*, Studies in Systems, Decision and Control 245,
https://doi.org/10.1007/978-3-030-23593-2_2

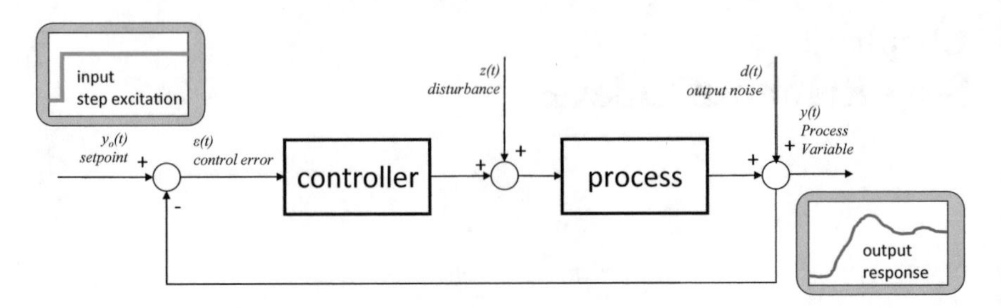

Fig. 2.1 Hierarchical functional levels of the control system (*denotes levels, where CPA is used)

(2.1)) and we may measure the loop output (PV, CV), like presented in the Fig. 2.1. This approach seems to be easy, however it has limitations, which may appear very difficult to be met in industrial reality.

$$\vartheta(t) = \begin{cases} 1 & \text{for } t \geq 0 \\ 0 & \text{for } t < 0 \end{cases} \tag{2.1}$$

First of all, we have to be able to perform such an experiment. We have to be aware that generally any intervention into the process (adding of the step excitation into the process) is considered as a disturbance or threat by plant personnel. Despite our explanations that it is only a minor disturbance, installation owner argues that it disturbs regular operation, product quality, production throughput or plant emissions. Any kind of such experiments need to be conducted cautiously according to the plant safety regulations and in close cooperation with plant operational and control personnel.

The fact that we may perform the experiment does not mean that obtained curves will give reliable information. Time-based identification or parameter estimation using step response works properly only in situations when the noise-to-signal ratio is relatively small. High noises may significantly impede proper analysis of the step response. Figure 2.2 presents the step response of the same process with different noise to signal ratios. We see that the first case (a) with low noise enables the shape analysis and the identification of the step response parameters, while in the second one (b) any decisions might be highly relative.

The noise-to-signal ratio is one of the limitations during the design and conduction of the excitation experiment. We should require to have the step magnitude as high as it is possible. However, its value might be constrained by the other loop properties. Process nonlinearities, actuators' characteristics, installation operating regimes, technological limitations or plant personnel concerns may force the use of smaller step magnitude vales. Thus, the experiment setup is a compromise between noise issues and other plant technological considerations.

Different situation appears once we cannot perform the step test experiment. In such a situation we have to rely on the normal operation loop data. We have to start the analysis from the selection of appropriate historical data. They should be relevant

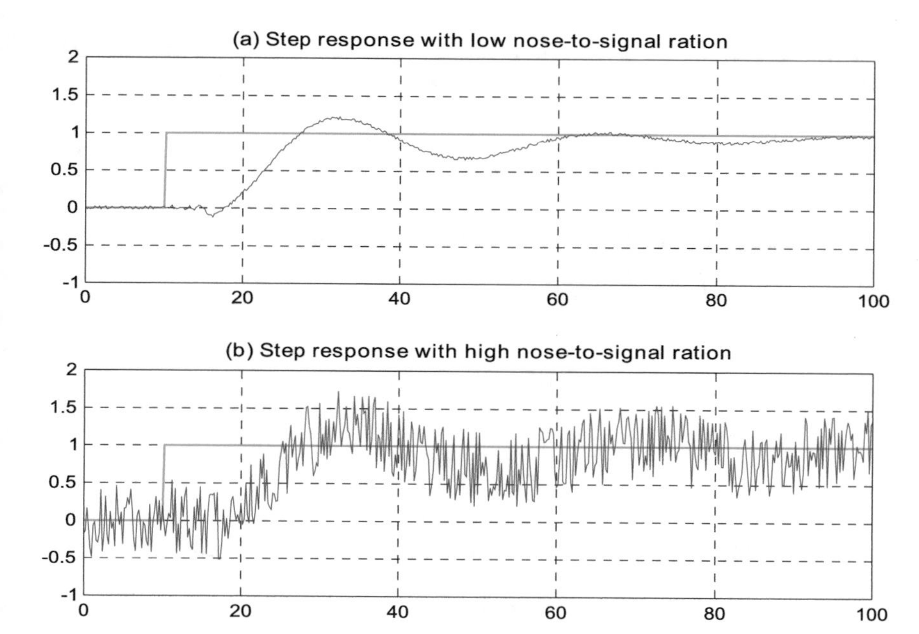

Fig. 2.2 Typical step responses with different noise-to-signal ratios

Fig. 2.3 Close loop model with input y_0 and output y

$y_0(t)$ $y_0(j\omega)$ → $g(t)$ $h(t)$ $G(j\omega)$ → $y(t)$ $y(j\omega)$

and reliable. Industrial life hardly meets our wishful expectations. We have to select normal operating period, i.e. without process or I&C infrastructure failures. Next, we have to secure proper conditions for data collection and export. One has to remember that majority of the control system historians or SCADA/MES software stores only selected and filtered samples. The historical data acquirement should be prepared and agreed with plant control and IT personnel to avoid incomplete, inconsistent or corrupted time series.

Once we have correct time series of the loop input and output variables at our disposal, we may start the process of the step response estimation. The process uses relations between the most prominent non-parametric models of time-invariant, linear processes, i.e. the impulse response, the step response and the frequency response. Figure 2.3 reflects this basic relationship.

The impulse response is denoted by $g(t)$ and is defined as the output of the process (loop) excited by Dirac's impulse function $\delta(t)$ (2.2).

$$\delta(t) = \begin{cases} \infty & \text{for } t = 0 \\ 0 & \text{for } t \neq 0 \end{cases} \tag{2.2}$$

The output of the linear process to any arbitrary input can be determined using the convolution integral by means of the impulse response, as in the case (2.3).

$$y(t) = \int_{o}^{t} g(\tau)\, y_0(t-\tau)\, d\tau \tag{2.3}$$

As one can see, the impulse and step responses are closely related. The step response may be obtained using integration of the impulse response with respect to time (2.4), while the impulse is thus the time derivative of the step response (2.5).

$$h(t) = \int_{-\infty}^{t} g(\tau)\, d\tau \tag{2.4}$$

$$g(t) = \frac{dh(t)}{dt} \tag{2.5}$$

The frequency response is connected with the time responses through the relation (2.6).

$$G(j\omega) = \frac{y(j\omega)}{y_0(j\omega)} = \int_{0}^{\infty} g(t)\, e^{-j\omega t} dt \tag{2.6}$$

There exist many methods to identify step response non-parametric model, as for instance [4] spectral analysis methods, approaches using frequency response measurement, correlation analysis, prediction error methods, Laguerre orthonormal functions, wavelets and others [1]. Each method has its scope of applicability and no single and universal approach exists. Correlation analysis is often suggested as a classical method, especially due to the fact that it works properly with highly noisy data. Nonetheless, we have to be prepared that the resulting step response is often highly disturbed, hampering proper thresholding and extraction of the measure.

Although step test measures form the majority of the assessment methods, we may find in the literature measures derived from very similar approaches. Some authors consider step disturbance [3, 7, 12] or impulse response [2]. The issues associated with obtaining of such responses are similar to the close loop step response applied to the setpoint signal and have been already discussed. Relevant measures are presented in the paragraphs below.

2.2 Loop Step Response Measures

Step response indexes may be evaluated according to well known definitions [10]. Let us consider an exemplary step response sketched in Fig. 2.4.

One of the main parameters describing the control quality is the steady state, i.e. the settled value of the step response $\overline{y}$ (2.7). We need to remember that it may differ from the setpoint step magnitude Y_0.

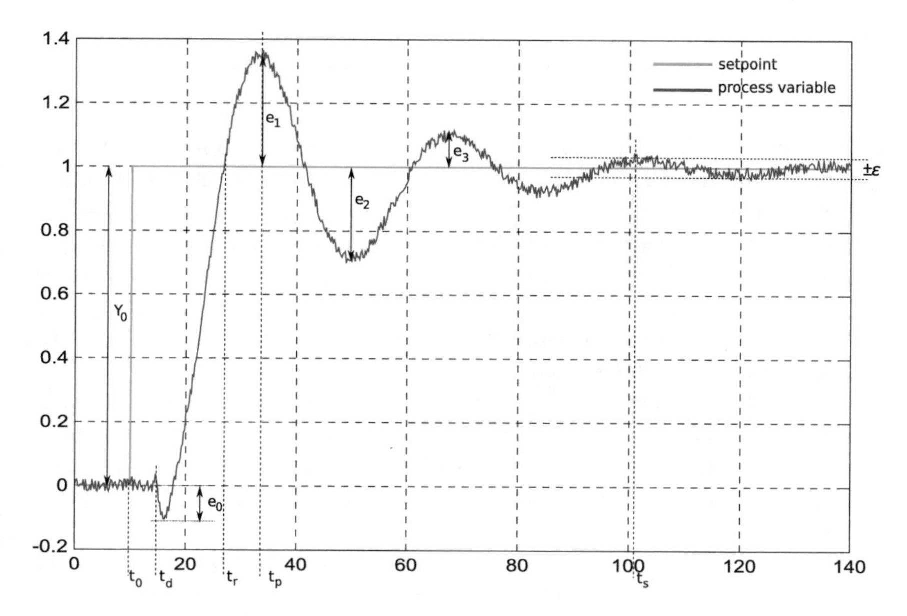

Fig. 2.4 Typical control loop step response with its parameters

$$\overline{y} = \lim_{t \to \infty} y(t) \tag{2.7}$$

Researchers consider the following measures that can be derived out of the response:

- **Offset (steady state error)** of the step response is the difference between the achieved and desired process variable values as $t \to \infty$ (2.8).

$$\varepsilon_{\infty} = Y_0 - \overline{y} \tag{2.8}$$

  The ultimate requirement for the control system, which is often not even formulated explicitly, is to have zero steady state value $\varepsilon_{\infty} = 0$. However, in some situations the non-zero steady state error may appear, as for instance in case of lack of the integration inside of the feedback loop or due to the nonlinearities and inappropriate limitations compliance (control signal constraints).
- **Overshoot** κ (2.9) is the common measure describing quality of the loop tuning. It reflects the loop oscillatory behavior of the under damped response. It is often desired to obtain the zero overshoot system (over-damped response), however it is not always achievable. The κ limitation is often considered as the loop tuning requirement. In an ideal case we should tend towards zero value, so the higher the overshoot, the worse the tuning and loop performance is. A high overshoot control is called *aggressive*.

$$\kappa = 100 \cdot \sup_{t \in (0,\infty)} \left\{ \frac{y(t)}{\overline{y}} \right\} = 100 \cdot \frac{e_1}{\overline{y}} \; [\%] \tag{2.9}$$

- **Settling time** t_s is the second common loop quality measure. It is often taken into consideration together with the overshoot. It informs how fast the system settles down within some band around the steady state value $\pm\varepsilon$. The best case is to minimize the settling time as much as possible, to achieve the fastest and more accurate setpoint changes tracking or disturbance rejection by the control system. On the other hand high values reflect very long response and setpoint reaching. Such a control is called *sluggish*. Minimization of settling time is often considered as the control loop tuning requirement.
  Actually, the meaning of settling time is opposite to the overshoot. Unfortunately, it is not possible to achieve extremely good overshoot and settling time simultaneously. Higher overshoot means lower settling time and vice-versa. Tuning is **the art of compromise** and a kind of multi-criteria optimization as far as the tuning process is considered.
- **Rise time** t_r is a very simplified measure, which informs the engineer how fast the system reaches setpoint after the change in loop input. The usage of the rise time is very limited as it does not incorporate any information about possible oscillations. Its application is only considered in fusion with other measures, like overshoot.
- **Peak time** t_p informs how fast the step response reaches its maximum value (peak). This index may be used to inform about the aggressiveness of control although its use is very limited. One has to remember that in case of no overshoot response $t_p \to \infty$. If selected, it should be used in specific cases and be accompanied by other measures.
- **Decay Ratio** DR has been discussed thoroughly by Seborg et al. [8] and proposed to be used together with overshoot and settling time. It is defined in a similar way to the overshoot (2.10), being defined as the ratio of successive peaks and troughs above (or below) the steady state value. It describes the factor by which oscillation is reduced during one complete cycle.

$$DR = 100 \cdot \frac{e_3}{e_1} \; [\%] \tag{2.10}$$

  The desired value of the decay ratio is $DR = 25\%$, which seems to be a good compromise between short rise time t_r and the short convergence (described by the settling time t_s) towards the desired value [6]. It has to be noted that similarly to the peak time Decay Ratio loses its informativeness, once the loop does not feature the oscillations.
- Relative **undershoot** r_{us} in the step response reflects real unstable zero (RHP) (2.11). The value of the undershoot increases as zero approaches origin. We have to know that once we witness nonminimumphase zero the fast settling time t_s requirement is incompatible with small undershoot r_{us}.

$$r_{us} = -100 \cdot \inf_{t\in(0,\infty)} \left\{ \frac{y(t)}{\overline{y}} \right\} = 100 \cdot \frac{e_0}{\overline{y}} \; [\%] \tag{2.11}$$

- **Peak value** measure is the same as the overshoot. It simply reflects the maximum value achieved by the loop step response (2.12).

$$PV = \overline{y} + e_1 \tag{2.12}$$

It should be noted that in an exact case the step response does not have all the features enlisted above. Minimum-phase plants do no exhibit undershoot. Processes without delay do not have delayed response. Many systems are aperiodic, i.e. their natural oscillations are impossible because of excessive dissipation of energy. In such case the majority of the above indicators does not apply and the settling time becomes the major step response performance index.

2.3 Loop Disturbance Step Response Measures

Typical oscillatory disturbance step response is sketched in Fig. 2.5. Researchers consider the following measures that can be derived from the disturbance step response:

- **Idle Index** II has been proposed by Hägglund [3]. The index takes into consideration the relationship between gradients of the controlled variable and the manipulated variable. We denote t_{pos} as the total time when both CV and MV signals have the same sign of gradient and t_{neg}, once they have opposite gradients. The value of the index is calculated according to the formula (2.13).

$$II = \frac{t_{pos} - t_{neg}}{t_{pos} + t_{neg}} \tag{2.13}$$

  The author proposes that large values of the index, i.e. $II > 0.4$ indicate sluggish controller tuning. Mid values $-0.4 < RI < 0.4$ should reflect well tuned loop. Larger negative values are unclear as they can originate from both aggressive or a well tuned loop. It should be noted that the method is sensitive to noises, as it requires accurate signal gradients estimations for index evaluation.
- **Area Index** AI of the disturbance step response reflects the way how the process output signal decays after the occurrence of the disturbance. According to the notation sketched in Fig. 2.5, the index is defined as the ratio between the area below the first peak A_0 and the summed areas of the following peaks (2.14).

$$AI = \frac{A_0}{A_0 + \sum_{i=1}^{\infty}(A_i^n + A_i^p)} \tag{2.14}$$

The index uses areas so the noises do not affect its calculation significantly. The author has proposed that the range of $0.3 < AI < 0.7$ indicates well-tuned loops, with lower values reflecting aggressive performance and higher ones sluggish ones.

- **Output Index** OI of the disturbance step response has been proposed by Visioli [12]. It is defined to reflect the asymmetry of the disturbance characteristics oscillation around zero (2.15). This measure takes into consideration oscillations and as such it applies to aggressively tuned controllers. However, it is based on areas and thus it is relatively robust compared to the noise-to-signal ratio.

$$OI = \frac{\sum_{i=1}^{\infty} A_i^n}{\sum_{i=1}^{\infty}(A_i^n + A_i^p)} \tag{2.15}$$

  The author has proposed that the value of the index $OI < 0.35$ applies to the loops with the rise time relatively too high compared to its optimal value. It has been suggested not to use OI measure separately, but in conjunction with the other ones.

- **R-Index** RI of the disturbance step response, according to its authors [7], depends on the shape of the first peak. It is intended that the ratio between its rising and falling parts (2.16) reflects the controller tuning.

$$RI = \frac{A_0^r}{A_0^d} \tag{2.16}$$

  The validation of the index performance shows that the values $RI \approx 0.7 \div 0.85$ indicate sluggish loop tuning, while small values for $RI < 0,4$ show aggressive tuning. Mid values of $0.5 < RI < 0.6$ reflect well tuned loop. As the index uses areas and the noises do not affect its calculation significantly.

Additionally, there are several works in the literature analyzing the relationships between this measures and loop tuning indications [9, 11, 12]. They also suggest fusion of the measures and the use of specific hybrid approaches consisting of the analysis with different indexes.

2.4 Loop Impulse Response Measures

Impulse response has been proposed by Goradia et al. [2] as the characteristics to evaluate control performance index. The method uses nine various curves and a pattern recognition method to classify which of them is better fitted by the real impulse response. The response is approximated using AR(20) regressive model. Thus the normal operation data can be used and the response is approximated one. The visual matching of the coefficients has been proposed originally. However, this method has been improved with the use of neural networks [5].

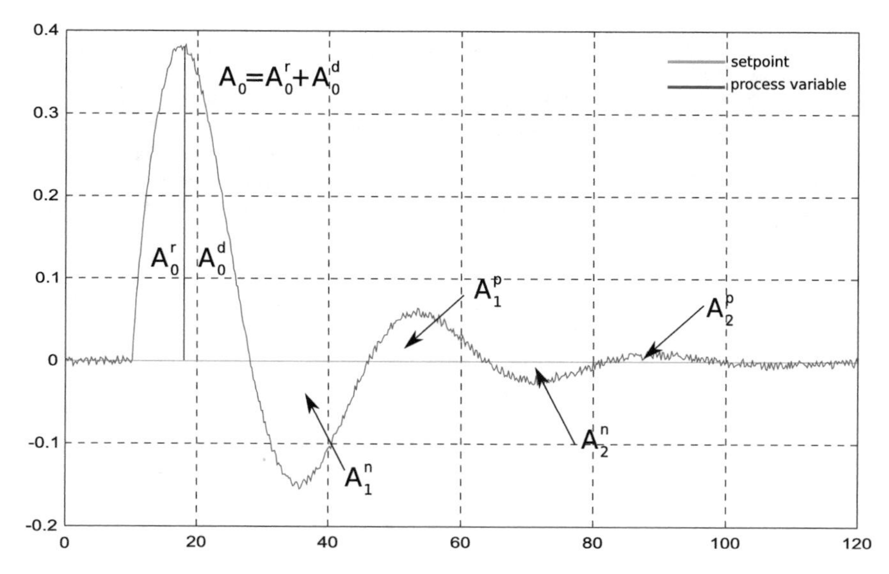

Fig. 2.5 Typical control loop disturbance step response with its parameters

The method takes into account active human decision impact factor. It is somehow a novelty as previously all approaches have tried to be autonomous. The indication that human impact is important forms a valuable contribution of this approach as it will be further shown.

References

1. Björklund, S.: A survey and comparison of time-delay estimation methods in linear systems. Technical Report Linköping Studies in Science and Technology, Thesis No. 1061, Linköping Universitet, Department of Electrical Engineering, Division of Automatic Control, Sweden (2003)
2. Goradia, D.B., Lakshminarayanan, S., Rangaiah, G.P.: Attainment of PI achievable performance for linear SISO processes with deadtime by iterative tuning. Can. J. Chem. Eng. **83**(4), 723–736 (2005)
3. Hägglund, T.: Automatic detection of sluggish control loops. Control Eng. Prac. **7**(12), 1505–1511 (1999)
4. Isermann, R., Münchhof, M.: Identification of Dynamic Systems: An Introduction with Applications. Springer Publishing Company, Incorporated (2014)
5. Jelali, M.: Control system performance monitoring assessment, diagnosis and improvement of control loop performance in industrial automation. Ph.D. thesis, Habilitationsschrift of the Universität Duisburg-Essen (2010)
6. Liptak, B.G.: Instrument Engineers' Handbook, Fourth Edition, Volume Two: Process Control and Optimization. CRC Press, Taylor & Francis Group, Boca Raton, FL (2005)
7. Salsbury, T.I.: A practical method for assessing the performance of control loops subject to random load changes. J. Process Control **15**(4), 393–405 (2005)
8. Seborg, D., Edgar, T.F., Mellichamp, D.: Process Dynamics & Control. Wiley, New York (2006)

9. Spinner, T.: Performance assessment of multivariate control systems. Ph.D. thesis, Texas Tech University (2014)
10. Swanda, A.P., Seborg, D.E.: Controller performance assessment based on setpoint response data. In: Proceedings of the 1999 American Control Conference, vol. 6, pp. 3863–3867 (1999)
11. Veronesi, M., Visioli, A.: A technique for abrupt load disturbance detection in process control systems. IFAC Proc. **41**(2), 14900–14905 (2008). 17th IFAC World Congress
12. Visioli, A.: Method for proportional-integral controller tuning assessment. Ind. Eng. Chem. Res. **45**(8), 2741–2747 (2006)

Chapter 3
Basic Data-Based Measures

God always takes the simplest way.

– Albert Einstein

The measures using loop step/impulse response have been considered in the previous paragraph. The site experiment is required practically to obtain proper characteristics. The analytical derivation of the response is seldom. Analysis of the methods shows that even in such a case the loop should be excited within the wide range of frequencies. This is hardly unachievable in normal close loop operation. Concerns about specific identification experiments remain the same as the ones enlisted for the step test.

Above discussion points out that the industrial CPA method should depend on the normal operation loop data. Historically, the methods using integral indexes were developed and applied. Due to their flexibility, noise ratio robustness, informativeness, easiness and common understanding they are widely used. Their comprehensive discussion is presented in the sections below.

The layout of this section resembles a simple control loop assessment procedure (see Fig. 3.1) performed in the time domain. It assumes no a priori knowledge about the case. The loop review and discussion about the selection of the variable for the analysis starts the procedure (Sect. 3.1). Next, the practical aspects on the data acquirement and pre-processing are presented (Sect. 3.2), being followed by the review of data time trends (Sect. 3.3) and evaluation of the time indexes (Sect. 3.4). The method for signal analysis concludes with a discussion about the use of static characteristics through the X-Y plots (Sect. 3.5).

P. D. Domański, *Control Performance Assessment: Theoretical Analyses and Industrial Practice*, Studies in Systems, Decision and Control 245,
https://doi.org/10.1007/978-3-030-23593-2_3

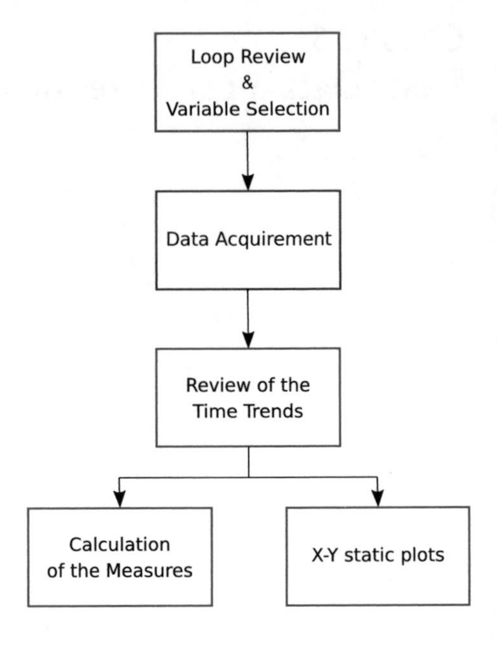

Fig. 3.1 Procedure for the CPA using simple data analysis

3.1 Control Loop Review and Variable Selection

The simple control loop performance assessment should start from the review of the considered loop, i.e. its structure and arrangements. One has to remember that the loops configurations may significantly differ. Each control system, PLC or DCS, uses different methodology to construct the loop. The system may use different and specific PID control configurations. The reviewer has to verify the structure of the loop and the algorithm.

The purpose of the control has to be learned first. We need to find out what process is controlled and what the main process variables are. Especially, the ultimate objective of the control, like for instance tracking versus disturbance rejection needs to be discovered. We need to identify the main loop actors, specifically to recognize potential measured and unmeasured disturbances. The conversation and the interview with plant personnel (operators and control engineers) is inevitable at this stage.

The type of the loop arrangement has to be assessed further. We need to check, whether it is a raw SISO loop, feedforward, cascaded control or three-element control. We need to check whether any types of the feedforward modules are used, how they are embedded into the loop template and how they are constructed. The feedforward block may be utilized for disturbance decoupling or for filtering and we have to verify these options. Furthermore, we need to assess whether any additional feedback blocks are used, as for instance compensator, filter, Smith predictor or nonlinear function block (e.q. actuator linearization). We also have to verify whether any switching, multiplexing, overlapping control or gain scheduling is employed. It is also important to verify the source and the character of the setpoint signal. The eventual human interventions frequency and their character need to be checked.

Furthermore, the type of the control algorithm which is applied within the loop has to be identified. PID loops form the majority, however the other algorithm might be used as well. There exist plenty of versions and modifications to the base PID form [1]. We may find an ideal algorithm, parallel (ISA standard) or series (interactive) form, PID control with derivative filter in parallel and series version, algorithm with the differentiation of the (filtered) output, control with setpoint weighting, non-interactive form of the algorithm, two degree of freedom algorithm, etc. It should be checked whether any anti-windup mechanism is employed eventually.

We need to verify the type of the controller action, i.e. direct or indirect. There might be as well some system dependent biases, by-passes, tracking mechanisms, M/A stations, interlocking. At this stage some process orientation, control infrastructure knowledge and the ability to read SAMA diagrams is very helpful. This background interview should never be omitted, because as in medicine it speeds up the diagnosis, clearing the path for real CPA activities.

Finally, we have to select a proper control loop variable for further evaluation of the measures. There are three main loop variables in general: setpoint $y_0(t)$, MV (controller output) $m(t)$ and CV (process output) $y(t)$. We often take into account the control error signal $\varepsilon(t) = y_0(t) - y(t)$ as well, however it does not have to be collected directly, as it may be simply calculated as the difference between the setpoint and process output. The control MAN/AUTO status should be considered as the auxiliary binary signal.

The process output variable is often considered as the main signal being further evaluated in order to calculate the measures. Virtually all the considered indexes and methods can use it. It is the loop output signal and as such it keeps information about the loop properties. It is indisputable in case of the constant setpoint control. It can be directly used in such a situation. However, once the setpoint signal changes the process output tracks it. Thus the signal itself consists of two elements: the background trend in line with setpoint fluctuations and the loop dynamical contribution depending on loop performance. The CV time series requires detrending first to discover loop properties in such a case.

There are many detrending algorithms, however their performance depends on a priori assumptions about the trend character. They are never perfect and ideal in cleaning of the time series. The remnants of the detrending may remain and unfortunately they will affect further analysis. Additionally, the detrending quality is not objective as it depends on human choices and/or loop environment, like disturbances. Detrending requires extreme caution from engineers. Once we consider the MV variable as the assessment signal similar discussion applies and the same considerations need to be taken into account.

Luckily, the control error signal is not affected by the setpoint trend and it keeps direct information about loop performance. It should be considered as the main signal for the loop investigation from that perspective, however in many situations it is worth to increase assessment ability adding information from the CV or MV variable.

The auxiliary loop signal of the MAN/AUTO status is important as well. First of all it enables to select periods of automatic operation. Industry likes to use it as the separate measure in form of the manual MAN mode share in the total time of operation as everybody wishes to use all the control loops in the automatic mode all the time. It is a practical measure of the loop tuning goodness and the operators' confidence in the daily operation basis.

3.2 Data Collection

Data collection is the next step of the process as the acquired datasets are further used for the evaluation of the CPA indexes. The objective is to collect true time series data of the selected variables. Although it looks simple, the industrial reality poses some issues. We may distinguish two groups of problems, which we must be aware of, i.e. process technological constraints and I&C infrastructure limitations.

Industrial process often behaves in a non-stationary way being affected by several impacts that influence normal operation. Collecting data for the CPA we should be aware of it. We should select appropriate operation periods for further analysis. The process should operate in normal regime without any emergency situations. It should not be affected by human interventions.

These disturbing impacts may have two natures: external or internal. As the external disturbances we may consider changes in technology, process reconfiguration, equipment switches, input product quality variations or plant operation regime exchanges conducted by the plant personnel. Weather conditions may also affect the process. Impacts of the internal origin are due to the technological or equipment failures, human interventions, operators biases, etc. All possible effects should be known a priori to select proper conditions for the definitions of the plant operation control assessment periods.

During the data collection process we also need to take into account the possible limitations imposed by historian software constraints. Databases often have limited storage and use custom specific algorithms for a data compression. Extraction of such compressed data may change its character, as it differs from the original raw values. We may witness holes in data files, step-wise profiles or linear interpolations. Thus, the historian setup has to be cross-checked and reconfigured if needed.

3.3 Observation of the Time Trends

Once data is selected and collected, we can start doing the analysis. It is a good practice to start with the visual trends inspection. Although the ultimate goal is to have autonomous assessment, visual data review of time trends is always helpful. We simply get the first impression on data quality and loop performance. Additionally, we have to take into account the fact that the collected data still does not have to be free of errors. There always may appear some lacking data or strange values, which are not generated through the process. Often these holes are filled with some texts values as for instance "`BAD DATA`" or "`N/D`". Thus the text dataset should be reviewed for such situations.

Even a small hole in data consisting of a few lacking samples can affect the whole process of the analysis. We need to be prepared to handle such instances. Sometimes we may just delete the records including such lacking data samples (static analysis). In case of the approaches based on models the better practice is just to interpolate lacking data, especially when the hole is small. The lack of 3 samples can be freely interpolated linearly, if the dataset includes 10000 samples. We should not deteriorate the process. On the other hand, it might be questionable if the amount of the lacking data is relatively visible within the entire time series.

More challenging situations may appear if the whole dataset is complete, however there appear samples with wrong values. Such anomalous values are called *artifacts* or *outliers*. They are not created through the process or the loop, but they can appear due to measurement system failures, communication losses or just I&C system (historian or interface) bad events. Quite often they are not even signaled by the system with bad quality tag. Their detection and isolation may not be straightforward [8].

Some measuring devices require calibration periods and during such moments the signal is lost or receives unexpectedly constant (last) value. Any cleaning or air blowing may also affect the real value. Such issues may be handled with the help of the site personnel which mostly has knowledge about such situations or in more serious cases the sensor validation and recovery solutions might be applicable [3, 4].

The other *artifacts* just accidentally appear and no clear reason might be determined. Sometimes this strange and unexpected peaks show up. The first choice method to deal with this is filtering. Unfortunately, filtering adds additional dynamics into data that may affect further assessment. The simplest way would be just to delete such single value peaks. Unfortunately, such an operation is subjective. The best way is to always confront it with plant personnel. Typical loop data are sketched in Fig. 3.2.

It is worth to apply sensor validation and recovery modules for large scale applications or autonomous systems. Despite the fact that there exist tools dealing with such situations, the task is always challenging and an assessment engineer should be aware of them and prepared.

3.4 Integral Measures

Once we come into possession of clear and reliable loop data we may calculate integral indexes. In literature we may find several different indexes, as for instance: integral square error (ISE), integral of the absolute error (IAE), integral of square time derivative of the control input (ISTC), integral of the absolute error multiplied by time (ITAE), integral of multiplied absolute error (ITNAE), quadratic error (QE) and total square variation (TSV). All the integral indexes have a lower bound of zero

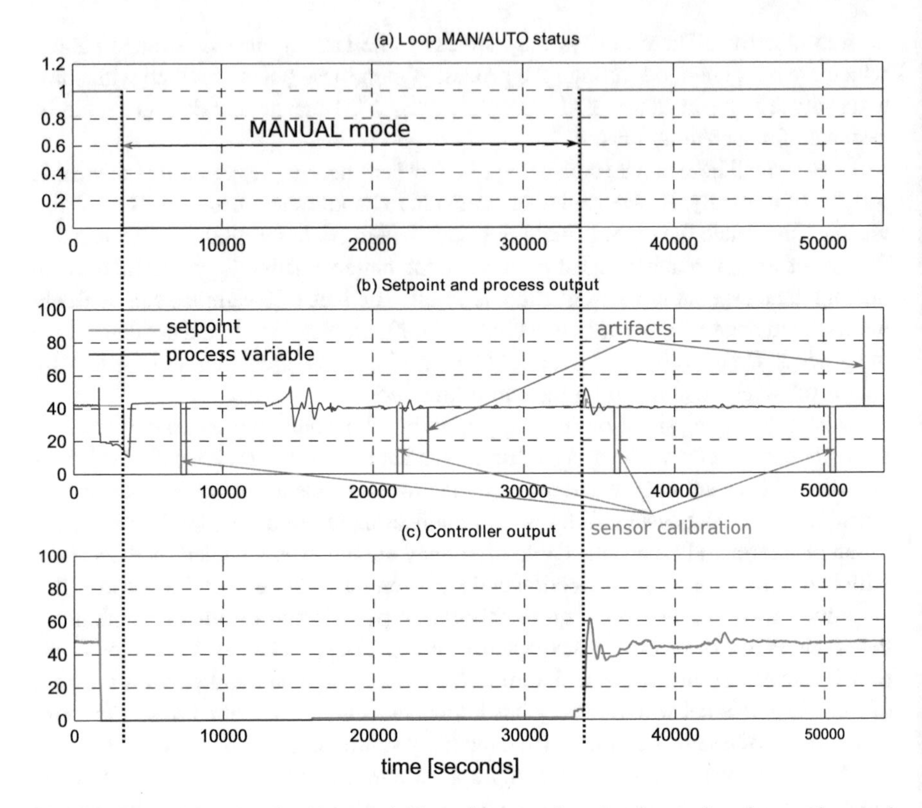

Fig. 3.2 Typical time trends of loop variables, with manual mode of operation, data artifacts and sensor calibrations

(ideal control showing exact tracking and no variability), but they also have no upper bounds. All specified measures are presented and discussed below.

The indexes are presented in the basic forms and in the literature one may find their variations, as for example time-weighted [10] modifications, normalized measures [6] or fractional order [15] versions. It should be noted that if the system has steady state error ε_∞, the $\varepsilon(t)$ has to be replaced by the $\varepsilon(t) - \varepsilon_\infty$.

3.4.1 ISE—Integral Square Error

Integral Square Error is the integral value of the squared control errors over specified period of time $t \in (t_1, t_2)$ (3.1). As we almost always consider digital data is calculated in the form of the summation of the N points collected with some sampling period (time decrement) where Δt. For easier interpretation of the results the index is evaluated as the mean value and exists under the name of the Mean Square Error (MSE) (3.2).

$$ISE = \int_{t_1}^{t_2} \varepsilon(t)^2 dt = \int_{t_1}^{t_2} \left[y_0(t) - y(t)\right]^2 dt \tag{3.1}$$

$$MSE = \frac{1}{N} \sum_{k=1}^{k=N} \varepsilon(k)^2 = \frac{1}{N} \sum_{k=1}^{k=N} \left[y_0(k) - y(k)\right]^2 \tag{3.2}$$

The ISE/MSE index highly penalizes errors with large values in opposition to smaller errors. Large values of the control error usually occur immediately after a disturbance and can be observed as the overshoot, as for instance in case of the step response. Thus this index is mostly used to indicate overshoot and aggressive control. It was shown [12] that tuning minimizing the ISE/MSE punishes large set-point deviations and generates aggressive control. It has been shown that MSE has no relationship to the economic performance of a loop [13], especially in case of disturbances [2], although it might be easier to trace mathematically.

3.4.2 IAE—Integral Absolute Error

Integral Absolute Error is the integral of the control error absolute values over specified period of time $t \in (t_1, t_2)$ (3.3). In the discrete time version it is frequently used as the mean sum of the absolute errors as in (3.4).

$$IAE = \int_{t_1}^{t_2} |\varepsilon(t)|\, dt = \int_{t_1}^{t_2} |y_0(t) - y(t)|\, dt \tag{3.3}$$

$$IAE = \frac{1}{N} \sum_{k=1}^{k=N} |\varepsilon(k)| = \frac{1}{N} \sum_{k=1}^{k=N} |y_0(k) - y(k)| \tag{3.4}$$

The IAE index does not distinguish between positive and negative contributions to the error. The index is less conservative and it is often used for an on-line controller tuning. IAE has the closest relationship to economic considerations [13]. It penalizes continued cycling. This index is appropriate for non-monotonic step responses and all kind of normal operation data.

3.4.3 ITAE—Integral of the Time Weighted Absolute Error

Integral Absolute Error is the integral of the control error absolute values multiplied by time over specified period of time $t \in (t_1, t_2)$ (3.5). As in the previous cases we use in practice discrete time version (3.4) evaluated with summation of N data samples.

$$ITAE = \int_{t_1}^{t_2} t \cdot |\varepsilon(t)|\, dt = \int_{t_1}^{t_2} t \cdot |y_0(t) - y(t)|\, dt \qquad (3.5)$$

$$ITAE = \sum_{k=1}^{k=N} k \cdot \Delta t \cdot |\varepsilon(k)| = \sum_{k=1}^{k=N} k \cdot \Delta t \cdot |y_0(k) - y(k)| \qquad (3.6)$$

ITAE is a very conservative performance index [7, 11]. Embedded multiplication by time strengthen the errors that persist a long time. It strongly weights larger errors that occur late in time, while less emphasis is placed on the initial control errors. A large value reflects bigger loop deviations. This index is often used in case of small datasets or data obtained using the step response. The ITAE index trades-off between the error magnitude and its settling time, i.e larger errors along part of the response become *good* if the time of response drops by an inverse proportion.

3.4.4 ITNAE—Integral of Multiplied Absolute Error

Integral Absolute Error is the integral of the control error absolute values multiplied by time to some power n over specified period of time $t \in (t_1, t_2)$ (3.7). As in the previous cases we use in practice discrete time version (3.8) evaluated with summation of N data samples.

$$ITNAE = \int_{t_1}^{t_2} t^n \cdot |\varepsilon(t)|\, dt = \int_{t_1}^{t_2} t^n \cdot |y_0(t) - y(t)|\, dt \qquad (3.7)$$

$$ITNAE = \sum_{k=1}^{k=N} (k \cdot \Delta t)^n \cdot |\varepsilon(k)| = \sum_{k=1}^{k=N} (k \cdot \Delta t)^n \cdot |y_0(k) - y(k)| \qquad (3.8)$$

ITNAE is the most conservative of the considered error criteria [6, 7, 11], because the multiplication by time (strongly powered by n) gives greater weighting to the errors that persists over a longer period of time. Its properties are quite similar to the ITAE. This index is often used in case of small data sets or data obtained with the step response.

3.4.5 QE—Quadratic Error

Quadratic Error is the mean of the control error absolute values summed with the weighted quadratic cost of the controller output over specified period of time powered to the power $n\, t \in (t_1, t_2)$ (3.9). As in the previous cases we use in practice discrete time version (3.8) evaluated with summation of N data samples averaged over the N.

$$QE = \int_{t_1}^{t_2} \left(\varepsilon(t)^2 + \lambda \cdot m(t)^2\right) dt = \int_{t_1}^{t_2} \left\{\left[y_0(t) - y(t)\right]^2 + \lambda \cdot m(t)^2\right\} dt \tag{3.9}$$

$$QE = \frac{1}{N}\sum_{k=1}^{k=N} \left(\varepsilon(k)^2 + \lambda \cdot m(k)^2\right) = \frac{1}{N}\sum_{k=1}^{k=N} \left\{\left[y_0(k) - y(k)\right]^2 + \lambda \cdot m(k)^2\right\} \tag{3.10}$$

QE is the standard criterion used for controller design with λ being a weighting factor. It is due to the contribution of both factors describing the control performance. i.e. control error (quality of control) and controller output (energy cost effect put on actuators). One has to be aware that the selection of the weighting factor δ is subjective [16] and in general it is case-dependent.

3.4.6 ISTC—Integral of Square Time Derivative of the Control Input

Integral of Square Time derivative of the Control input index is the variation of the ISE (MSE) index, however the controller output derivative is squared (3.11). Practical discrete time version is used to facilitate calculations (3.12). In some publications this index is denoted as Total Variation (TV) [14, 18] or Total Square Variation (TSV) [5, 17].

$$ISTC = TSV = \int_{t_1}^{t_2} \left(\frac{dm(t)}{dt}\right)^2 dt \tag{3.11}$$

$$ISTC = TSV = \frac{1}{\Delta t}\sum_{k=1}^{k=N-1} \left(m(k+1) - m(k)\right)^2 \tag{3.12}$$

Oscillations of the control input $m(t)$ can cause fatigue or damage of actuator equipment. The ISTC takes into consideration control input change. It is a good measure of the *smoothness* of a signal. A larger index value indicates the oscillations in the manipulated variable, which can result in unnecessary wear of the actuator or lost energy, while a smaller value reflects smooth operation and better loop performance.

This index is frequently used together with the ISE index in a similar way to the QE. Then, it is named the ISE-TSV measure $J_{ISE-TSV}$ (3.13). By varying weighting factor λ we obtain trade-off between the optimal ISE and TSV, however it must be noted that its selection is subjective [16].

$$J_{ISE-TSV} = (1-\lambda) \cdot ISE + \lambda \cdot TSV \tag{3.13}$$

3.5 Quality Evaluation Using X-Y Plots

Using of the X-Y plots during the Control Performance Assessment is a good practice, although it is seldom addressed in the research and publications. It is probably due to marginal scientific content. On the other hand, these plots can be frequently met in the industrial control assessment studies and reports.

Apart from the time trend plots, the X-Y or two dimensional (2D) diagrams are very informative. Human perception is perfect in recognition of patterns and classifications using drawings, unlike an automatic or autonomous systems. This fact has been already noticed in the area as Mitchell et al. [9] used a modified English language-idiom "*a picture is worth a thousand words*" in the title of their publication. Discussion about the subject of the X-Y plots should cover two aspects:

1. what signals should be subject to the plotting and analysis,
2. and how these plots should be prepared, what and how should be highlighted in drawings.

The X-Y plots are a static drawing and as such they should be used to analyze and reveal static properties of the loop. The most straightforward type of a drawing would be the characteristics of the process output as the function of the controller output, i.e. $CV = f(MV)$. This characteristic shows static properties of the controlled process. The best exemplification of such results can be a situation where some flow is controlled by the valve. In such a situation we obtain the valve K_v curve in form of the $flow = f(valve\_position)$.

Exemplary sketches of the X-Y plots are presented in Fig. 3.3. There are four different scattered point plots presenting process variable as the function of the controller output. A second order polynomial interpolation curve added to catch eventual background relationship that might exist in data.

It is clearly noticed that it is seldom to get a very clear and exact curve fitting. The data often consists of point clouds or sets of point clusters. Review of such simple plots enables us to make an initial diagnosis of the loop. First of all we can try to detect nonlinearity, as it is clearly recognizable in all four curves.

The characteristics visible in subfigure (a) show quite well a loop with insignificant nonlinearity and the points being relatively not scattered. The polynomial interpolation of the second order forms a good representation of the embedded relationship. Example (b) is quite similar to the previous one. However, the low values of the MV variable there are visible oscillations. They are visibly reinforced against other regions of the domain. They might originate with the stiction effect or other deadband type nonlinearity connected with unfitted controller setup.

Other two subfigures show evident data clustering in the form of horizontal brands (c) or slanting ones in case (d). This effect might be explained by the fact that the loop properties also depend on some other variable not taken into consideration. The varying pressure might be the reason in case of the valve flow characteristics, as the flow through the valve depends on both valve opening and the flowing fluid pressure. In such cases three dimensional (3D) plots should be considered. These X-Y-Z plots

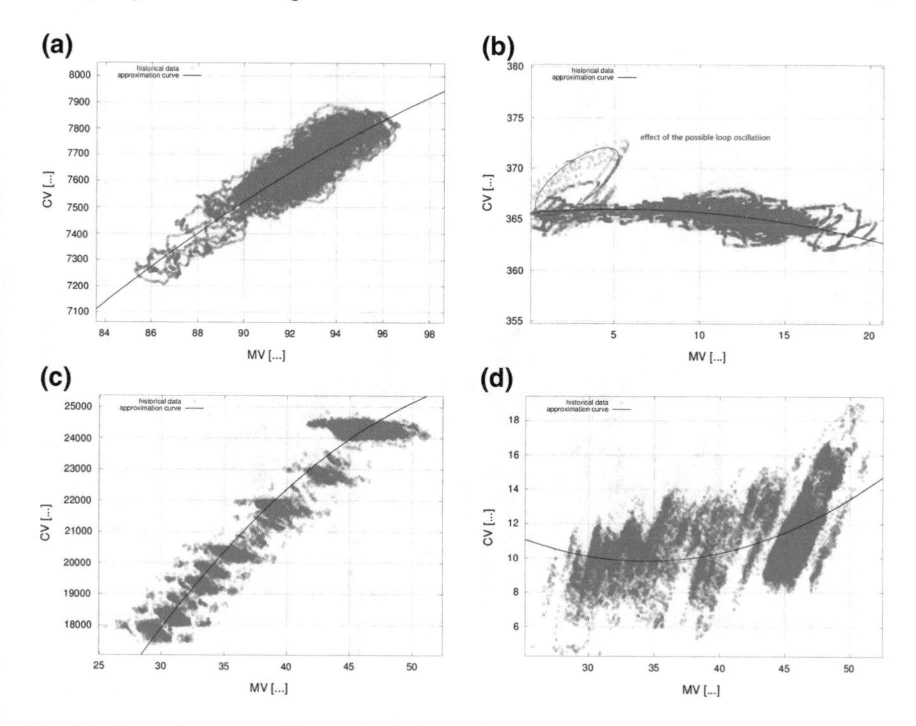

Fig. 3.3 Examples of the X-Y plots for the industrial loop data

might be presented in two formats, i.e. in form of the classical 3D plot or the surface contour map.

The K_v characteristics for some industrial valve are presented below. Classical three-dimensional plot is sketched in Fig. 3.4 and surface contour map is presented in Fig. 3.5. In both cases the plots present real historical data and some approximated surface. These data represent a strongly nonlinear valve. Such visualization tools enable rapid visual interpretation of the plots.

In case of actuators, especially valves, there might be possible even more precise insight into the equipment. In modern plants the actuators often have internal feedback signal about its position. Thus we obtain two-staged flow of information (see Fig. 3.6). The *position demand* signal $m_{dm}(t)$ is sent from the control system (actually the controller output). Next, the demand is being realized by the driver and positioner unit. The currently achieved real position of the positioner (valve) is sent back to the control system as the *position feedback* $m_{feed}(t)$. Next, the process reacts and we measure the *controlled variable* $y(t)$.

This internal position feedback signal enables us to decompose the X-Y static loop analysis into two parts. First we may assess the actuator drive performance through the characteristic $m_{feed}(t) = f(m_{dm}(t))$. In an ideal case we should always obtain the exact $y = x$ curve. If not, it means that we have equipment driver problems.

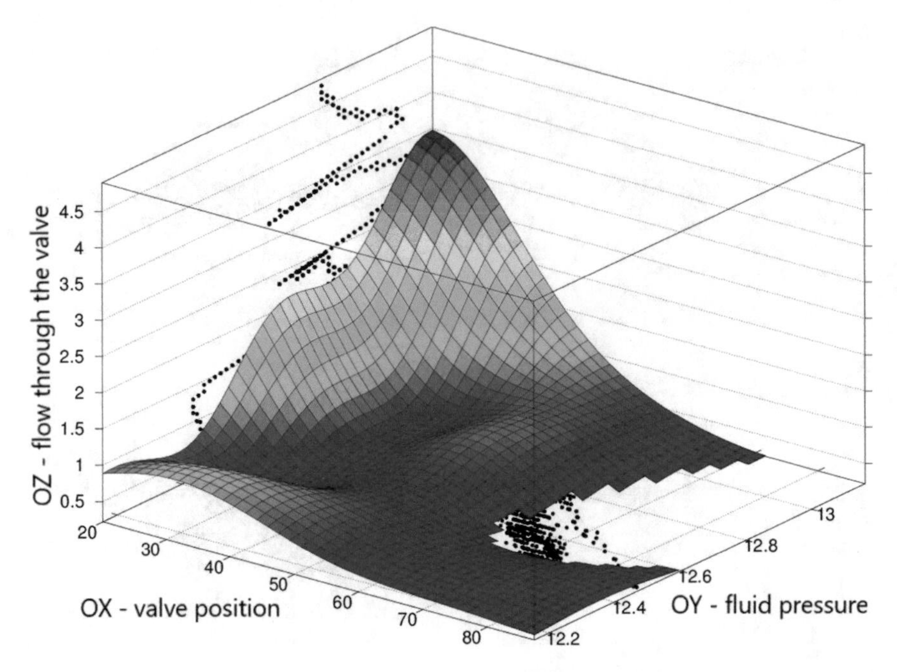

Fig. 3.4 Industrial example of the 3D characteristics of the valve with the scattered data (black dots) and approximated surface

The next characteristic $y(t) = f\left(m_{feed}(t)\right)$ reflects the real MC-CV relationship. The industrial example of such an approach is sketched in Fig. 3.7.

In some situations, especially for actuators driven by electric drivers, there might be more degrees of freedom as the actuator characteristics may consist of two design curves: one for the current realization and another for the drive positioner Fig. 3.8.

As one can see, the usage of the static X-Y characteristics brings important contribution to the loop analysis, especially associated with the actuator performance. One subject is to plot the static scattered points plots. Another issue is their interpretation. The valuable observations and conclusions may be derived once we are able to extract background relationships from the scattered points. The data interpolation methods are providing useful tools for this.

The subject of data interpolation both in 2D and 3D is another broad area of research. There are many various approaches and solutions solving different issues of the robust regression. However, the methods are not the goal from the Control Performance Assessment perspective. We require robust and relatively easy methodology in a CPA task.

The first choice is to use polynomial approximation, mostly using the least squares method. It is a reliable, fast and very popular approach. In fact, in most of the cases approximation with the *2nd* order polynomial (or the *3rd* order at most) is satis-

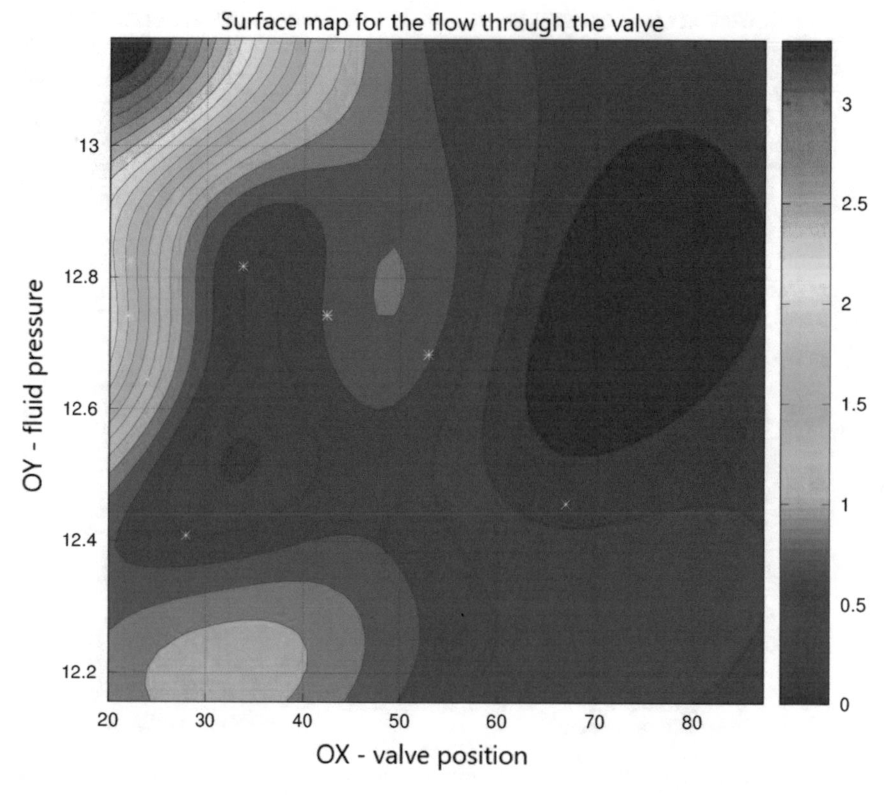

Fig. 3.5 Industrial example of the 3D contour surface plot of the valve with the scattered data (white stars) and approximated surface

Fig. 3.6 Signals associated with the actuator

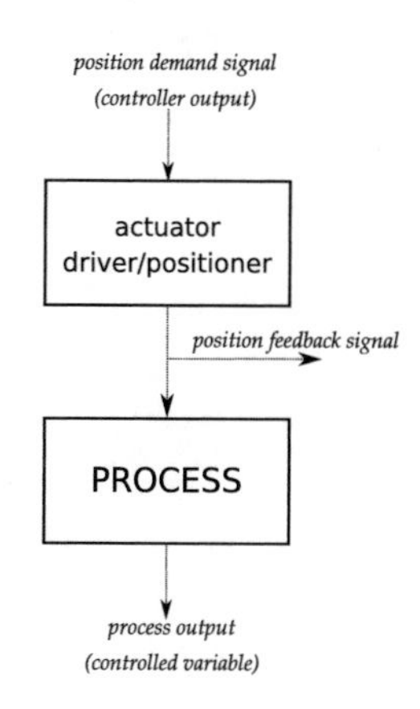

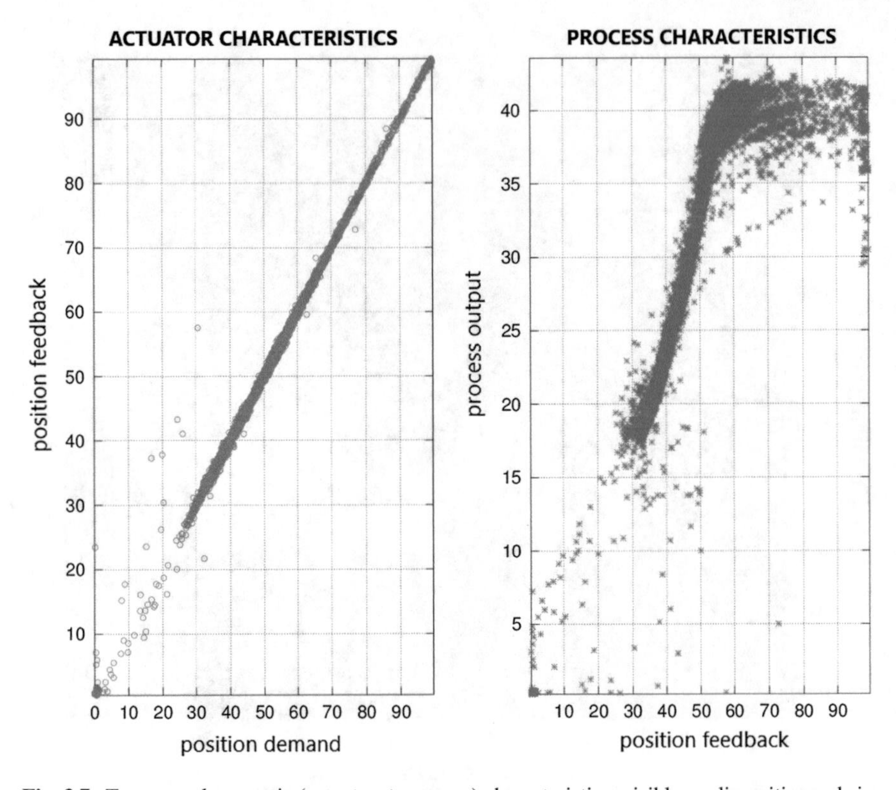

Fig. 3.7 Two-stage loop static (actuator + process) characteristics–visible nonlinearities and significant saturation of the valve

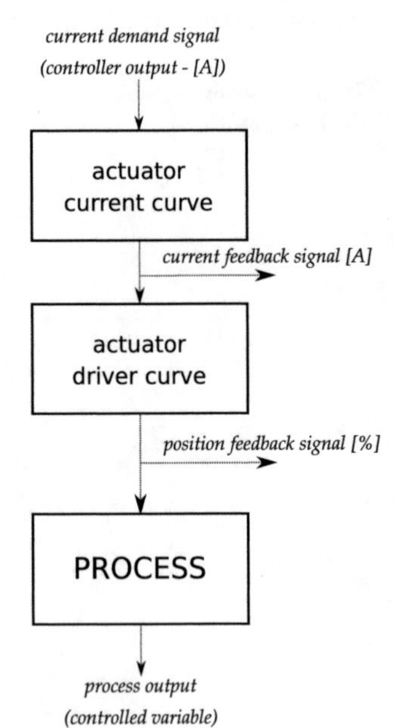

Fig. 3.8 Signals associated with the electrically driven actuator

factory. One may consider any other static interpolation methods as artificial neural networks, kernel regression, orthonormal functions or others in more challenging situations.

References

1. Åström, K.J., Hägglund, T.: PID Controllers: Theory, Design, and Tuning, 2nd edn. International Society for Measurement and Control, Research Triangle Park, N.C, ISA (1995)
2. Domański, P.D.: Statistical measures for proportional-integral-derivative control quality: simulations and industrial data. Proc. Inst. Mech. Eng. Part I J. Syst. Control Eng. **232**(4), 428–441 (2018)
3. Domański, P.D., Plamowski, S., Warchoł, M., Świrski, K.: (2004) Sensor validation and recovery. In: Proceedings of International Conference on Complex Systems Intelligence and Modern Technological Applications CSIMTA 2004, Cherbourg, France, pp. 766–771 (2004)
4. Fortuna, L., Graziani, S., Rizzo, A., Xibilia, M.G.: Soft Sensors for Monitoring and Control of Industrial Processes (Advances in Industrial Control). Springer-Verlag, New York (2006)
5. Gao, X., Yang, F., Shang, C., Huang, D.: A novel data-driven method for simultaneous performance assessment and retuning of PID controllers. Ind. Eng. Chem. Res. **56**(8), 2127–2139 (2017)
6. Jelali, M.: Control Performance Management in Industrial Automation: Assessment. Diagnosis and Improvement of Control Loop Performance. Springer-Verlag, London (2013)
7. Levine, W.S.: The Control Handbook. Jaico Publishing House (1996)
8. Mehrotra, K.G., Mohan, C.K., Huang, H.M.: Anomaly Detection Principles and Algorithms, 1st edn. Terrorism, Security, and Computation. Springer Publishing Company, Inc (2017)
9. Mitchell, W., Shook, D., Shah, S.L.: A picture worth a thousand control loops: an innovative way of visualizing controller performance data. Invited Plenary Presentation, Control Systems (2004)
10. Nishikawa, Y., Sannomiya, N., Ohta, T., Tanaka, H.: A method for auto-tuning of PID control parameters. Automatica **20**(3), 321–332 (1984)
11. Seborg, D., Edgar, T.F., Mellichamp, D.: Process Dynamics & Control. Wiley, New York (2006)
12. Seborg, D.E., Mellichamp, D.A., Edgar, T.F., Doyle, F.J.: Process Dynamics and Control. Wiley (2010)
13. Shinskey, F.G.: Process control: as taught vs as practiced. Ind. Eng. Chem. Res. **41**, 3745–3750 (2002)
14. Skogestad, S.: Simple analytic rules for model reduction and PID controller tuning. J. Process Control **13**(4), 291–309 (2003)
15. Tavazoei, M.S.: Notes on integral performance indices in fractional-order control systems. J. Process Control **20**(3), 285–291 (2010)
16. Unbehauen, H.: Controller design in time domain. In: Unbehauen, H. (ed.) Encyclopedia of Life Support Systems (EOLSS). Eolss Publishers, Oxford (2009)
17. Yu, Z., Wang, J.: Performance assessment of static lead-lag feedforward controllers for disturbance rejection in PID control loops. ISA Trans. **64**, 67–76 (2016)
18. Yu, Z., Wang, J., Huang, B., Li, J., Bi, Z.: Design and performance assessment of setpoint feedforward controllers to break tradeoffs in univariate control loops. In: 19th IFAC World Congress, IFAC Proceedings vol. 47(3), pp. 5740–5745 (2014)

Chapter 4
Statistical Measures

Sanity is not statistical.

–George Orwell

4.1 Time Series Histogram

Histograms are traditionally used by statisticians to discover and present the underlying distribution shape for some selected continuous dataset. Such plot enables the analysis and assessment of the data statistical properties, like for instance the shape of the probabilistic distribution, the type of its approximation with the Probabilistic Density Function (PDF), and characteristics factors (mean, standard deviation, skewness, outliers, etc.).

Construction of the histogram starts from the splitting of the data into the intervals, called bins. After that each bin will include some number of data samples. Basing on such obtained bins we plot the bar chart representing a data histogram. Three exemplary time series analysis plots are presented in the figures below. The upper plot shows the control error time trend and the lower one presents histogram of the considered control error. Additionally, the data mean x_0 and standard deviation $St Dev$ values are shown. A nice bell-shaped histogram is shown in Fig. 4.1. Figure 4.2 shows the shape with significant outliers and fat-tails. An asymmetric and scattered shape is sketched in Fig. 4.3.

It is very important to set proper bins for histogram evaluation. An example is presented in Fig. 4.4. It shows two histograms prepared with the same data as in Fig. 4.3. The left situation presents the bin width too small because, it shows too much individual data and does not reveal the underlying pattern (frequency distribution)

P. D. Domański, *Control Performance Assessment: Theoretical Analyses and Industrial Practice*, Studies in Systems, Decision and Control 245,
https://doi.org/10.1007/978-3-030-23593-2_4

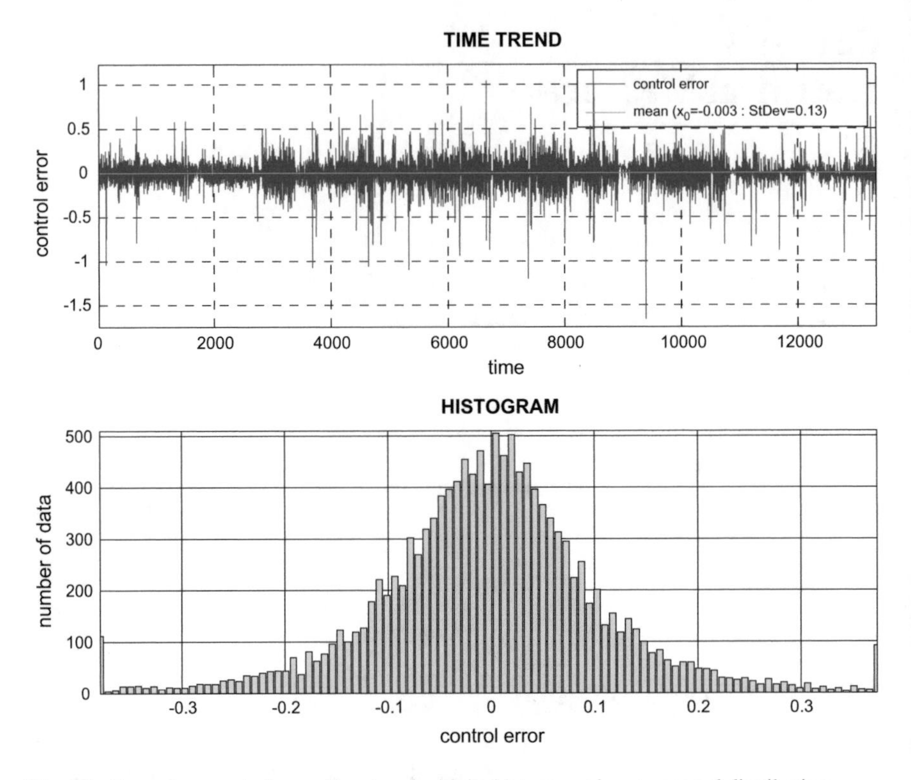

Fig. 4.1 Exemplary control error time trend with its histogram close to normal distribution

of the data. On the other end of the scale is a plot presented on the right, where the bins are too large, and again, we are unable to find the underlying trend in the data.

It is the area of the bar that reflects the frequency of occurrences for each bin in a histogram. The height of the bar does not have to indicate, how many number of data points really fit into each individual bin. The frequency of occurrences within the selected bin is indicated by the product of the height multiplied by the width of the bin. One of the reasons that the height of the bars is often taken seriously into consideration is because many histograms often have equally spaced bars (bins). The height of the bin does reflect the frequency only under these circumstances.

The shape of the histogram is a non parametric model of the time series distribution. Its maximum value indicates the measure of the mostly occurring value. The shape of the decaying slopes and its broadness reflects the variable fluctuations character. There are several industrial studies and a good practice to use and assess histograms itself in the CPA projects and reports.

The further step in the time series probabilistic properties assessment is to evaluate and analyze a PDF distribution relevant to the data. The review of the possible

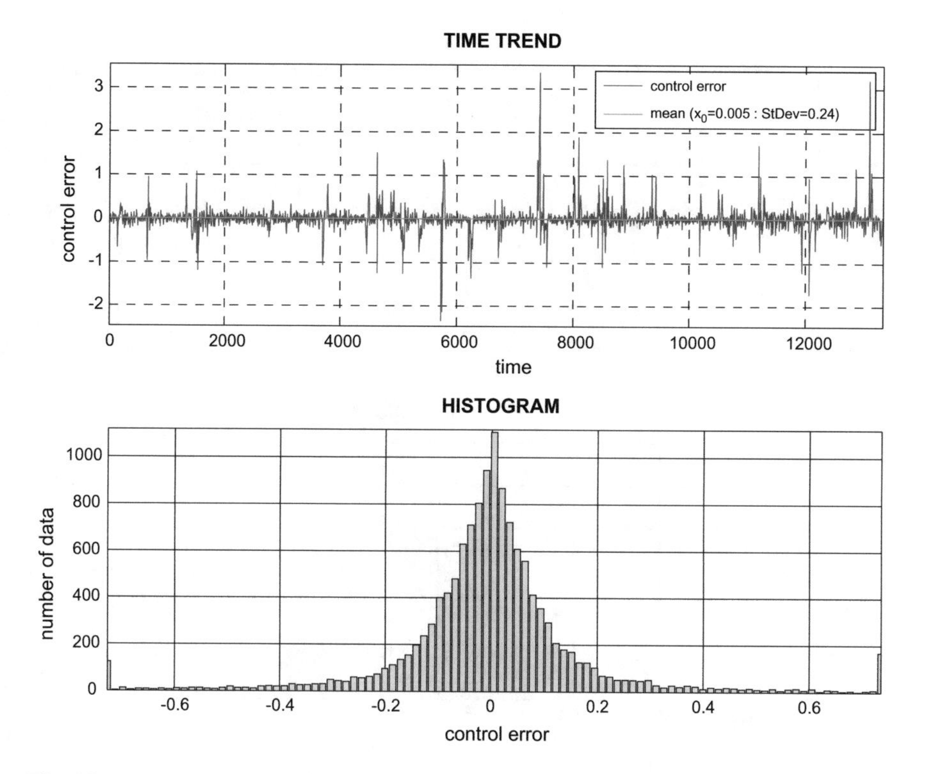

Fig. 4.2 Exemplary control error time trend with its histogram exhibiting fat tails

functions will start with a normal Gaussian distribution (Sect. 4.2) being followed by the fat-tailed functions (Sect. 4.3) and other rarely used ones (Sect. 4.4). The statistical approach concludes in Sect. 4.5 with the description of the robust statistics.

4.2 Gaussian Basic Statistics

Normal Gaussian probabilistic distribution function is described as the function of the variable x and is characterized by two parameters: mean x_o and standard deviation σ (4.1). The function is symmetrical and the mean is responsible for the position and the standard deviation for the broadness.

$$F^{Gauss}(x)_{x_o,\sigma} = \frac{1}{\sqrt{2\pi\sigma^2}} e^{-\frac{(x-x_0)^2}{2\sigma^2}} \tag{4.1}$$

Examples of the PDF shapes for normal distribution are sketched on Fig. 4.5. Both factors, i.e. x_o and σ exist and can be evaluated analytically. The Eqs. (4.2) and (4.3) presents the results in a discrete time case $(x_1, \ldots, x_N)$, where N is number of data points.

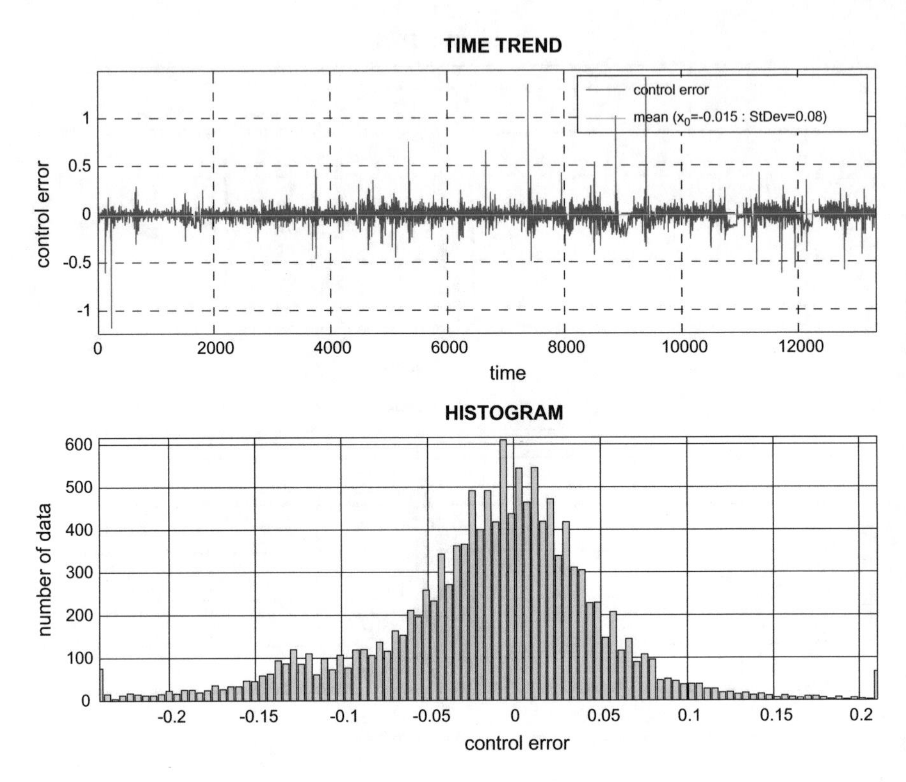

Fig. 4.3 Exemplary control error time trend with asymmetrical and tailed histogram

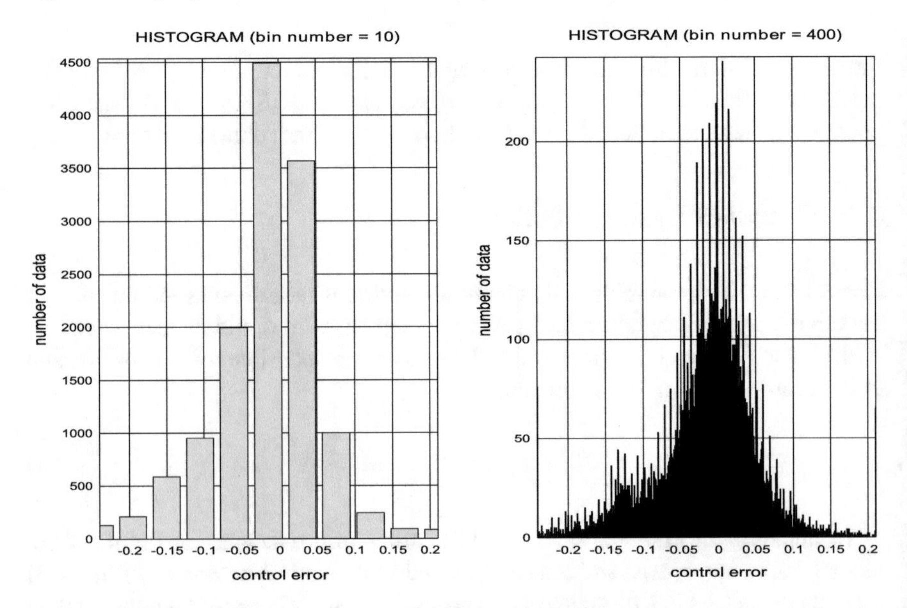

Fig. 4.4 Control error histogram shapes with different bin numbers

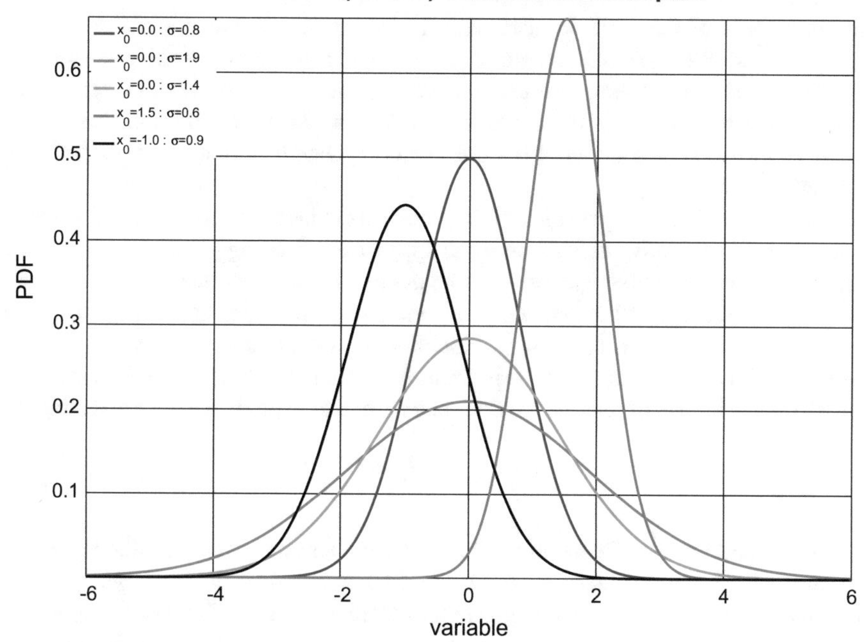

Fig. 4.5 Examples of normal (Gauss) probabilistic distribution functions

$$mean\,(x) = x_o = \frac{1}{N}\sum_{i=1}^{N} x_i \tag{4.2}$$

$$std\,(x) = \sigma = \sqrt{\frac{\sum_{i=1}^{N}\left(x_i - x_0\right)^2}{N-1}} \tag{4.3}$$

Statistical properties of the Gaussian normal distribution are used in the definition of several very popular control performance KPIs. Mean x_o and standard deviation of the signal σ are the most common ones. In some cases the standard deviation is exchanged with the variance $Var(x)$ (4.4). Both indexes may be used alternatively.

$$Var\,(x) = \sigma^2 \tag{4.4}$$

Mean x_0 reflects variable average value, i.e. steady state error in case of control error analysis. Its desired value is zero thus it is relatively easy to measure the distance from its optimal value. The non-zero value may be caused by one of the following reasons: the feature due to the lack of the integration within the feedback loop, nonlinear process behavior or the constraints that do not permit to reach the desired operating point (setpoint).

Standard deviation σ informs about signal variability, as it is directly related to the broadness of the variable histogram. Higher σ means larger variations, while small values reflect fewer fluctuations. As we apply the control quality assessment the variability of the control error is considered. Thus, it informs about variations of the setpoint tracking or in the disturbance rejection. Both directly reflect dynamic quality of the control loop. Higher values associated with large fluctuations indicate poorer control.

Unfortunately, the target value of the standard deviation is unknown and we may only observe its relative change. In some approaches, it is suggested to estimate the optimal achievable value σ_{opt} using some assumed control strategy, as for instance minimum variance or PID control. Such a information enables to build constructive relative measure σ_{rel} in form (4.5). Nonetheless, such approaches always require some a priori knowledge about the process. There exists a conflict between the constructiveness in case of relative measure and ease in case of plain standard deviation.

$$\sigma_{rel} = \frac{\sigma_{opt}}{\sigma} \tag{4.5}$$

These measures are frequently followed by higher order statistics like skewness or Kurtosis. Skewness (4.6) is a measure for PDF asymmetry and its shape. It brings information about positive and negative bias from the mean value. For the symmetrical distribution it is equal to zero. If it is negative, there are more negative items in the data. While if it is positive, we face the opposite situation.

$$skewness\,(x) = \frac{1}{N\sigma^3}\sum_{i=1}^{N}(x_i - x_0)^3 \tag{4.6}$$

Kurtosis (4.7) is a descriptor of the shape of a probability distribution. It is a measure for data concentration. The higher kurtosis is, the more scattered data is and the distribution function shape is more flat. Small values result in slender PDF.

$$kurtosis\,(x) = \frac{1}{N\sigma^4}\sum_{i=1}^{N}(x_i - x_0)^4 - 3 \tag{4.7}$$

The importance of these measures and their common acceptance is unquestionable and the majority of researchers and control engineers use them. However, we have to be aware that they are valid, while the signal histogram has Gaussian properties and there are no *outliers*. The first may be performed graphically. The approach is based on fitting of the PDF function to the data histogram. The fitting mostly minimizes the error between the area under the PDF function and the area of the histogram, i.e. the product of the height multiplied by the width of the bins (see Fig. 4.6). Visual inspection of such plots is an informal way of assessment, but also some measures of fitting can be used as for instance the mean square error.

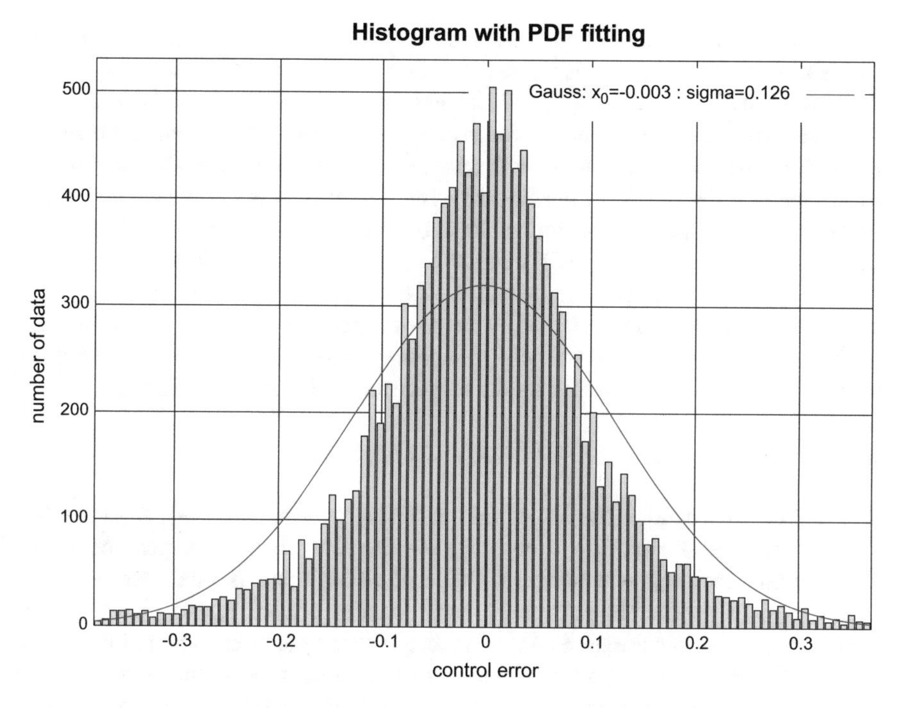

Fig. 4.6 Control error variable histogram with Gauss normal PDF fitted

These properties may be validated formally with specific normal distribution hypothesis tests. One of them is Kolmogorov-Smirnov normality test (4.8). In statistics, the Kolmogorov-Smirnov test (KS) is a nonparametric test for the equality of continuous, one-dimensional probability distributions that can be used to compare a sample with a reference PDF. The Kolmogorov-Smirnov statistic quantifies a distance between the empirical distribution function of the sample and the cumulative distribution function of the reference distribution. The Kolmogorov-Smirnov serves as a goodness of fit test. In the special case of testing for normality of the distribution or fitting with other probabilistic distribution functions.

$$KS = \max_{x} \left| \hat{F}_1(x) - \hat{F}_2(x) \right| \tag{4.8}$$

Another approach is to check normality hypothesis through skewness or α stability. Skewness test is based on the difference between the data's skewness and value of zero, with some selected p-value (value $p = 0.05$ is used in the paper). In a very similar way the α stability hypothesis is tested, when distance from value of 2 of α stability index is checked with some selected p-value ($p = 0.05$). Besides, there are also many other tests, like Smirnov-Cramer-von Mises [6], Anderson-Darling [2], Shapiro-Wilk [30], Lilliefors [23]. Experience shows that visual inspection and the histogram analysis enables better assessment, however it is more time consuming than simple normal hypothesis validation with some of the above tests.

Apart from the approaches described above there may be found analyzes of more advanced statistical approaches as for instance cumulants, bispectrum and bicoherence. The first and second order statistics can be used to describe linear process. Processes that deviate from the linearity can be conventionally considered using higher order statistics. These notions have been used to develop new indexes, like the Non-Gaussianity Index (NGI) and the NonLinearity Index (NLI) [5].

The approach uses the interpretation in the frequency domain. Second order moment in time domain is reflected by the power spectrum in the frequency domain. It shows the decomposition of the signal energy across the frequencies. The third order cumulants are represented by the bispectrum in the frequency domain (4.9)

$$B(f_1, f_2) \triangleq E\left[X(f_1) X(f_2) X^*(f_1 + f_2)\right], \tag{4.9}$$

where $B(f_1, f_2)$ denotes the bispectrum in the bi-frequency (f_1, f_2), $X(.)$ the discrete Fourier transform of the time series and the $X^*(.)$ is the complex conjugate. Bispectrum is the conjugate number with both the margin and the phase. The analysis of the $B(f_1, f_2)$ should be performed as the function of both f_1 and f_2. Thus three dimensional 3D plots are used. As the bispectrum measured at some point reflects the interaction between two frequencies it can be used as the as the relation to the nonlinearities characterizing the assessed signal. In practice, the bispectrum is normalized and we obtain the measure called the bicoherence (4.10), which is limited to the $bic^2(f_1, f_2) \in (0, 1)$

$$bic^2(f_1, f_2) \triangleq \frac{|B(f_1, f_2)|^2}{E\left[|X(f_1) X(f_2)|^2\right] \cdot E\left[|X(f_1 + f_2)|^2\right]}. \tag{4.10}$$

It can be shown that the squared bicoherence can be expressed for linear time series x_k as (4.11). If it is equal to zero $bic^2(f_1, f_2) = 0$ the original time series is Gaussian.

$$bic^2(f_1, f_2) = \frac{E\left[x_k^3\right]^2}{E\left[x_k^2\right]^3} = \frac{\mu_3^2}{\sigma^6}. \tag{4.11}$$

Two tests are developed to validate such a hypothesis.

- $NGI \leqslant 0$ meaning that the x_k is Gaussian.
  Test validates whether the squared bicoherence is zero or within some threshold.
- $NLI = 0$ meaning that the x_k is linear.
  Test validates whether for the squared bicoherence being non-zero the underlying process generating signal x_k is linear (also with the threshold applied).

The best practice estimates for the hypotheses and the thresholds are proposed by the authors of the method. The approach is applied to the analysis of the process and controller output signals in the CPA task. It is shown that the method may help in the assessment of the actuators, mostly the valves. It correctly detects the presence of significant nonlinearities enabling to detect valve stiction, slipping and its static characteristics shape.

Importance of Gaussian measures and their common acceptance is unquestionable. Majority of researchers and control engineers use them. However, we have to be aware that they are valid, while the signals have Gaussian properties. It was shown that for large sets of industrial control error histograms only minority (percentage share $\approx 6\%$ of the loops) has normal distribution [9]. Majority of them has fat tails. Mostly α-stable distribution is the best fitted (more then 60%), while the rest holds Cauchy PDF. Thus the other probabilistic density functions and the associated measures will be discussed in the following paragraphs.

4.3 Fat-Tailed Distributions

Non-Gaussian properties may be noticed through observation of the industrial time series histograms. If data has been described by unbiased Brownian motion then the histogram would have had a classical Gaussian bell shape. However, practice shows that often Gaussian distribution does not exactly fit into histogram long tails. Even average value is not perfectly represented. It seems that it may be better approximated by specific fat tailed distribution. Review of the available literature did not give any clear hint. It seems that this subject has not been popular in control community. Analysis of data from other contexts and domains, like financial or atmospheric phenomenon data analysis shows frequent uses of Cauchy and α-stable distributions [12, 31]. To confirm this hypothesis more control data than addressed in this paper was validated. Results have initially shown that it is not bad hypothesis. Nonetheless, it would be interesting to explore this research area.

Let us start the non-Gaussian discussion from the industrial examples. The presented plots show control error histograms together with the fitted two PDFs: normal Gauss and fat-tailed α-stable distribution. Observing the sketched plots one may see that in some cases Gaussian hypothesis is fully justified (see Fig. 4.7). However, in mots cases fat tail functions form better alternative, even in Gaussian case (Fig. 4.7), but fat-tailed (Fig. 4.8) and other ones (Figs. 4.9 and 4.10) as well. Review of the industrial data show only the minority of the signals well exhibiting the Gaussian properties [9].

The discussion on the origin of the fat tails in the distribution is not evident [16, 34]. It is well known from practice that Gaussian function tends to underestimate the probability of the events far from the mean (called *outliers*). They are responsible for the tails. There are known several reasons for their existence and the discussion can be done using the similarity with other contexts where the subject has been thoroughly analyzed (like physics or economics). On the one hand, they can be due to the certain behavioral patterns, i.e underlying certain degree of persistence in the control error time series [24]. On the other hand, fat tails can also appear due to long-range correlations in a “hidden variable” [26] such as volatility. Finally, the fat

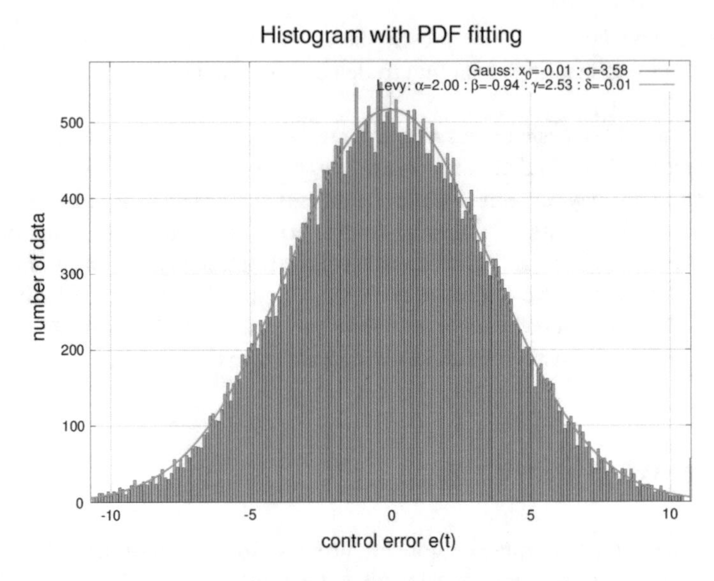

Fig. 4.7 Properly designed and well-tuned loop without disturbances–Gaussian character

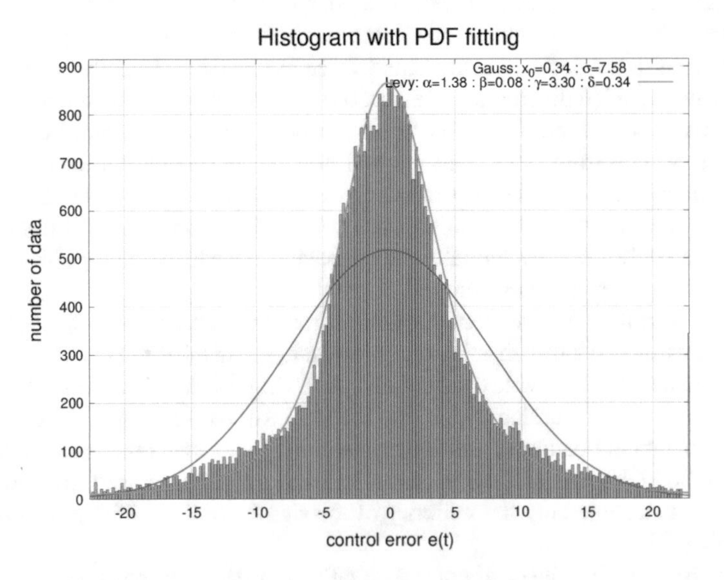

Fig. 4.8 Fat-tailed histogram for the correlated loop–uncoupled disturbances

tails may be generated when the signal involves various origins with more than one normal distribution [14].

None of the above potential explanations can be excluded from the real industrial control loop data. Persistence behavior may just occur due to the process complexity existing in multivariate and multi-loop systems nonlinear systems with frequent

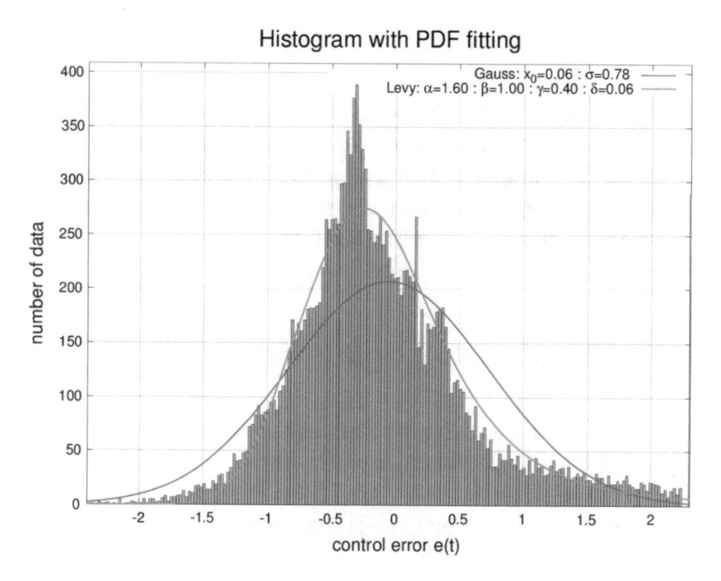

Fig. 4.9 Nonlinear loop behavior reflected in the asymmetric histogram

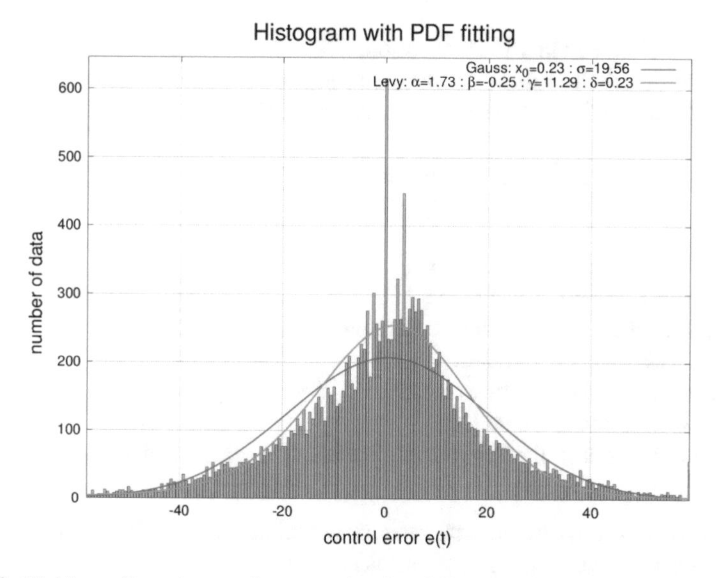

Fig. 4.10 Highly nonlinear loop performance showing shifted and skewed histogram

failures and disruptions and evident human impact due to the operators' interventions, biasing, manual operation, etc. Feedback itself may also contribute to the overall picture. Correlations appear simply because of the internal and external disturbances influencing the loop with various and varying delays. As the system is multivariate and exposed to the various disturbances and noises of different origins, it is highly probable that they have different normal distributions.

Thus, the existence of fat tails is not rare and should not be neglected. Two main fat-tail models of the stable distributions are discussed in the following subsections to present potential of using non-Gaussian probabilistic density functions factors as the CPA measures. At first the general form of the α-stable distribution is considered being followed by the special case of the Cauchy function.

4.3.1 α-Stable Probabilistic Density Function

An interesting alternative as a source of the statistical measures are α-stable functions. Given independent and identically distributed random variables $X_1, X_2, \ldots, X_n$ and X, the X is said to follow an α-stable distribution, if there exists a positive constant C_n and a real number D_n such that the relation holds (4.12)

$$X_1 + X_2 + \cdots + X_n \stackrel{d}{=} C_n \cdot X + D_n, \tag{4.12}$$

where $\stackrel{d}{=}$ denotes equality in distribution.

α-stable distribution does not have closed probabilistic density function and is expressed through the characteristics equation (4.13).

$$F^{stab}_{\alpha,\beta,\delta,\gamma}(x) = \exp\left\{i\delta x - |\gamma x|^{\alpha}\left(1 - i\beta l(x)\right)\right\}, \tag{4.13}$$

where
$l(x) = \begin{cases} sgn(x)\tan\left(\frac{\pi\alpha}{2}\right) & \text{for } \alpha \neq 1 \\ -sgn(x)\frac{2}{\pi}\ln|x| & \text{for } \alpha = 1 \end{cases},$
$0 < \alpha \leq 2$ called *stability* index or characteristic exponent,
$|\beta| \leq 1$ is a *skewness* parameter,
$\delta \in \mathbb{R}$ is a distribution *location* (mean),
$\gamma > 0$ is distribution *scale* factor.

The function is described by four parameters. There are special cases with a closed form of the PDF:

- $\alpha = 2$ reflects independent realizations, especially for $\alpha = 2$, $\beta = 0$, $\gamma = 1$ and $\delta = 0$ we get exact normal distribution equation,
- $\alpha = 1$ and $\beta = 0$ denote the Cauchy case that is considered in details in the following paragraph and
- $\alpha = 0.5$ and $\beta = \pm 1$ denote the Lévy case, which is not considered in the analysis.

General form of the α-stable distribution can be considered as the potential source of the factors that may be used as the CPA measures. Some examples of the PDF shapes are sketched in Fig. 4.11.

The function has more degrees of freedom, as it is parametrized by four parameters. It should be also stated that in general form there is no analytical solution to

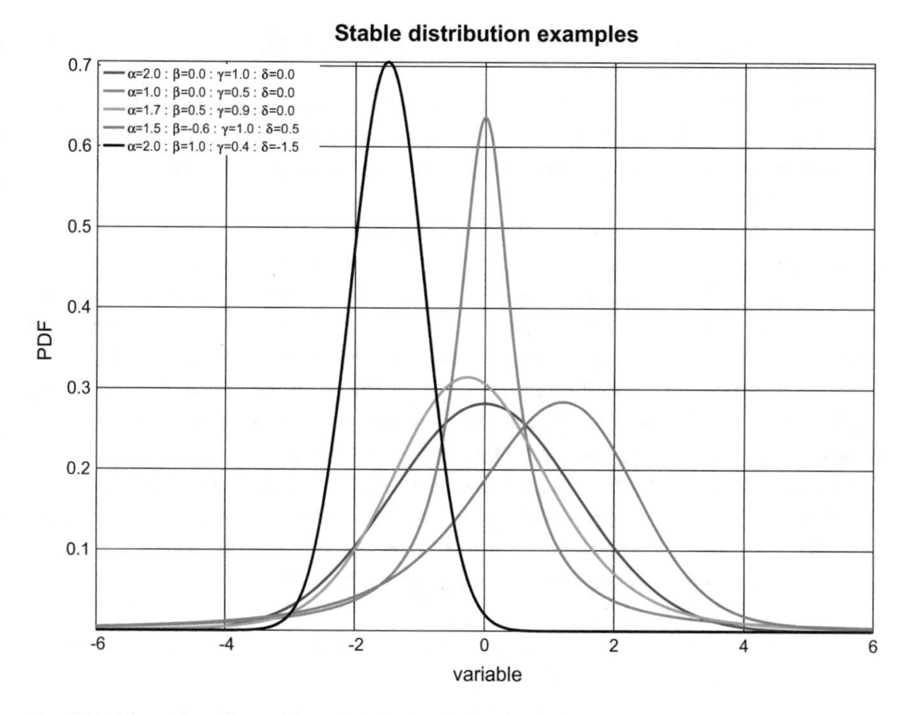

Fig. 4.11 Examples of α-stable probabilistic distribution functions

the evaluation of mean value and standard deviation. Industrial data analysis shows that it is also well suited for loop variables characterized by the fat tails. Stability parameter α is responsible for the fatness of the tails. Thus it may help to indicate the same phenomena that cause fat tails, i.e. auto-correlations, complexity or embedded persistence.

Location parameter δ keeps some information of the function position, but it should not be considered identical to the PDF mean. Nonetheless it may be considered as the measure of the steady state error with the detection possibilities to the normal mean value.

Additionally α-stable distribution delivers two more shaping parameters. β informs us about distribution skewness. This factor may have the similar meaning to the general (normal distribution) skewness coefficient. It indicates loop asymmetric performance, which is not desired. The desired value should be $\beta = 0$. Any variations from the zero value may be caused by loop/actuator nonlinearities or inadequate operating point, mostly caused by wrong control philosophy or instrumentation limitations.

The scaling factor γ has the same meaning as the γ parameter of Cauchy PDF and a very similar one to the normal standard deviation σ. It reflects the PDF broadness, thus it indicates loop fluctuations. The smaller the value, the better dynamical control, however there exists lower limitation associated with the best achievable control that may be evaluated with different senses, as for instance minimum variance, LQG or PID control.

Although the variance may be infinite for stable distribution it should not eliminate fat tail distributions from the considerations. Several statistical processes have "infinite variance", however may exhibit finite, and often very well behaved, mean deviations [32].

The research over these parameters has shown that in several cases they can be even better than the Gaussian "predecessors". Analysis of the loops controlled by the PID [10] and GPC predictive controller [13] has shown their robustness to various noises and disturbances having different statistical properties. Also their industrial validation has confirmed the results obtained with computer simulation [11].

Following above interpretations perspective for good control loop, one would expect to have zero value for location and skewness, minimum value of γ and $\alpha \approx 2$ as this is the case for independent control error realizations.

Evaluation of the parameters for the α-stable probabilistic density function may be achieved with different methods [27] as for instance fast but not very accurate quantiles method [25], iterative Koutrouvelis approach based on the characteristics function estimation [20], logarithmic moment method [21] Maximum Likelihood (ML) approach [4] being the most accurate but at the highest calculation cost. The Koutrouvelis approach has been used in further simulations considered to be the good compromise between the accuracy and the calculation effort.

Concluding, α-stable probabilistic density function offers four possible indexes:

- the **position factor** δ measuring loop steady state error value,
- the **stability factor** α reflecting loop persistence,
- the **skewness factor** β indicating loop asymmetric performance,
- and the **scaling coefficient** γ reflecting control loop tuning dynamic goodness.

4.3.2 Cauchy Probabilistic Density Function

Cauchy PDF is an example of the fat-tail distribution. It is also known, especially among physicists, as the Lorentz or Cauchy-Lorentz PDF, Lorentz(-ian) function or Breit-Wigner distribution. The distribution is described by the density function (4.14)

$$F_{\delta,\gamma}^{Cauchy}(x) = \frac{1}{\pi\gamma}\left(\frac{\gamma^2}{(x-\delta)^2+\gamma^2}\right), \tag{4.14}$$

where
$\delta \in \mathbb{R}$ is a distribution *location* (mean) parameter,
$\gamma > 0$ is distribution *scale* factor.

The Cauchy function is symmetrical. The shape for values further from mean does not decay so fast as it is with normal distribution. It is described, similarly to Gauss, by two parameters. They have similar meaning to the ones of normal distribution.

Location factor δ informs about function mean value, so in case of control error one would wish to have it equal to zero. The reasons for the non-zero value are exactly the same as for the normal distribution. Scale factor informs about function slenderness and as such it indicates signal variability. Similarly to normal standard deviation optimal value of γ is unknown and we have to more rely on its relative changes. Exemplary shapes of Cauchy PDF with different coefficients are presented in Fig. 4.12. The research over these parameters has shown that they are also robust against various loop disturbances [10, 11, 13].

There are several methods dedicated towards finding of the proper Cauchy distribution for the certain time series [29]. The fitting of the PDF shape to the histogram with the Maximum Likelihood Estimation (ML) is one of the most popular and robust ones [15]. The method implemented in Octave successive quadratic programming solver [3] has been applied in all the considered examples and simulations.

Concluding, similarly to the normal Gauss measures the Cauchy approach offers two indexes:

- the **position factor** δ measuring the loop steady state error value,
- and the **scaling coefficient** γ reflecting control loop tuning dynamic goodness.

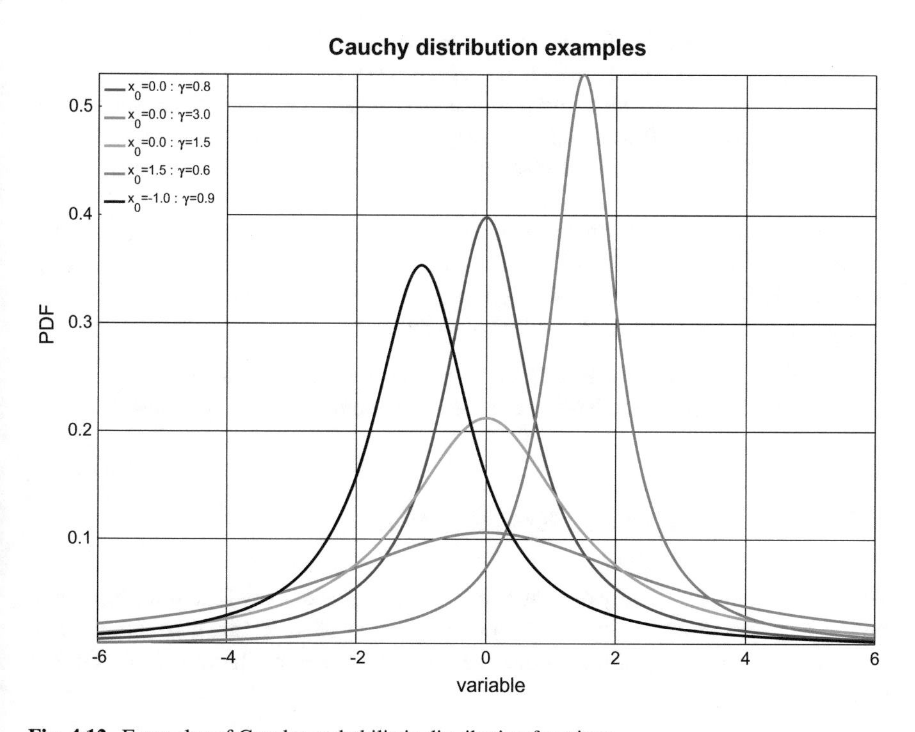

Fig. 4.12 Examples of Cauchy probabilistic distribution functions

4.4 Other Non-Gaussian Functions

Apart from the Gaussian and fat-tails stable distributions some other density functions can be met in the considerations such as Laplace or GEV. They are covered in the further presentations. In some research also gamma PDF is taken into consideration. However, its character does not fit potential shapes existing in the control error analysis. Thus it will not be further analyzed.

4.4.1 Laplace Double Exponential Distribution

The Laplace probabilistic density function is sometimes called the double exponential distribution. It is formed as a function of differences between two independent variables with identical exponential distributions. Its probability density function is given by formula (4.15).

$$F_{\mu,b}^{Lap}(x) = \frac{1}{2b} e^{-\frac{|x-\mu|}{b}}, \tag{4.15}$$

where $\mu \in \mathbb{R}$ is a *location* factor and $b > 0$ is a scale parameter.

Its shape decays exponentially and is characterized by parameter b. As it is fully described by two parameters, one would like to expect zero value of location μ and minimal value of scale factor. Unfortunately its optimal value is unknown and we may only rely on its relative changes. Exemplary shapes of Laplace PDF are presented in Fig. 4.13.

Laplace distribution parameter estimation can be achieved with various approaches, however maximum likelihood [19] or Bayesian approaches [22] are the most common. In this work the MLE approach implemented in Matlab is used.

Concluding, the Laplace PDF offers two indexes:

- the **position factor** μ measuring the loop steady state error value,
- and the **scale factor** b reflecting control loop tuning dynamic goodness.

4.4.2 GEV–Generalized Extreme Value

GEV distribution is a family of continuous probabilistic density functions developed within extreme value theory to combine properties of different distributions, like Gumbel, Fréchet and Weibull. Its distribution is described by the following formula (4.16).

$$PDF_{\mu,\sigma,\xi} = \frac{1}{\sigma} t(x)^{\xi+1} e^{-t(x)}, \tag{4.16}$$

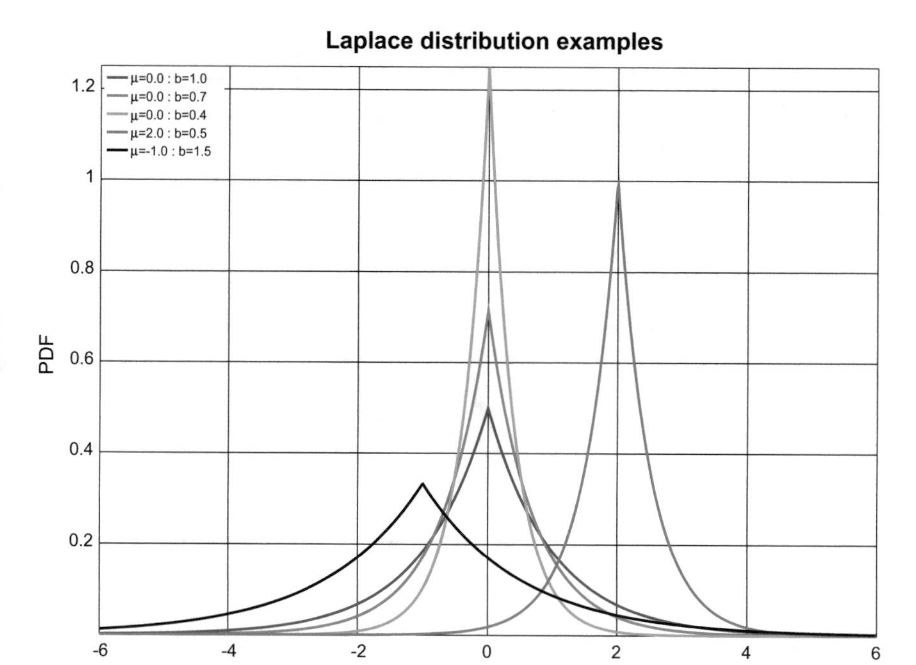

Fig. 4.13 Examples of Laplace probabilistic distribution functions

where

$$t(x) = \begin{cases} \left(1 + \left(\frac{x-\mu}{\sigma}\right)\xi\right)^{-\frac{1}{\xi}} & \text{if } \xi \neq 0 \\ e^{-\frac{x-\mu}{\sigma}} & \text{if } \xi = 0 \end{cases},$$

$\mu \in \mathbb{R}$ is a PDF *location* parameter,
$\sigma > 0$ is distribution *scale* parameter,
and $\xi \in \mathbb{R}$ is distribution *shape* factor.

GEV distribution function examples are sketched on Fig. 4.14. GEV parameters does not reflect exactly statistics mean value and standard deviation. For that purpose mean value is calculated according to (4.17) and (4.18), respectively.

$$mean_{GEV} = \mu - \frac{\sigma}{\xi} + \frac{\sigma}{\xi} \cdot \Gamma(1 - \xi) \tag{4.17}$$

$$stdev_{GEV} = \sqrt{\frac{\sigma^2}{\xi^2}\left(\Gamma(1 - 2\xi) - \Gamma^2(1 - \xi)\right)} \tag{4.18}$$

Thus from the perspective of steady-state error detection perspective one should consider location factor μ or PDF mean value $mean_{GEV}$. As the shape is considered

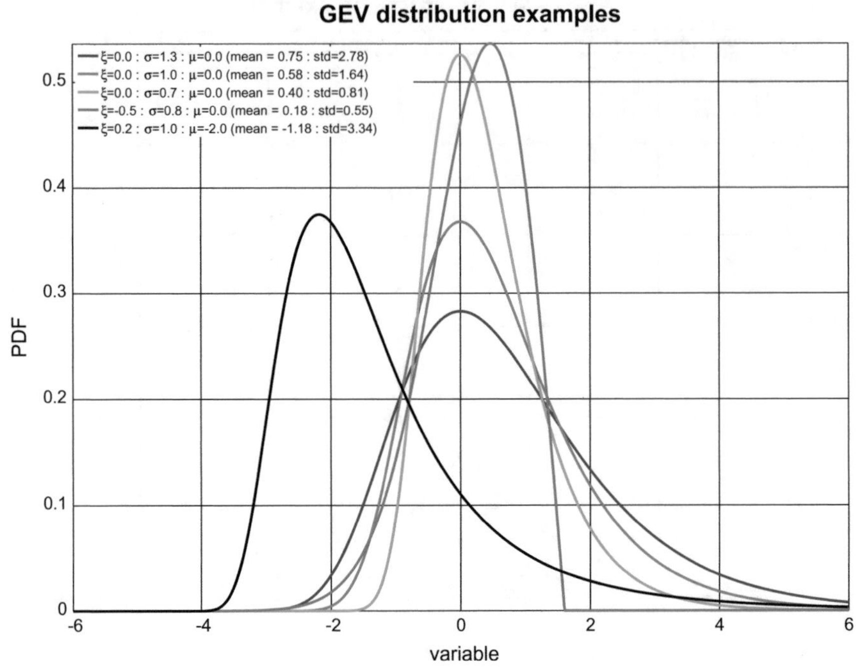

Fig. 4.14 Examples of GEV probabilistic distribution functions

there are more options, like scale factor σ and shape parameter ξ and additionally GEV standard deviation $stdev_{GEV}$ may be also taken into consideration.

GEV distribution parameter estimation is also done with ML algorithm [28], however other approaches as for instance maximum product of spacings [35] or the mixed maximum likelihood and L-moments estimation procedure [1] have been also proposed. The ML approach implemented in Matlab is used in the simulations.

Concluding, the GEV approach offers five possible indexes:

- the **location parameter** μ and the mean value $\boldsymbol{mean_{GEV}}$ measuring the loop steady state error value,
- the **GEV standard deviation** $\boldsymbol{stdev_{GEV}}$ reflecting control loop tuning quality,
- and the **GEV shape** $\boldsymbol{\xi}$ and **GEV scale** $\boldsymbol{\sigma}$ potentially indicating control loop tuning dynamic goodness.

4.5 Robust Statistics

The existence of outliers in data implies fat tails in their distributions. This feature generates serious challenge in the non-Gaussian data analysis. We have to decide how to consider the outliers, i.e. accept them or decline. In the first one the outliers are considered to sustain an important information about the loop. This approach leads to the incorporation of the fat tailed probabilistic distribution functions and the measures using their factors. This potential has been presented in the above chapters.

On the other hand we may adopt the opposite assumption. The outliers are irrelevant, affect the application of Gaussian approaches and as such should be somehow neglected (removed) [8]. Once data is *clean* and free of the outliers one can apply canonical approach with the use of normal Gaussian function. This methodology is the underlying idea behind the proposal and evaluation of **robust statistics**.

They were introduced several years ago [17], but Huber's research [18] gave them a new application feedback. They achieve good performance for data having various probability distributions, especially for normal ones. Robust methods may be used to estimate location, scale, and other regression parameters for time series affected by outliers. This feature is the most interesting in our case. The fact that the robust approach is not sufficiently used in control is probably connected with the fact that they have appeared and are mostly investigated in the research context of the chemical engineering. It is clearly reflected by the ratio between the number of publications in chemistry and control engineering.

The average value of the data (4.2) is the best-known estimate of a true value of a random variable x_k. It is the location estimator of the general position of the x_k. This estimator looses its reliability in presence of the outliers in the data. Thus it is called non-robust estimator. The robust ones will aim to describe well the time series properties regardless of the data content. From that perspective the parametric location estimators of the α-stable or Laplace distribution offer better estimates than the Gaussian one (4.2). In contrary we may not to reject outliers, but to transform the data. Such the estimators are named the semi-non-parametric ones. The median value is considered as the reasonable alternative. It tells, where the middle of a data set of the length N is (4.19)

$$median^N(x) = x_m^N = \begin{cases} \frac{x_{\frac{N}{2}:N} + x_{\frac{N}{2}+1:N}}{2} & \text{if } N \text{ is even,} \\ x_{\frac{N+1}{2}:N} & \text{if } N \text{ is odd,} \end{cases} \tag{4.19}$$

where $x_{1:N} \leqslant x_{2:N} \leqslant \cdots \leqslant x_{2:N}$ are the ordered observations. There are many robust location and scale estimators in the literature apart from the simplest median. M-estimators with Huber ψ-function or logistic function may be used [7].

The ML-estimators can be generalized to the class of M-estimators. An M-estimator of location $x^M H_0$ is defined as the solution of the Eq. (4.20)

$$\sum_{i=1}^{N} \psi\left(\frac{x_i - \mu}{\hat{\sigma_0}}\right) = 0, \tag{4.20}$$

where $\psi(.)$ is any non-decreasing odd function, μ is a location estimator and $\hat{\sigma_0}$ is a preliminary scale estimator, like for instance the Median Absolute Deviation or any other highly robust scale estimator. M-estimators are affine equivariant and the Eq. (4.20) can be solved with the Newton-Raphson algorithm using the sample median as a starting value. The Huber ψ function is defined as (4.21)

$$\psi_H(x, b) = \max(-b, \min(x, b)), \tag{4.21}$$

where b is a user defined constant. Its limit is the median as $b \to 0$ and the mean with $b \to \infty$. The value $b = 1.345$ gives 95% efficiency at the normal model. The efficiency of Huber's estimator increases as the constant b increases, while the robustness is decreases with the increasing constant b.

On the other hand the logistic smooth ψ function is defined as (4.22)

$$\psi_{log}(x) = \frac{e^x - 1}{e^x + 1}, \tag{4.22}$$

which may be rewritten as $\psi_{log}(x) = 2F(x) - 1$, where $F(x) = 1/(1 + e^{-x})$ is the cumulative distribution function of the logistic PDF, also known as the sigmoid function.

Analogously to the location estimators the scale M-estimator is defined as the solution to the Eq. (4.23)

$$\frac{1}{N}\sum_{i=1}^{N}\rho\left(\frac{x_i - \hat{\mu_0}}{\sigma}\right) = \kappa, \tag{4.23}$$

where $0 < \kappa < \rho(\infty)$, $\rho(.)$ is is even, differentiable and non-decreasing on the positive numbers loss function, σ is a location estimator and $\hat{\mu_0}$ is a preliminary location estimator, like for instance the highly robust sample median. While the logistic ψ function (4.22) is taken as $\rho(.)$ we obtain the logistic ψ scale estimator.

The methods implemented in the LIBRA toolbox [33] have been used in all the following analyzes. In a similar way to the normal Gauss measures, the robust approach offers two indexes:

- **robust position estimators**, like M position estimator with Huber ψ or logistic function measuring the loop steady state error value,
- and **robust scale estimators** as for instance the logistic ψ scale M-estimator indicating control loop dynamic tuning quality.

References

1. Ailliot, P., Thompson, C., Thomson, P.: Mixed methods for fitting the GEV distribution. Water Resour. Res. **47**(5), w05551 (2011)
2. Anderson, T.W., Darling, D.A.: A test of goodness-of-fit. J. Am. Stat. Assoc. **49**, 765–769 (1954)
3. Axensten, P.: Cauchy CDF, PDF, inverse CDF, parameter fit and random generator. http://www.mathworks.com/matlabcentral/fileexchange/11749-cauchy/ (2006)

4. Borak, S., Misiorek, A., Weron, R.: Models for heavy-tailed asset returns. In: Cizek, P., Härdle, K.W., Weron, R. (eds.) Statistical Tools for Finance and Insurance, 2nd edn, pp. 21–56. Springer, Berlin (2011)
5. Choudhury, M.A.A.S., Shah, S.L., Thornhill, N.F.: Diagnosis of poor control-loop performance using higher-order statistics. Automatica **40**(10), 1719–1728 (2004)
6. Cramer, H.: On the composition of elementary errors. Scand. Actuar. J. **1**, 13–74 (1928)
7. Croux, C., Dehon, C.: Robust estimation of location and scale. Wiley StatsRef: Statistics Reference Online (2014)
8. Daszykowski, M., Kaczmarek, K., Heyden, Y.V., Walczak, B.: Robust statistics in data analysis–a review: basic concepts. Chemom. Intell. Lab. Syst. **85**(2), 203–219 (2007)
9. Domański, P.D.: Non-gaussian properties of the real industrial control error in SISO loops. In: Proceedings of the 19th International Conference on System Theory, Control and Computing, pp. 877–882 (2015)
10. Domański, P.D.: Non-Gaussian statistical measures of control performance. Control Cybern. **46**(3), 259–290 (2017)
11. Domański, P.D.: Statistical measures for proportional-integral-derivative control quality: simulations and industrial data. Proc. Inst. Mech. Eng. Part I J. Syst. Control Eng. **232**(4), 428–441 (2018)
12. Domański, P.D., Gintrowski, M.: Alternative approaches to the prediction of electricity prices. Int. J. Energy Sect. Manag. **11**(1), 3–27 (2017)
13. Domański, P.D., Ławryńczuk, M.: Assessment of the GPC control quality using non-Gaussian statistical measures. Int. J. Appl. Math. Comput. Sci. **27**(2), 291–307 (2017)
14. Engle, R.: Risk and volatility: econometric models and financial practice. Am. Econ. Rev. **94**(3), 405–420 (2004)
15. Ferguson, T.S.: Maximum likelihood estimates of the parameters of the cauchy distribution for samples of size 3 and 4. J. Am. Stat. Assoc. **73**(361), 211–213 (1978)
16. Gremm, M.: The origin of fat tails. https://ssrn.com/abstract=2337830 (2013). SSRN
17. Hawkins, D.M.: Identification of Outliers. Chapman and Hall, London, New York (1980)
18. Huber, P.J., Ronchetti, E.M.: Robust Statistics, 2nd edn. Wiley (2009)
19. Johnson, N., Kotz, S., Balakrishnan, N.: Continuous univariate distributions. No. t. 2 in Wiley series. In: Probability and Mathematical Statistics: Applied Probability and Statistics. Wiley (1995)
20. Koutrouvelis, I.A.: Regression-type estimation of the parameters of stable laws. J. Am. Stat. Assoc. **75**(372), 918–928 (1980)
21. Kuruoglu, E.E.: Density parameter estimation of skewed alpha-stable distributions. IEEE Trans. Signal Process. **49**(10), 2192–2201 (2001)
22. Li, L.: Bayes estimation of parameter of laplace distribution under a new LINEX-based loss function. Int. J. Data Sci. Anal. **3**(6), 85–89 (2017)
23. Lilliefors, H.: On the Kolmogorov-Smirnov test for normality with mean and variance unknown. J. Am. Stat. Assoc. **62**, 399–402 (1967)
24. Mantegna, R.N., Stanley, H.E.: Stochastic process with ultraslow convergence to a Gaussian: the truncated Lévy flight. Phys. Rev. Lett. **73**, 2946–2949 (1994)
25. McCulloch, J.H.: Simple consistent estimators of stable distribution parameters. Commun. Stat. Simul. Comput. **15**(4), 1109–1136 (1986)
26. Pasquini, M., Serva, M.: Clustering of volatility as a multiscale phenomenon. Eur. Phys. J. B Condens. Matter Complex Syst. **16**(1), 195–201 (2000)
27. Royuela-del Val, J., Simmross-Wattenberg, F., Alberola-Lopez, C.: Libstable: fast, parallel, and high-precision computation of α-stable distributions in R, C/C++, and MATLAB. J. Stat. Softw. Art. **78**(1), 1–25 (2017)
28. Scarf, P.A.: Estimation for a four parameter generalized extreme value distribution. Commun. Stat. Theory Methods **21**(8), 2185–2201 (1992)
29. Schuster, S.: Parameter estimation for the Cauchy distribution. In: 2012 19th International Conference on Systems, Signals and Image Processing (IWSSIP), pp. 350–353 (2012)

30. Shapiro, S.S., Wilk, M.B.: An analysis of variance test for normality (complete samples). Biometrika **52**(3–4), 591–611 (1965)
31. Shen, X., Xu, Zhang Y.H., Meng, S.: Observation of alpha-stable noise in the laser gyroscope data. IEEE Sens. J. **16**(7), 1998–2003 (2016)
32. Taleb, N.N.: Real-world Statistical Consequences of Fat Tails: Papers and Commentary. STEM Academic Press, Technical Incerto Collection (2018)
33. Verboven, S., Hubert, M.: LIBRA: a Matlab library for robust analysis. Chemom. Intell. Lab. Syst. **75**, 127–136 (2005)
34. Viswanathan, G.M., Fulco, U.L., Lyra, M.L., Serva, M.: The origin of fat tailed distributions in financial time series, Cornell University Library (2002). arXiv:cond-mat/0112484v4
35. Wong, T.S.T., Li, W.K.: A note on the estimation of extreme value distributions using maximum product of spacings. Lect. Notes Monogr. Ser. **52**, 272–283 (2006)

Chapter 5
Model-Based Measures

Can we make a mechanical model of it?

–Lord Kelvin

The measures summarized in the previous sections are defined in time domain. They use loop time series data and do not require any a priori knowledge about the loop or background process. They are fully data-driven. They all share the similar shortcut. None of them offers any distance from the measured index value to the optimal one. Thus, apart from the actual measured index value J_{act} one would require to estimate the lowest (the best achievable) limit of performance index J_{opt}. It is clear that such an estimation requires more knowledge on the process and this set of the approaches is named model-driven.

Starting from the first paper by Harris in 1989 [24], many authors (T.J. Harris, L. Desborough, B.S. Ko, T.F. Edgar, B. Hang, S.L. Shah, N.F. Thornhill, A. Horch, A.J. Isaksson or M.J. Grimble.) have proposed dozens of the approaches based on the underlying benchmarking idea for various control configuration and performance features. The subject is very broad and one may find extensive descriptions of these methods in the several overviews, books and monographes [17, 26, 35, 36, 47]. During long period of time these methods have been applied in many industries and large practical experience has been collected [1, 61].

This chapter is organized as follows. At first the basic minimum variance CPA methodology (Sect. 5.1) is introduced, as it is the basic for any further control benchmarking approach. It is followed by the review of the approaches addressing other than minimum variance benchmarking (Sect. 5.2). Finally, the overview of other general model-based approaches (Sect. 5.3) is presented.

P. D. Domański, *Control Performance Assessment: Theoretical Analyses and Industrial Practice*, Studies in Systems, Decision and Control 245,
https://doi.org/10.1007/978-3-030-23593-2_5

5.1 Minimum Variance Approach

The following process as in Fig. 5.1 is considered in the underlying works [10, 24, 29, 40] introducing minimum variance approach. It is described by the discrete time linear characteristic equation (5.1).

$$y(k) = G(z^{-1})z^{-l}m(k) + d(k) = \frac{B(z^{-1})}{A(z^{-1})}z^{-l}m(k) + d(k), \tag{5.1}$$

where $G(z^{-1})$ is the process transfer function without a delay described by the polynomials $B(z^{-1})$ and $A(z^{-1})$ and l is pure discrete time delay. The disturbance signal $d(k)$ is considered to be generated by Auto Regressive Integrated Moving Average (ARIMA) process (5.2) driven by the white noise signal $\upsilon(k)$ having zero mean and variance σ_υ^2. This is a non-stationary noise, being characterized by the the variance increasing with time, i.e. $\sigma_d^2(t) = t \cdot \sigma_\upsilon^2$.

$$d(k) = G_d(z^{-1})\upsilon(k) = \frac{D(z^{-1})}{C(z^{-1})(1 - z^{-1})}\upsilon(k), \tag{5.2}$$

We assume that the controller $G_c(z^{-1})$ is linear with the control signal (MV) defined as (5.3)

$$m(k) = G_c(z^{-1})\varepsilon(k) = G_c(z^{-1})\left(y_0(k) - y(k)\right). \tag{5.3}$$

Putting together all of the equations with the assumption of the zero setpoint ($y_0(k) = 0$), one may derive the close loop equation (5.4)

$$y(k) = \frac{G_d(z^{-1})}{1 + G(z^{-1})G_c(z^{-1})z^{-l}}\upsilon(k). \tag{5.4}$$

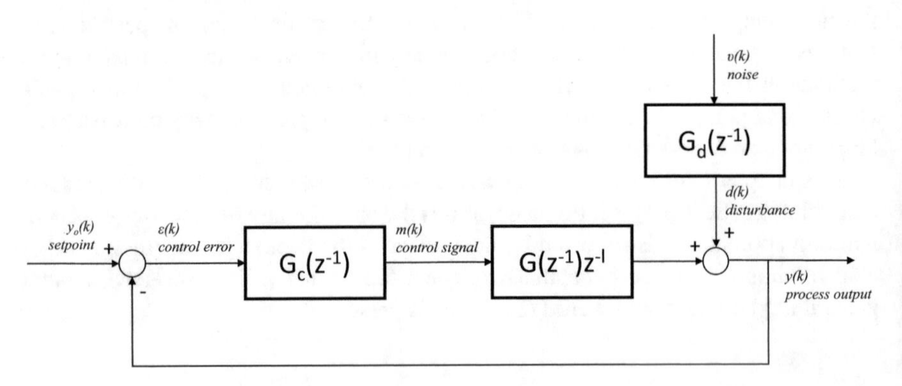

Fig. 5.1 Control loop with *feedback* control

We assume the numerator and denominator polynomials of the close loop system by $P(z^{-1})$ and $Q(z^{-1})$, respectively. Then we may solve the Diophantine equation and obtain (5.5):

$$y(k) = \frac{P(z^{-1})}{Q(z^{-1})}\upsilon(k) = g(z^{-1})\upsilon(k), \tag{5.5}$$

where a polynomial $g(z^{-1})$ is the impulse response of the closed loop system (5.6):

$$g(z^{-1}) = g_0 + g_1 z^{-1} + g_2 z^{-2} + g_3 z^{-3} + \cdots . \tag{5.6}$$

Now the minimum quality limit in the sense of the minimum variance controller may be derived. We use the fact that some part of the output variance in the close loop is invariable and may be estimated [24]. It means that the coefficients g_i of the impulse response higher than the discrete delay, i.e. $k = l, l+1, \ldots$ are equal to zero. Using the close loop system (5.5) and the noise description (5.2) we obtain the relation (5.7):

$$G_d(z^{-1}) = \underbrace{f_0 + f_1 z^{-1} + \cdots + f_{l-1} z^{l-1}}_{F(z^{-1})} + R(z^{-1})z^{-1}, \tag{5.7}$$

where the coefficients f_k (for $k = 1, \ldots, l-1$) are constant and $R(z^{-1})$ is some transfer function.

Now we may reformulate the relation (5.5) into the general equation (5.8):

$$\begin{aligned} y(k) &= \frac{F(z^{-1}) + R(z^{-1})z^{-1}}{1 + G(z^{-1})G_c(z^{-1})z^{-l}}\upsilon(k) \\ &= \left[F(z^{-1}) + \underbrace{\frac{R(z^{-1}) - F(z^{-1})G(z^{-1})G_c(z^{-1})}{1 + G(z^{-1})G_c(z^{-1})z^{-l}}}_{L(z^{-1})} z^{-1} \right] \upsilon(k) \\ &= F(z^{-1})\upsilon(k) + L(z^{-1})\upsilon(k-1). \end{aligned} \tag{5.8}$$

Two elements in the right side of the Eq. (5.8) are independent, due to the fact that $F(z^{-1})\upsilon(k) = f_0\upsilon(k) + \cdots f_{k-l}\upsilon(k - l + 1)$. As the result we estimate the output variance (5.10)

$$Var\,(y(k)) = Var\,\big(F(z^{-1})\upsilon(k)\big) + Var\,\big(L(z^{-1})\upsilon(k-1)\big) \tag{5.9}$$

and

$$Var\,(y(k)) \geqslant Var\,\big(F(z^{-1})\upsilon(k)\big). \tag{5.10}$$

Following the above relations (5.10) the equity $L = 0$, i.e. $R - FGG_c = 0$ holds and lead to the minimum variance control rule

$$G_c(z^{-1}) = \frac{R(z^{-1})}{F(z^{-1})G(z^{-1})}. \tag{5.11}$$

The variance of the loop (5.11) output variable $y(k)$ is equal to:

$$\sigma_y^2 = \sigma_{MV}^2 = \left(g_0^2 + g_1^2 + \cdots + g_{l-1}^2\right)\sigma_\upsilon^2 = \left(f_0^2 + f_1^2 + \cdots + f_{l-1}^2\right)\sigma_\upsilon^2. \tag{5.12}$$

The variance σ_{MV}^2 is an impassable lower constraint of the assumed control quality. It may be achieved with the Åström minimum variance controller and depends only on the process delay l.

The above relations allowed to define several closely connected control quality indexes, as for instance:

- $\eta_0 = \frac{\sigma_y^2}{\sigma_{MV}^2}$, $\eta_0 \geqslant 1$,
- $\eta_1 = \frac{\sigma_{MV}^2}{\sigma_y^2}$, $0 \leqslant \eta_1 \leqslant 1$,
- $\eta_2 = 1 - \frac{\sigma_{MV}^2}{\sigma_y^2}$, $0 \leqslant \eta_2 \leqslant 1$,
- $\eta_3 = \frac{\sigma_y^2 - \sigma_{MV}^2}{\sigma_{max}^2 - \sigma_y^2}$ [46].

The control assessment procedure is organized according to the following steps:

1. estimate process time delay l,
2. identify the close loop model relating disturbance $\upsilon(t)$ and the process output $y(t)$, like for instance ARMA regression model or the other,
3. estimate the minimum variance σ_{MV}^2,
4. estimate actual variance of the process output signal σ_y^2,
5. evaluate the respective control performance measure η and asses the loop,
6. alternately, one may calculate the autocorrelation function for the process output $y(t)$ to look for the significant correlations beyond the estimated time delay.

The proposed indexes do not depend on the noise $\upsilon(k)$ amplitude, however the accurate knowledge of the process delay is required and crucial. This fact may be considered as the main drawback of the approach as the delay estimation is a challenging task in the industrial reality [5].

It is crucial for the benchmarking procedure to estimate the lower quality constraint. It leads to the definition of the algorithm, which evaluates the highest achievable control quality. Filtering and CORrelation analysis (FCOR) algorithm proposed in [30] is based on filtering and correlation analysis. Evaluation of the cross-correlation function between the output and the noise enables to estimate σ_{MV}^2 and the indexes.

We consider stable close loop system, described by the impulse response or the infinite order MA model (5.5). Multiplication of the Eq. (5.5) by $\upsilon(k), \upsilon(k-1), \ldots, \upsilon(k-l+1)$ and evaluation of the expected value for both sides give:

$$
\begin{aligned}
r_{yv}(0) &= E\left[y(0)v(0)\right] = g_0 \cdot \sigma_v^2 \\
r_{yv}(1) &= E\left[y(1)v(1)\right] = g_1 \cdot \sigma_v^2 \\
r_{yv}(2) &= E\left[y(2)v(2)\right] = g_2 \cdot \sigma_v^2 \\
&\vdots \\
r_{yv}(l-1) &= E\left[y(l-1)v(l-1)\right] = g_{l-1} \cdot \sigma_v^2.
\end{aligned}
\tag{5.13}
$$

Thus, the constant part of the output signal variation (achievable with the Åström minimum variance control) is:

$$
\begin{aligned}
\sigma_{MV} &= \left(g_0^2 + g_0^2 + \cdots + g_{l-1}^2\right) \cdot \sigma_v^2 \\
&= \left[\left(\frac{r_{yv}(0)}{\sigma_v^2}\right)^2 + \left(\frac{r_{yv}(1)}{\sigma_v^2}\right)^2 + \cdots + \left(\frac{r_{yv}(l-1)}{\sigma_v^2}\right)^2\right] \cdot \sigma_v^2 \\
&= \left[r_{yv}(0)^2 + r_{yv}(1) + \cdots + r_{yv}(l-1)\right] / \sigma_v^2.
\end{aligned}
\tag{5.14}
$$

Substituting Eq. (5.14) to the indexes' definitions we obtain:

$$
\begin{aligned}
\eta_1(l) &= \left[r_{yv}(0)^2 + r_{yv}(1) + \cdots + r_{yv}(l-1)\right] / \sigma_y^2 \sigma_v^2 \\
&= \rho_{vy}(0)^2 + \rho_{vy}(1) + \cdots + \rho_{vy}(l-1) \\
&= ZZ^T
\end{aligned}
\tag{5.15}
$$

where Z is a vector of the cross correlation coefficients between $y(k)$ and $v(k)$ for $k = 0, 1, \ldots, l-1$

$$
Z = \left[\rho_{vy}(0), \rho_{vy}(0), \ldots \rho_{vy}(l-1)\right].
\tag{5.16}
$$

Estimates $\hat{\rho}_{vy}(k)$ for the cross correlation coefficients $\rho_{vy}(k)$:

$$
\hat{\rho}_{vy}(k) = \frac{\frac{1}{N}\sum_{i=1}^{N} y(i)v(i-k)}{\sqrt{\frac{1}{N}\sum_{i=1}^{N} y^2(i)\frac{1}{N} \cdot \sum_{i=1}^{N} v^2(i)}}.
\tag{5.17}
$$

The values for the $v(i)$ are approximated by the innovations $\hat{v}(i)$ obtained with the whitening filters received with the estimation of the coefficients of the polynomials for the standard regression models. This model-based assessment methodology uses above presented benchmark based on the minimum variance optimal controller and the estimates of the innovations exciting stochastic model of the controlled process.

In theory this approach requires knowledge of the process delay and the loop variables. Actually, it is required to identify the complex system that consist of the controller, the process and the disturbance model. This task is difficult due to high orders and complex structure [3]. Delay existing in the control loop can be considered simply as the transportation delay. The task becomes complicated and depends on the optimization procedure, once it starts being associated with the dynamics. The other limitation contributes to the fact that delay identification is not easy and may be strongly biased by loop disturbances and other external correlations [5].

Practical implementations of the algorithm may vary. They often include further modifications. Time series analysis of the data is performed to obtain an appropriate model of the system [7, 8]. It often includes correlation analysis, detrending or filtering of process variables, which introduces further human decisions into the index evaluation.

The original formulation has been initially proposed for the SISO feedback controllers. Further research has been addressing other configurations of the control system, like univariate feedforward [10], unstable and nonminimumphase systems [65], MIMO configurations [31], varying setpoint [48], cascaded control [38]. The researchers have introduced several improvements to the method as for instance Variability Matrix [12], filtering [34], optimization based approaches [54], multi-model mixing time-variant minimum variance [44] and many others [27, 52, 71].

Literature [25] identifies shortcuts of the minimum variance approach, like complexity of non-linear behavior, representation for the considered system and the disturbance and finally efficiency of the model fitting.

5.2 Other Optimal Benchmarking Controllers

The benchmarking uses in the original formulation Åström minimum variance optimal control. It has been found to be not very practical. Additionally, it is not the only controller that can be considered as the best achievable algorithm. The index compares actual performance J_a versus the optimal benchmark one J_o and is defined as

$$\eta = \frac{J_o}{J_a}. \tag{5.18}$$

Process industry is based on the PID algorithm. More than 90% of the controllers uses PID-based loops. Thus, it has been natural for the research to address PID benchmarks as well. Once we consider stochastic control objective, we address the ability to reject unmeasured stochastic disturbances, which may be assessed through the output variance

$$J_o^{PID} = \sigma_{PID}^2 = \min_{PID} \sigma_y^2, \tag{5.19}$$

Solution to the above problem may be obtained with the non-convex optimization [40, 54, 56].

The deterministic performance is connected with the setpoint tracking quality and measured disturbance rejecting ability. One of the proposed approaches uses the lower bound of the IAE index for input load disturbances based on the direct synthesis method [70]. Next, there has been established empirical formulation of the lower bound for IAE and the rising time using first order and second order with time delay models for setpoint step response [32].

The widespread use of the PID control algorithm in process industry motivated the researchers to extend the approach towards optimal PID benchmarking [11, 40]. The best achievable PID controller is used to determine its variance, which is subsequently used as the performance limit. The method also delivers settings of the evaluated optimal PID algorithm, which may be further used. As an example the Multiscale Principal Component Analysis (MSPCA) for performance assessment and diagnosis of PID controllers in steel rolling processes has been proposed in [50] with an optimal PID algorithm utilized to evaluate the performance benchmark.

The PID-based minimum variance approaches to the CPA task possess two important limitations. The first one is associated with the potential mismatch between theoretical PID algorithms and the practical PID loop structure, which is implemented inside of the DCS/PLC logics. One should remember that industrial loops are equipped with many different features (various algorithm versions, tracking, filtering, linearization, dead-bands, tracking, etc.) that cannot be exactly reflected in the theoretical evaluations and thus the optimal PID control may be unachievable in practice. The second limitation is associated with the joint effect of the feedback-feedforward action. These two actions are separated in practical implementations and are decomposed into two separate actions. In the CPA framework, they are treated jointly.

The IMC (Internal Model Control) [13, 15] algorithms form the class of model-based approaches extending PID complexity towards the processes with delays and parameter uncertainties. The main feature of IMC control is an internal model, which enables to generate the prediction for a one single moment in a future associated with the model delay. There are several variants of the IMC approach in the literature. Typical version of the algorithm is sketched in Fig. 5.2.

The IMC algorithm has been used as the minimum variance control benchmark due to its good trade-off between robustness and performance (especially for the process delays). Although the design of the IMC algorithm requires full knowledge

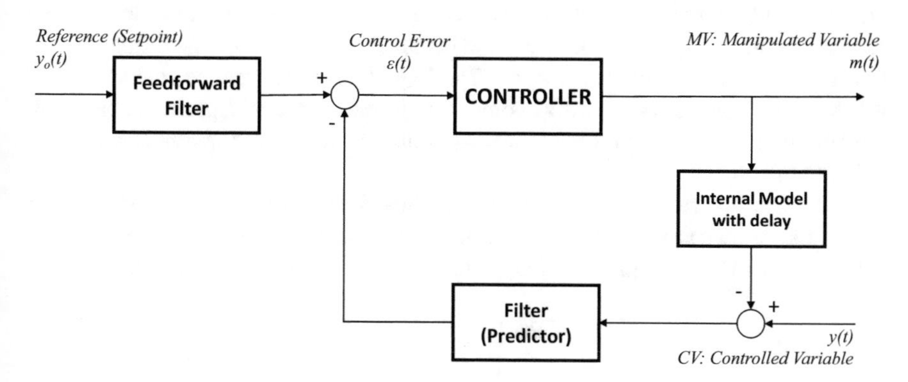

Fig. 5.2 IMC controller structure

of the process model and disturbance, it is not needed for the benchmark evaluation. The IMC-based approach allows to achieve relatively good performance, however far from the performance indicated by the Harris minimum variance index [36]. Consecutive research has started to investigate more complicated controllers that might be applied as the benchmarks, i.e. APC structures.

The advanced version of the minimum variance approach has been proposed in [19] and has been called the GMV (Generalized Minimum Variance). It allows to penalize the cost of the control actions. The method aims at the minimization of the quadratic performance index:

$$J_o^{GMV} = E\left\{[P_c\varepsilon(k) + F_c m(k)]^2\right\}, \tag{5.20}$$

where the P_c and F_c denote the control error and manipulating variable weighting function, respectively. The P_c element often includes integration, while the F_c function incorporates process delay. Unfortunately these functions cannot be selected arbitrary. There is restricting assumption that the closed-loop system has to be stable. Finally we obtain the GMV quality index defined in a standard form:

$$\eta^{GMV} = \frac{J_o^{GMV}}{J_a}. \tag{5.21}$$

The procedure to obtain the minimum variance bound is generally very similar to the standard formulation by Harris. Nonetheless, the computational effort is larger and more process knowledge is required. The evaluation uses both control error signal and the manipulating variable. There is also a need to select the appropriate value for the weighting factor, which sets the trade-off between tracking efficiency and the control actions effort. There are several discussion on the way, how this weighting should be performed (see [18, 22]).

Going beyond the above approaches has inspired researcher to consider Linear Quadratic Gaussian controller as the reference benchmark [4, 30, 37, 49]. The method does not force the user to apply the LQG controller. It only supports the user with the achievable bound for the linear controller. This approach also delivers the second degree of freedom as it enables to penalize control actions. Additionally it is not restricted in the weights selection. The LQG benchmark can be evaluated with much higher computational effort comparing to the previous approaches based on the minimum variance controller. The next limitation factor of the method is the need to know the process model.

The performance index for the LQG benchmark is evaluated using the trade-off curve, also known as the performance limit curve. The solution is generated through solving of the H2/LQG problem [21], where the objective function is defined by

$$J_o^{LQG} = var\left\{y(k)\right\} + \lambda var\left\{m(k)\right\}, \tag{5.22}$$

where λ is a positive penalty coefficient and denotes the trade-off coefficient between tracking and control cost. Different optimal solutions of $Var\,(y(k))$ and $Var\,(m(k))$

may be evaluated by altering its value. The trade-off between the controlled variable performance and the manipulated variable cost provides various achievable optimal control regions. It is responsible for the solution position on the trade-off curve. This curve determines the performance limitation for all possible linear controllers, which might be applied to the considered linear time-invariant process. These solutions include the minimum variance controller as well. The LQG approach is generally applicable to the task of disturbance rejection, despite the fact that the LQG design can handle tracking problems as well.

In a general form the evaluation of the benchmark is done through the selected steps:

1. model identification (ARMAX, ARIMAX),
2. Kalman filter design,
3. LQG design.

As the result two possible measures may be generated:

$$\eta_y^{LQG} = \frac{\sigma^2_{y,LQG}}{\sigma^2_y}, \tag{5.23}$$

$$\eta_m^{LQG} = \frac{\sigma^2_{m,LQG}}{\sigma^2_m}, \tag{5.24}$$

The first measure (5.23) defines optimal performance with respect to the controlled variable variance, while the second one (5.24) with respect to the manipulated variable variance.

The LQG benchmark still is characterized by the disadvantages similar to other minimum variance measures. First of all, more process information is required, i.e. the knowledge of the manipulated and controlled variables. Additionally the construction of the trade-off curve introduces further uncertainties. The calculation effort is even larger as the state estimator has to be developed followed by the solution of the Riccati equations. All these aspects reduce practical applicability of the LQG benchmark [41].

Similarly to the development of the advanced control, the continuation of the research towards Model Predictive Control (MPC), has formed the natural step. Model predictive control technology can now be found widely in a variety of applications including petroleum, chemical and pulp and paper industries. The LQG-type constrained benchmark has been proposed for the assessment of MPC [39]. This approach uses the knowledge of the process model, which is used for the evaluation of the constrained minimum variance controller. Stable inverse of the process model is the additional methodology assumption.

Other methods being proposed for the MPC benchmarking are the design-case approach [55], which uses the MPC controller criterion as the measure performance index J^{MPC}, the infinite-horizon MPC [36]. Further research continues in the direction of the constraint benchmarking, economic performance assessment [67] and on-line model predictive control performance benchmarking and monitoring [31, 73] or the orthogonal projection of the current output onto the space spanned by past outputs, inputs or setpoint using normal routine close loop data [62]. Different MPC approaches can be applied, mostly concentrating on the DMC and GPC algorithms. The final measure performance index has similar formulation to the other minimum variance benchmarks.

$$\eta^{MPC} = \frac{J_o^{MPC}}{J_a}. \tag{5.25}$$

Control reality of process industry shows that the most important and challenging controller structure is multivariate. Thus, the CPA research should also address multivariable configuration. The simplest approach is to decompose the MIMO case into several MISO cases. It might be used in many cases although it is restricted to the situations when there is no coupling between the decomposed loops.

Minimum variance MIMO approaches use the so called interactor matrix (evaluated with filtering and correlation analysis FCOR algorithm). It is used to evaluate the multivariable minimum variance control problem and the the performance assessment index for MIMO processes [69]. Further approaches have been using the historical and user-specified benchmark as two parallel indexes [33, 72] or the procedure using simplified closed-loop model identification and verification if the plant model lies in a performance set that entails the acceptable plant changes for the existing controller under the original disturbance levels [6].

Still multivariate scenarios keep similar shortcuts of other minimum variance approaches increased with the need to build and evaluate the interactor matrix. Approach may be improved with the simplifications as for instance due to evaluation of the lower and upper bands, without the minimum variance itself.

There are other approaches, apart from above mentioned the most popular benchmarks, as for instance:

(a) Extended Horizon Performance Index (EHPI) [9, 64], which considers user specifications to avoid the need for the loop time delay knowledge.
(b) The utilization of the spectral based approach in the minimum variance benchmark has been proposed and evaluated in [66].
(c) Performance index based on desired pole locations [28] that compares the actual variance to the one obtained, when placing all closed-loop poles with one at the origin.
(d) Relative Variance Index (RVI) [2, 71], which compares actual control performance to both minimum variance control and open-loop control, demonstrating the benefits of using RVI compared to Harris index.
(e) The state space controllers has been also considered [20], as the possible benchmark formulations.
(f) Time scale selection method with the Laguerre stochastic model in the minimum variance control has been proposed in [74], but the delay selection problem still remains unsolved.

(g) The method based on multi-model mixing time-variant minimum variance [44].
(h) Various user defined benchmarks based on the historical, best achievable loop performance. The proposed approaches exist in the literatures under different names as for instance baseline [16], historical data benchmark, reference data set benchmark [14] or reference distribution [43].
(i) CPA for a class of nonlinear systems modeled by auto-regressive second order Volterra series with a general linear additive disturbance using the nonlinear GMV concept [45].

5.3 Other Model-Based Approaches

Apart from the mainstream approach, which uses the minimum variance benchmarking, some other model-based methods have been also developed. Very often proposed methodologies address control different strategies:

(a) The relative performance index has been proposed by Li [42]. The approach uses reference model that reflects the required closed-loop behavior and enables to evaluate the performance index defined as the ratio between the reference and actual value of a measure.
(b) The approach proposed by Gupta et al. in [23] uses the model of the process accompanied with the estimation of the input and output disturbances with the Auto Regressive (AR) process.
(c) Close loop model identification with the use of the state space filters with the recursive least squares procedure further used to calculate the aggressiveness and sluggishness indexes [51].
(d) Model-based approaches using frequency domain analysis of closed loop systems to detect the stiction [60].
(e) Evaluation of the model residuals that further may be used to design the performance assessment index for model predictive control [63].
(f) The frequency approach using nonparametric models in form of the Bode curves has been proposed in [53] and further developed for the fractional order models in [57, 58].
(g) Factorization methods to estimate the generalized interactor matrix used a reference behavior estimation [59].
(h) In the recent approach Yan et al. [68] has proposed the oscillation detection algorithm, which uses Hidden Markov Models (HMM) to recognize the oscillatory behavior.

References

1. Bauer, M., Horch, A., Xie, L., Jelali, M., Thornhill, N.: The current state of control loop performance monitoring—a survey of application in industry. J. Process Control **38**, 1–10 (2016)
2. Bezergianni, S., Georgakis, C.: Controller performance assessment based on minimum and open-loop output variance. Control Eng. Pract. **8**(7), 791–797 (2000)
3. Bialic, G.: Methods of control performance assessment for sampld data systems working under stationary stoachastics disturbances. Ph.D. thesis, Dissertation of Technical University of Opole, Poland (2006)
4. Bialic, G., Błachuta, M.J.: Performance assessment of control loops with PID controllers based on correlation and spectral analysis. In: Proceedings of the 12th IEEE International Conference on Methods and Models in Automation and Robotics, MMAR'2006 (2006)
5. Björklund, S.: A survey and comparison of time-delay estimation methods in linear systems. Tech. Rep. Linköping Studies in Science and Technology, Thesis No. 1061, Linköping Universitet, Department of Electrical Engineering, Division of Automatic Control, Sweden (2003)
6. Bombois, X., Potters, M., Mesbah, A.: Closed-loop performance diagnosis for model predictive control systems. In: 2014 European Control Conference (ECC), pp. 264–269 (2014)
7. Cinar, A., Marlin, T.E., MacGregor, J.F.: Automated monitoring and assessment of controller performance. In: IFAC Proceedings, iFAC Symposium on On-line Fault Detection and Supervision in the Chemical Process Industries, Newark, Delaware, 22–24 April, vol. 24, no. 4, pp. 163–167 (1992)
8. CPC: Univariate Controller Performance Assessment. University of Alberta, Computer Process Control Group, Limited Trial Version (2010)
9. Desborough, L., Harris, T.J.: Performance assessment measures for univariate feedback control. Can. J. Chem. Eng. **70**(6), 1186–1197 (1992)
10. Desborough, L., Harris, T.J.: Performance assessment measures for univariate feedforward/feedback control. Can. J. Chem. Eng. **71**(4), 605–616 (1993)
11. Eriksson, P.G., Isaksson, A.J.: Some aspects of control loop performance monitoring. In: 1994 Proceedings of IEEE International Conference on Control and Applications, vol. 2, pp. 1029–1034 (1994)
12. Farenzena, M.: Novel methodologies for assessment and diagnostics in control loop management. Ph.D. thesis, Dissertation of Universidade Federal do Rio Grande do Sul, Porto Alegre, Brazil (2008)
13. Frank, P.M.: Entwurf von Regelkreisen mit vorgeschriebenen Verhalten. Wissenschaft + Technik: Taschenausgaben, Braun (1974)
14. Gao, J., Patwardhan, R., Akamatsu, K., Hashimoto, Y., Emoto, G., Shah, S.L., Huang, B.: Performance evaluation of two industrial MPC controllers. Control Eng. Pract. **11**(12), 1371–1387. In: 2002 IFAC World Congress (2003)
15. Garcia, C.E., Morari, M.: Internal model control. A unifying review and some new results. Ind. Eng. Chem. Process Des. Dev. **21**(2), 308–323 (1982)
16. Gerry, J.P.: Real-time performance assessment. In: Liptak, B.G. (ed.) Instrument Engineers' Handbook: Process Control and Optimization, vol. 2, 4th edn, pp. 311–317. CRC Press (2005)
17. Gomez, D., Moya, E.J., Baeyens, E.: Control performance assessment: a general survey. In: L F de Carvalho, A.P., Rodriguez-Gonzalez, S., De Paz Santana, J.F., Rodriguez, J.M.C. (eds.) Distributed Computing and Artificial Intelligence: 7th International Symposium, Springer, Berlin, Heidelberg, pp. 621–628 (2010)
18. Grimble, M., Majecki, P.: Weighting selection for controller benchmarking and tuning. Tech. Rep. PAM-12-TN-1-V1, Industrial Control Centre, University of Strathclyde, Glasgow, UK (2004)
19. Grimble, M.J.: Controller performance benchmarking and tuning using generalised minimum variance control. Automatica **38**(12), 2111–2119 (2002)
20. Grimble, M.J.: Restricted structure control loop performance assessment for PID controllers and state-space systems. Asian J. Control **5**(1), 39–57 (2003)

21. Grimble, M.J.: Robust Industrial Control Systems: Optimal Design Approach for Polynomial Systems. Wiley (2006)
22. Grimble, M.J., Uduehi, D.: Process control loop benchmarking and revenue optimization. In: Proceedings of the 2001 American Control Conference, vol. 6, pp. 4313–4327 (2001)
23. Gupta, A., Mathur, T., Stadler, K.S., Gallestey, E.: A pragmatic approach for performance assessment of advanced process control. In: 2013 IEEE International Conference on Control Applications (CCA), pp. 754–759 (2013)
24. Harris, T.J.: Assessment of closed loop performance. Can. J. Chem. Eng. **67**, 856–861 (1989)
25. Harris, T.J., Yu, W.: Controller assessment for a class of non-linear systems. J. Process Control **17**(7), 607–619 (2007)
26. Hoo, K.A., Piovoso, M.J., Schnelle, P.D., Rowan, D.A.: Process and controller performance monitoring: overview with industrial applications. Int. J. Adapt. Control Signal Process. **17**(7–9), 635–662 (2003)
27. Horch, A., Isaksson, A.J.: A modified index for control performance assessment. In: Proceedings of the 1998 American Control Conference, vol. 6, pp. 3430–3434 (1998)
28. Horch, A., Isaksson, A.J.: A modified index for control performance assessment. J. Process Control **9**(6), 475–483 (1999)
29. Huang, B., Shah, S.L.: Limits of control loop performance: practical measures of control loop performance assessment. In: AIChE Annual Meeting (1996)
30. Huang, B., Shah, S.L.: Performance Assessment of Control Loops: Theory and Applications, 1st edn. Springer, Berlin (1999)
31. Huang, B., Shah, S.L., Kwok, E.K.: On-line control performance monitoring of MIMO processes. In: Proceedings of the 1995 American Control Conference, vol. 2, pp. 1250–1254 (1995)
32. Huang, H.P., Jeng, J.C.: Monitoring and assessment of control performance for single loop systems. Ind. Eng. Chem. Res. **41**(5), 1297–1309 (2002)
33. Huang, Q., Zhang, Q.: Research on multivariable control performance assessment techniques. In: 2011 International Symposium on Advanced Control of Industrial Processes (ADCONIP), IEEE, pp. 508–511 (2011)
34. Jain, M., Lakshminarayanan, S.: A filter-based approach for performance assessment and enhancement of SISO control systems. Ind. Eng. Chem. Res. **44**(22), 8260–8276 (2005)
35. Jelali, M.: An overview of control performance assessment technology and industrial applications. Control Eng. Pract. **14**(5), 441–466 (2006)
36. Jelali, M.: Control Performance Management in Industrial Automation: Assessment, Diagnosis and Improvement of Control Loop Performance. Springer, Berlin (2013)
37. Kadali, R., Huang, B.: Controller performance analysis with LQG benchmark obtained under closed loop conditions. ISA Trans. **41**(4), 521–537 (2002)
38. Ko, B.S., Edgar, T.F.: Performance assessment of cascade control loops. AIChE J. **46**(2), 281–291 (2000)
39. Ko, B.S., Edgar, T.F.: Performance assessment of constrained model predictive control systems. AIChE J. **47**(6), 1363–1371 (2001)
40. Ko, B.S., Edgar, T.F.: PID control performance assessment: the single-loop case. AIChE J. **50**(6), 1211–1218 (2004)
41. Kozub, D.J.: Controller performance monitoring and diagnosis. Industrial perspective. In: IFAC Proceedings, 15th IFAC World Congress, vol. 35, no. 1, pp. 405–410 (2002)
42. Li, Q., Whiteley, J.R., Rhinehart, R.R.: A relative performance monitor for process controllers. Int. J. Adapt. Control Signal Process. **17**(7–9), 685–708 (2003)
43. Li, Q., Whiteley, J.R., Rhinehart, R.R.: An automated performance monitor for process controllers. Control Eng. Pract. **12**(5), 537–553, Fuzzy System Applications in Control (2004)
44. Liu, M.C.P., Wang, X., Wang, Z.L.: Performance assessment of control loop with multiple time-variant disturbances based on multi-model mixing time-variant minimum variance control. In: Proceeding of the 11th World Congress on Intelligent Control and Automation, pp. 4755–4759 (2014)

45. Maboodi, M., Khaki-Sedigh, A., Camacho, E.F.: Control performance assessment for a class of nonlinear systems using second-order volterra series models based on nonlinear generalised minimum variance control. Int. J. Control **88**(8), 1565–1575 (2015)
46. Meng, Q.W., Gu, J.Q., Zhong, Z.F., Ch, S., Niu, Y.G.: Control performance assessment and improvement with a new performance index. In: 2013 25th Chinese Control and Decision Conference (CCDC), pp. 4081–4084 (2013)
47. Ordys, A., Uduehi, D., Johnson, M.A.: Process Control Performance Assessment—From Theory to Implementation. Springer, Berlin (2007)
48. Perrier, M., Roche, A.A.: Towards mill-wide evaluation of control loop performance. In: Proceedings of the Control Systems, pp. 205–209 (1992)
49. Pour, N.D., Huang, B., Shah, S.: Performance assessment of advanced supervisory-regulatory control systems with subspace LQG benchmark. Automatica **46**(8), 1363–1368 (2010)
50. Recalde, L.F., Katebi, R., Tauro, H.: PID based control performance assessment for rolling mills: a multiscale PCA approach. In: 2013 IEEE International Conference on Control Applications (CCA), pp. 1075–1080 (2013)
51. Salsbury, T.I.: Continuous-time model identification for closed loop control performance assessment. Control Eng. Pract. **15**(1), 109–121 (2007)
52. Sarabi, B.K., Maghade, D.K., Malwatkar, G.M.: An empirical data driven based control loop performance assessment of multi-variate systems. In: 2012 Annual IEEE India Conference (INDICON), pp. 070–074 (2012)
53. Schlegel, M., Skarda, R., Cech, M.: Running discrete Fourier transform and its applications in control loop performance assessment. In: 2013 International Conference on Process Control (PC), pp. 113–118 (2013)
54. Sendjaja, A.Y., Kariwala, V.: Achievable PID performance using sums of squares programming. J. Process Control **19**(6), 1061–1065 (2009)
55. Shah, S.L., Patwardhan, R., Huang, B.: Multivariate controller performance analysis: methods, applications and challenges. In: Proceedings of the Sixth International Conference on Chemical Process Control, pp. 199–207 (2001)
56. Shahni, F., Malwatkar, G.M.: Assessment minimum output variance with PID controllers. J. Process Control **21**(4), 678–681 (2011)
57. Skarda, R., Cech, M., Schlegel, M.: Bode-like control loop performance index evaluated for a class of fractional-order processes. In: IFAC Proceedings, 19th IFAC World Congress, vol. 47, no. 3, pp. 10622–10627 (2014)
58. Skarda, R., Cech, M., Schlegel, M.: Simultaneous control loop performance assessment and process identification based on fractional models. IFAC-PapersOnLine. In: 9th IFAC Symposium on Advanced Control of Chemical Processes ADCHEM 2015, vol. 48, no. 8, pp. 859–864 (2015)
59. Souza, D.L., Oliveira-Lopes, L.C.: A novel controller performance index using a local process model factorization. J. Adv. Res. Dyn. Control Syst. **6**(3), 8–23 (2014)
60. Srinivasan, B., Spinner, T., Rengaswamy, R.: A reliability measure for model based stiction detection approaches. In: IFAC Proceedings, 8th IFAC Symposium on Advanced Control of Chemical Processes, vol. 45, no. 15, pp. 750–755 (2012)
61. Starr, K.D., Petersen, H., Bauer, M.: Control loop performance monitoring—ABB's experience over two decades. IFAC-PapersOnLine. In: 11th IFAC Symposium on Dynamics and Control of Process Systems Including Biosystems DYCOPS-CAB 2016, vol. 49, no. 7, pp. 526–532 (2016)
62. Sun, Z., Qin, S.J., Singhal, A., Megan, L.: Control performance monitoring via model residual assessment. In: 2012 American Control Conference (ACC), pp. 2800–2805 (2012)
63. Sun, Z., Qin, S.J., Singhal, A., Megan, L.: Performance monitoring of model-predictive controllers via model residual assessment. J. Process Control **23**(4), 473–482 (2013)
64. Thornhill, N.F., Oettinger, M., Fedenczuk, P.: Refinery-wide control loop performance assessment. J. Process Control **9**(2), 109–124 (1999)
65. Tyler, M.L., Morari, M.: Performance assessment for unstable and nonminimum-phase systems. IFAC Proceedings, vol. 28, no. 12, pp. 187–192 (1995)

66. Vishnubhotla, A.: Frequency and time-domain techniques for control loop performance assessment. Ph.D. thesis, University of Alberta, Department of Chemicai and Materiah Engineering, Canada (1997)
67. Xu, F., Huang, B., Tamayo, E.C.: Assessment of economic performance of model predictive control through variance/constraint tuning. In: IFAC Proceedings, 6th IFAC Symposium on Advanced Control of Chemical Processes, vol. 39, no. 2, pp. 899–904 (2006)
68. Yan, Z., Chen, J., Zhang, Z.: Using hidden markov model to identify oscillation temporal pattern for control loops. Chem. Eng. Res. Des. **119**, 117–129 (2017)
69. Yu, J., Qin, S.J.: MIMO control performance monitoring using left/right diagonal interactors. J. Process Control **19**(8), 1267–1276, Special Section on Hybrid Systems: Modeling, Simulation and Optimization (2009)
70. Yu, Z., Wang, J.: Assessment of proportional–integral control loop performance for input load disturbance rejection. Ind. Eng. Chem. Res. **51**(36), 11744–11752 (2012)
71. Yuan, Q., Lennox, B.: A new framework for controller performance assessment. In: 2007 IEEE International Conference on Control and Automation, pp. 2679–2684 (2007)
72. Yuan, Q., Lennox, B., McEwan, M.: Analysis of multivariable control performance assessment techniques. J. Process Control **19**(5), 751–760 (2009)
73. Zhang, R., Zhang, Q.: Model predictive control performance assessment using a prediction error benchmark. In: 2011 International Symposium on Advanced Control of Industrial Processes (ADCONIP), pp. 571–574 (2011)
74. Zhong, Z., Meng, Q., Liu, J.: Control system comprehensive performance assessment. In: Proceedings of the 33rd Chinese Control Conference, pp. 6268–6272 (2014)

Chapter 6
Frequency Based Methods

We will proceed no further in this business.
–William Shakespeare

The utilization of the frequency based approach in the research on CPA is relatively rare. The frequency based analysis mostly focuses on the classical notions of Nyquist, Bode and Nichols charts or phase and gain margins. Bode plots present in the double logarithmic scale the relation of the transfer function magnitude $|G(j\omega)|$ and phase $\varphi(\omega)$ against the frequency Nyquist chart, called the polar plot presents the real versus the imaginary part of the frequency domain transfer function $G(j\omega)$ with tω being a parameter along the curve. The Nichols plot is a single curve in a coordinate system with phase angle as the abscissa and log modulus as the ordinate (ω is a parameter along the curve) [7].

Phase and gain margins are used as common control quality indexes. The gain margin is defined as the reciprocal of the intersection for open loop frequency domain transfer function polar plot on the negative real axis. Phase margin is defined as the angle between the negative real axis and a radial line drawn from the origin to the point where the open loop frequency domain transfer function intersects the unit circle. The lower the margin value is, the closer to the stability edge the close loop system is.

The maximum closed loop log modulus can be also considered as o potential performance quality measure in the frequency domain [5]. It directly measures the closeness of the open loop frequency domain transfer function to the $(-1, 0 \cdot j)$ point at all frequencies, and thus it is more robust than the margins.

As the frequency plots form the non-parametric models the control performance assessment mostly requires the process mathematical model. Furthermore, one has to be aware that due to the fact that we are estimating control loop quality, the identification has to be organized in the close loop with all the limitations impeded

P. D. Domański, *Control Performance Assessment: Theoretical Analyses and Industrial Practice*, Studies in Systems, Decision and Control 245,
https://doi.org/10.1007/978-3-030-23593-2_6

by active feedback. The next shortcut associated with these models is that they are non-parametric, mostly in form of the curves. Method requires in most cases human-in-the-loop visualization supporting the decision-making process, determining if loop quality is acceptable or not. The above issues significantly limit practical implementation of the frequency-based approaches in the [6, 9].

The first relevant research has been proposed by Kendra and Cinar in [4]. Performance assessment has been based on the comparison between the observed frequency response characteristics and the design specifications. It has been evaluated through the excitation of the setpoint with a zero mean, pseudo random binary sequence and observation of the control error controlled variable responses. In the same years Vishnubhotla [14] proposed the formulation of the minimum variance approach using spectral analysis of the close loop system.

The maximum closed-loop modulus has been successfully used to monitor the performance of the controlled process and can be extended to include an economic aspect to the control performance assessment algorithm [2, 6]. However, this method requires additionally to perform the dedicated testing experiment to gather appropriate data, which may limit method industrial applicability.

This shortcut is considered and omitted in the works of Schglegel and Skarda. Model estimation has been done using normal operation process data with the assumption that plant dynamics is sufficiently excited [8]. The authors use running Discrete Fourier Transform (DFT) to assess different aspects of the control loop, as for instance an increasing valve stiction may allow to reveal a cause of oscillations in the loop or can be also used for continuous monitoring of process changes at particular frequencies. Authors have proposed specific sensitivity function based on Bodes theorem and an assumption of process monotonicity and controller exhibiting integrating behavior at low frequencies (like PI/PID) as a reference model.

It has enabled to propose two dedicated indexes of the peak value and the available loop bandwidth. The key samples of actual sensitivity function are estimated by Fourier transform and compared to the ideal values. The proposed method has been applied to assess PID based control loops. The substantial advantage of the proposed approaches is that it provides information about controller performance, i.e. its sluggishness or aggressiveness.

The methodology has been extended in the consecutive research [11, 12]. The new approach a priori has assumed that the process can be described by multiple fractional-pole model [3], which covers the large number of essentially monotone real processes.

$$G\left(s\right)=\frac{K}{\prod_{i=1}^{p}\left(\tau_{i}s+1\right)^{n_{i}}}. \tag{6.1}$$

The assessment procedure consists of the three following steps:

1. closed loop identification of two frequency response points,
2. frequency response points validation by error estimation algorithm,
3. evaluation of the process model.

Some indirect references to the CPA task may be obtained with the use of the reference to disturbance ratio (RDR) [1]. The proposed methodology calculates the ratio of reference signal to disturbance signal energy at the system output and provides a quantitative measure of disturbance rejection performance of control systems on the bases of communication channel limitations. It provides a straightforward analytical method for the comparison and improvement of implicit disturbance rejection capacity of closed loop control systems. The proposed methodology has been tested with the simulations for the classical PID and fractional order PID algorithms.

Another assessment opportunity can be obtained with the utilization of the sensitivity function [13]. The approach can be used to design fractional-order PID controller controller and furthermore it might be applied to control quality measurement task. The assessment of the PI IMC controller using frequency domain defined phase and gain margins has been considered in [10]. Although these methods do not address directly the CPA task, their application as potential control quality measures is straightforward.

References

1. Alagoz, B.B., Tan, N., Deniz, F.N., Keles, C.: Implicit disturbance rejection performance analysis of closed loop control systems according to communication channel limitations. IET Control Theory Appl. **9**(17), 2522–2531 (2015)
2. Belanger, P.W., Luyben, W.L.: A new test for evaluation of the regulatory performance of controlled processes. Ind. Eng. Chem. Res. **35**(10), 3447–3457 (1996)
3. Charef, A., Sun, H.H., Tsao, Y.Y., Onaral, B.: Fractal system as represented by singularity function. IEEE Trans. Autom. Control **37**(9), 1465–1470 (1992)
4. Kendra, S.J., Cinar, A.: Controller performance assessment by frequency domain techniques. J. Process Control **7**(3), 181–194 (1997)
5. Luyben, W.L.: Process Modeling, Simulation, and Control for Chemical Engineers. Chemical Engineering Series. McGraw-Hill (1990)
6. O'Connor, N., O'Dwyer, A.: Control loop performance assessment: a classification of methods. In: Proceedings of the Irish Signals and Systems Conference, Queens University Belfast, pp. 530–535 (2002)
7. Ogata, K.: Modern Control Engineering, 4th edn. Prentice Hall PTR, Upper Saddle River, NJ, USA (2001)
8. Schlegel, M., Skarda, R., Cech, M.: Running discrete Fourier transform and its applications in control loop performance assessment. In: 2013 International Conference on Process Control (PC), pp. 113–118 (2013)
9. Shardt, Y., Zhao, Y., Qi, F., Lee, K., Yu, X., Huang, B., Shah, S.: Determining the state of a process control system: current trends and future challenges. Can. J. Chem. Eng. **90**(2), 217–245 (2012)
10. da Silva Moreira, L.J., Junior, G.A., Barros, P.R.: IMC PI control loops frequency and time domains performance assessment and retuning. IFAC-PapersOnLine. In: 3rd IFAC Conference on Advances in Proportional-Integral-Derivative Control PID 2018, vol. 51, no. 4, pp. 148–153 (2018)
11. Skarda, R., Cech, M., Schlegel, M.: Bode-like control loop performance index evaluated for a class of fractional-order processes. In: 19th IFAC World Congress, IFAC Proceedings, vol. 47, no. 3, pp. 10622–10627 (2014)

12. Skarda, R., Cech, M., Schlegel, M.: Simultaneous control loop performance assessment and process identification based on fractional models. IFAC-PapersOnLine. In: 9th IFAC Symposium on Advanced Control of Chemical Processes ADCHEM 2015, vol. 48, no. 8, pp. 859–864 (2015)
13. Tepljakov, A., Petlenkov, E., Belikov, J.: A flexible MATLAB tool for optimal fractional-order PID controller design subject to specifications. In: 2012 31st Chinese Control Conference (CCC) (2012)
14. Vishnubhotla, A.: Frequency and time-domain techniques for control loop performance assessrnent. Ph.D. thesis, University of Alberta, Department of Chemicai and Materiah Engineering, Canada (1997)

Chapter 7
Alternative Indexes

For a complex natural shape, dimension is relative. It varies with the observer. The same object can have more than one dimension, depending on how you measure it and what you want to do with it.

– Benoît B. Mandelbrot

Alternative indexes in the CPA task go beyond classical and commonly used assessment methodologies. They try to address practical aspects, which are frequently met in the process industry reality. One may distinguish two general classes of indexes from the perspective of possible implementations and practical use:

- *non-constructive* descriptive measures not giving any hints, what should be done to improve performance,
- and *constructive* ones not only monitoring but also enabling root-cause analysis and showing direction of performance improvement. Constructive indexes are much more informative as they help in further control improvement.

Another general property of any control performance assessment methodology is the source of data being further used during index evaluation. The ideal approach should not require any specifically or artificially generated data (step test, dedicated testing, open loop, etc.) for evaluation and should utilize signals from normal process operation. Further practical requirement assumes that the industrial method should not depend on any specific process knowledge or user defined subjective parametrization. One has to be aware that such dedicated knowledge may be a source of the errors or results falsification and/or may require the presence of the expert interventions in the method formulation.

The review of the previously described approaches shows that the minimum variance based methods constitutes the majority of the research, despite its shortcuts,

P. D. Domański, *Control Performance Assessment: Theoretical Analyses and Industrial Practice*, Studies in Systems, Decision and Control 245,
https://doi.org/10.1007/978-3-030-23593-2_7

i.e. the need for process knowledge or its delay at least and relatively high computational burden. The results of minimum variance based approach are based on the Gaussian disturbance properties assumption. It might be expected that the assumed benchmarks might not be achievable nor falsified, when applied to control loops witnessing non-Gaussian properties. Observation of industrial examples [12] shows that the majority of the loops in process industry generates data, which are characterized by the non-Gaussian, mostly fat-tail properties.

In last years we may find different attempts to the methods using nonlinear data analysis, which are not based on the assumption of Gaussian statistics. In 2012 Srinivasan et al. [79] has proposed the utilization of the Hurst exponent evaluated with Detrended Fluctuation Analysis (DFA). The approach has been tested for SISO control loop with PI and PID controllers. The same direction has been followed by Spinner [77, 78]. The authors have used DFA analysis to the process output. Reason why detrending has been applied was caused by large variability of the process output as it often has to track variations in setpoint.

Different approach has been prosed by Pillay in 2014 [70]. He suggested to apply DFA analysis and Hurst exponent evaluation to MSE of the control error signal. His examples are also based on the SISO loops controlled by PID controller.

Further research followed this direction investigating the area of the persistence [14], fractal [13, 16], multi-fractal [15] and fractional order measures for different control algorithms [18] proposing comprehensive procedures, obtaining promising results and being followed by the industrial applications [21].

Finally, an interesting approach to use minimum entropy index with Renyi's quadratic entropy [7, 39, 92–94] has been proposed. The index has been tested with simulated temperature cascade control industrial example of the disturbance tower. It has been extended from SISO loops to the cascaded control. Entropy can be used to measure many different properties of the disturbances, such as similarity, equality, disorder, and so on [20]. In fact this approach follows analysis applied in other domains like biological systems [25]. Alternative non-Gaussian approaches, like persistence, fractal, multi-fractal, fractional order or entropy based indexes are still not well established. The following sections bring them closer in details.

7.1 Fractal Approach

Fractal and persistence approach forms an extension of the non-Gaussian statistical analysis presented in Sect. 4. These methods are very popular in economic [43, 45, 48, 73, 84], natural and social sciences [6, 59, 60, 64, 69, 76, 80]. Fractals have also met acceptance in medicine [54, 67, 68] and some engineering areas, like telecommunication [3, 23, 49], informatics [50], wind energy [43, 44] or prediction [17]. In control engineering they are relatively rarely used and lie outside of the mainstream interest.

Fractal time series analysis has been initially developed to reflect and capture economical features. In economy there exists effectiveness hypothesis assuming that if prices reflect all publicly available information, new prices are only caused by new information. The prices should hold properties of the Brownian motion. It assumes that future is independent on the past and the present. However, practice does not reflect it. Information is neither complete, nor known *a priori*. We do not react immediately and simultaneously. Following that, we obtain rather fractional Brownian motion. Research shows that similar behavior and results are observed in many different areas. Fractal methods have found several applications in hydrology [30], geology [85], solar physics [62], meteorology, seismology, biology, medicine, telecommunication, networks, etc. Analysis of control engineering data in form of the time series originating from real complex industrial systems reveals similar properties.

Unfortunately, there are only a few reported publications about fractal approach in control engineering. Authors in [79] have addressed the subject of the Hurst exponent application in the assessment of PI and PID controllers. In [9] scaling (Hurst) exponent has been used to assess Kalman filter performance. In the recent works authors [10] have performed diagnosis of MIMO control loops with Hurts exponent evaluated using Detrended Fluctuation Analysis algorithm with the Mahalanobis distance. The same approach has been applied to the disturbed univariate and multivariate systems with disturbances [11] as well. Further research has addressed similar approaches using multiple scales and the notion of crossover [12, 13].

The notion of fractals (in Latin *fractus*—broken, fragmentary), self-similar or *infinitely subtle* has been introduced in *70's* by Benoit Mandelbrot [58]. Fractal analysis is especially useful when the data are characterized by power-law, self-similarity, scale-invariance or nonlinear properties [22, 72]. Following fractal theory the traditional concept of three-dimensional space has been be extended into the fractal (fractional) dimension (FD).

Self-similarity is the underlying concept behind fractals and power laws reflecting invariance against changes in scale or size [73]. It may be perceived as the symmetry shaping of our environment and the way one may interpret it. The idea of an alternative approach to the concept of randomness [80] has caused the development of fractal analysis, although the mathematicians have noticed and investigated similar behavior long before. Fractals have existed as the mathematical formulas much earlier. Researchers have found that there exist specific objects with a common property of non-integer Hausdorff number [73]. The previous research of Cantor, Peano, Hilbert, Koch, Sierpiński or Julia enabled Mandelbrot to develop the fundamentals of his theory. Simultaneously computers helped a lot enabling much easier visualization. The concept of fractal has become more friendly. Concluding, we may point out three main properties of fractals: self-similarity, Hausdorff dimension larger then topological one and simple recursive definition.

7.1.1 Hurst Exponent as a Persistence Measure

The first step in the verification whether the data under consideration hold fractal features and whether this approach is adequate is to check if the data satisfy the assumptions of the power law. The test is simple. We need to verify if frequency of an event, varies as a power of some attribute of that event. George K. Zipf, professor of the German philology at Harvard University, has noticed that the frequency of the words in a text is inversely proportional to its rank:

$$c(w) \approx \left\lfloor \frac{A}{r(w)} \right\rfloor. \tag{7.1}$$

where:

- A—number of different words in a text,
- $\lfloor x \rfloor$—integer value **x**.

The power law is often visualized with the double logarithmic plot, in which the above relation forms a line. Similar features have been also observed by Pareto in a distribution of wealth or by Gutenberg and Richter for the relation between the magnitude and a total number of earthquakes. The *log-log* relationship is characterized by its slope. One of the methods to asses such phenomena is to calculate a value of the so called Hurst (scaling) exponent denoted by H.

H. E. Hurst worked on Nile dams having history of *847* river overflows [34]. It has been assumed that the overflows are independent on history. He proved otherwise. Hurst exponent is defined as the asymptotic property of the rescaled range R/S plot.

$$E\left[\frac{R(n)}{S(n)}\right] = cn^{H}, n \to \infty \tag{7.2}$$

where n is a number of observations, R reflects data range of length n, S—standard deviations in time slot n, c is a positive constant and H a Hurst exponent. H is calculated using logarithms

$$ln\, E\left[\frac{R(n)}{S(n)}\right] = \ln\, c\, +\, H \ln\, n \tag{7.3}$$

plotted in double logarithmic scale $E\left[\frac{R(n)}{S(n)}\right]$ from n estimating H as the line slope. Exemplary R/S plot is sketched in Fig. 7.1.

Small value of n will make the result anti-persistent and the value of Hurst parameter will be deceiving. On the other side a large value of n together with short data set may produce too few samples for correct self similar process H estimation. Therefore, the number of observations selection should be judicious. Hurst exponent is a very important parameter in fractal theory. Its value reflects the following properties of the considered time series.

- $H < 0$ means that the white noise may be over differenced.
- $0 < H < 0.5$ means anti-persistent (ergodic) time series, i.e. data returning towards average. Data increase in the past suggests decrease in the future and vice versa.

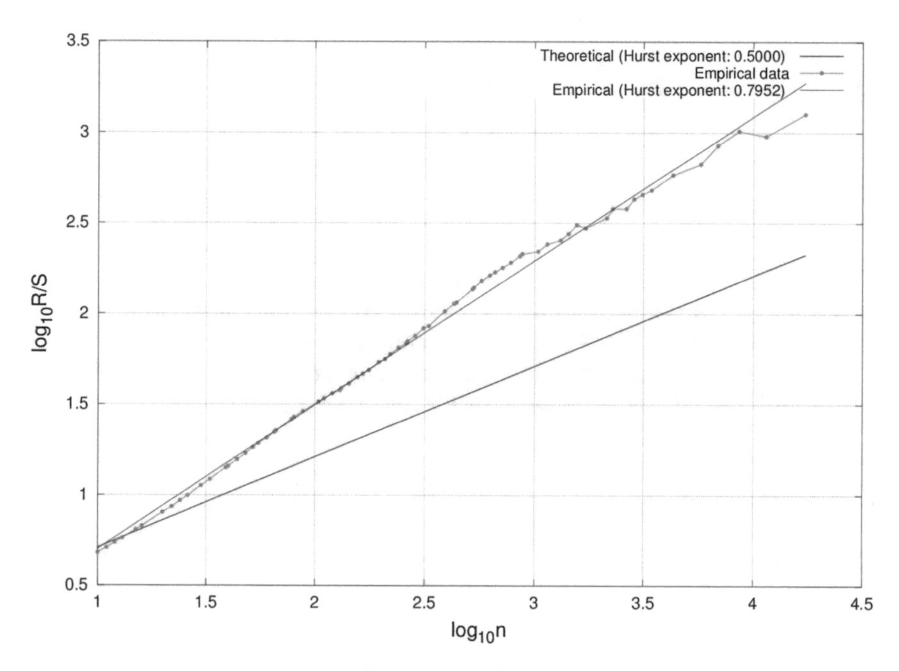

Fig. 7.1 The example of the persistent ($H \approx 0.8$) R/S plot

- If $H = 0.5$ then time series has properties of the Brownian motion. All observations are statistically independent. We get stochastic uncorrelated process. Gaussian process is an example of such data.
- $0.5 < H < 1$ means persistent process, i.e. data increase/decrease in the past implies increase/decrease in the future, respectively. It is said to be fractional Brownian motion—fractional white noise with Long Range Dependence (LRD).
- $1 \leq H < 1.5$ reflect fractional brown noise with no dependency in time domain, i.e. fractional integration of white noise, with $d \in [0.5, 1)$.
- $H = 1.5$ is a Brown noise.

Normally in the control engineering time series analysis, Hurst exponent is limited to a set $H \in (0, 1)$, however its measuring might be problematic. In some cases, relations in data are so complex and corruptions inside of data (trends, periodicity) so strong that conventional methods fail to properly reflect them. It may result in Hurst exponent estimates greater than 1.

It should be noted that there exists the hypothesis about the equivalence between parameters of α-stable distribution and the Hurst exponent. The research has shown that stability factor α of the α-stable distribution characteristics equation (4.12) may be equivalent to the inverse of Hurst exponent ($H = 1/\alpha$) [69]. There exists several assumptions on data characteristics limiting validity of this hypothesis. This relationship is often biased by the varying long-range correlations of the small and large fluctuations. Thus, this rule has a very limited applicability.

In literature we may find several further parameters originating from Hurst exponent, as for example fractal correlation

$$C = 2^{(2H-1)} - 1 \tag{7.4}$$

or fractal dimension

$$D = 2 - H. \tag{7.5}$$

R/S plot is the original method to estimate H. It enables to evaluate not only single scaling exponents, but also reveals additional features like multiple scalings, crossover points, etc. Apart from above Hurst exponent evaluation method, there are other algorithms [56], as for instance (in historical order).

7.1.1.1 Periodogram

The periodogram method [26] is defined by:

$$I(\xi) = \frac{1}{2\pi N} \left| \sum_{j=1}^{N} X_j e^{ij\xi} \right|^2 \tag{7.6}$$

where, x_i is the frequency and $i = \sqrt{-1}$. For a series with finite variance, $I(\xi)$ is an estimate of the spectral density of the series. A *log-log* plot of $I(\xi)$ should have a slope of $(1 - 2H)$ close to the origin.

7.1.1.2 Higuchi (Fractal Dimension Method) Algorithm

In the Higuchi algorithm [33], time series is grouped into n partitions. At first, the aggregated sums of the series are evaluated. In the next step absolute differences of these cumulative sums between the partitions are analyzed to find out the fractal dimension of the time series.

This process is repeated for n number of groups. The result changes as n increases and is related to the Hurst parameter H of the time series. A *log-log* plot of the statistic versus number of the partitions is ideally expected to be linear with a slope related to H.

7.1.1.3 Haslett and Raftery Estimate

The algorithm is based on the ARFIMA model and uses spatial autocorrelation [31].

$$\left(1-\sum_{i=1}^{p}\phi_i B^i\right)(1-B)^d X_t = \left(1-\sum_{i=1}^{q}\theta_i B^i\right)\varepsilon_t \tag{7.7}$$

For solving and H estimation,maximum likelihood estimation problem is solved usually faster than Cholesky decomposition or Durbin-Levinson algorithm.

7.1.1.4 Boxed or Modified Periodogram Method

The method is a modification of the Periodogram method [32]. The algorithm divides the frequency axis into logarithmically equally spaced boxes and averages the periodogram values corresponding to the frequencies inside the box.

7.1.1.5 The Haar Wavelet and Maximum Likelihood Estimation

A Hurst parameter estimation method is based on Haar wavelet and maximum likelihood estimation [41]. This method firstly uses discrete wavelet transform using the Haar basis to the increments of fractional Brownian motion, also known as discrete fractional Gaussian noise. The method uses likelihood estimation and yields coefficients which are weakly correlated and have a variance that is exponentially related to scale. They further enable to evaluate Hurst exponent [87].

7.1.1.6 Aggregated Variance Method

This method [5] considers variance of $\left(X_t^{(m)}\right)$, where $X_t^{(m)}$ is a time series obtained from X_t by aggregated over the m number of blocks

$$Var\left(X_t^{(m)}\right) = m^{2H-2},\ \frac{N}{m} \to \infty\ and\ m \to \infty \tag{7.8}$$

7.1.1.7 Variance of Residuals Method

In method described by Peng [66] and called variance of residuals method, the series is divided into blocks of size m. Within each block, the cumulated sums are computed up to t and a least-squares line $a + b * t$ is fitted to the cumulated sums. Then the sample variance of the residuals is computed which is proportional to m^{2*H}. The mean is computed over the blocks. The slope $\beta = 2 * H$ from the least square regression provides an estimate for the Hurst exponent H.

7.1.1.8 Absolute Value Method

In the absolute value method [82], the time series data X_i is divided into blocks of size m and an average within each block for successive values of m is given by

$$X^{(m)}(k) = \frac{1}{m} \sum_{i=(k-1)m+1}^{km} X(i), \; k = 1, 2, \ldots \tag{7.9}$$

Afterwards data $X_1, \ldots, X_n$ is divided into N/m blocks of size m and the sum of the absolute values of the aggregated series is computed as $\frac{1}{N/m}\sum_{k=1}^{N/m} \left|X^{(m)}(k)\right|$. After that calculation, the logarithm of this statistical value is plotted versus $\log(m)$. For long range dependent time series, the slope of the line is generally $(H - 1)$.

7.1.1.9 Differential Variance

The series is divided into n groups [83]. Within each partition, the variance, relative to mean of the total series, is evaluated. The difference of the variances is then calculated. A measure of the change as a variable parameter of these calculations between different partitions is calculated. The number of groups (n) is increased and the algorithm is repeated. The observed variability changes with increasing n and is related to the Hurst parameter H of the time series. This methodology is known as differenced variance approach. A *log-log* plot of variability changes with the number of partition is linear and related with a slope of H and can be estimated by linear regression.

7.1.1.10 Abry and Veitch's Wavelet-Based Analysis

The scale behavior in data can be estimated with the wavelet-based tool from the plot of function [1]

$$\log_2 \left(\frac{1}{nj} \sum_k |d_x(j, k)|^2 \right) \tag{7.10}$$

versus j, where $d_x(j, k) = \langle x, \psi_{j,k} \rangle$, and $\psi_{j,k}$ is the dual mother wavelet. This wavelet-based method can be implemented very efficiently allowing the direct analysis of very large data sets and is robust against the presence of deterministic trends.

7.1.1.11 Whittle's Maximum Likelihood Estimator

The Whittle estimation algorithm originates from periodogram as well [81]. It considers the function:

$$F(\alpha) = \int_{-\pi}^{+\pi} \frac{I(\lambda)}{g(\lambda, \alpha)} d\lambda \tag{7.11}$$

where $I(\lambda)$ is the periodogram, $g(\lambda, \alpha)$ is the spectral density at the frequency λ, and α is the vector of unknown parameters. The Whittle estimator is the value of α that minimizes the function $F(.)$. When dealing with white noise, a fGn process or ARFIMA processes, α is simply the parameter H. If the series is assumed to be $FARIMA(p, d, q)$, then α includes unknown coefficients in the autoregressive and moving average polynomials.

7.1.1.12 Diffusion Entropy Method

The purpose of Diffusion Entropy method [29] is to find the scaling factor without disturbing data with any form of detrending. The existence of scaling implies the existence of a PDF $p(x, t)$ that scales according to the equation

$$p(x, t) = \frac{1}{t^\delta} \cdot F\left(\frac{x}{t^\delta}\right), \tag{7.12}$$

where δ is the PDF scaling exponent. From the Shannon entropy

$$S(t) = -\int_{-\infty}^{+\infty} p(x, t) \cdot \ln p(x, t)\, dx \tag{7.13}$$

we get the following relation

$$S(\tau) = A + \delta\tau, \tag{7.14}$$

where

$$A = -\int_{-\infty}^{+\infty} F(y) \cdot \ln F(y)\, dy \tag{7.15}$$

and

$$\tau = \ln(t/t_0) \tag{7.16}$$

Equation (7.14) is the key relation for understanding detection of PDF scaling exponent δ.

7.1.1.13 Detrended Fluctuation Analysis

Fluctuation analysis was originally suggested by Peng [66]. According to the DFA method [86], for the initial time series $X(i)$ the cumulative time series $y(t) = \sum_{i=1}^{t} X(i)$ is constructed, which is then divided into N segments of length τ, and for each segment $y(t)$ the following fluctuation function is calculated:

$$F^2(\tau) = \frac{1}{\tau} \sum_{t=1}^{\tau} (y(t) - Y_m(t))^2 \tag{7.17}$$

where $Y_m(t)$ is a local m-polynomial trend within the given segment. The averaged on the whole of the time series $y(t)$ function $F(\tau)$ depends on the length of the segment: $F(\tau) \ltimes \tau^H$. The diagram of dependence of $\log F(\tau)$ on $\log \tau$ represents a line approximated by the least square method in some interval. Estimate of the exponent H is a slope of the line representing *log-log* dependence.

7.1.1.14 Koutsoyiannis' Method

Koutsoyiannis proposed an iterative method [48] to determine the standard deviation σ and the Hurst exponent H that minimize the error $e^2(\sigma, H)$:

$$e^2(\sigma, H) = \sum_{k=1}^{k'} \frac{\left[\ln \sigma + H \ln k + \ln c_k(H) - \ln s^{(k)}\right]^2}{k^p} \tag{7.18}$$

where $s^{(k)} \approx c_k(H) k^H \sigma$ and $c_k(H) = \sqrt{\frac{n/k - (n/k)^{2H-1}}{n/k - 1/2}}$. The minimization of $e^2(\sigma, H)$ is done numerically.

7.1.1.15 Kettani and Gubner Method

Let $X_1, X_2, \ldots, X_n$ be a realization of a Gaussian second-order self-similar process [46]. The estimated Hurst parameter can be calculated by

$$\widehat{H}_n = \frac{1}{2}\left[1 + \log_2 \left(1 + \widehat{\rho}_n(1)\right)\right], \tag{7.19}$$

where $\widehat{\rho}_n(k)$ denotes the sample autocorrelation. The 95% confidence interval of H is centered around the estimate $\widehat{H}_n$. For an FARIMA(0,d,0) process we get

$$\widehat{d}_n = \frac{\widehat{\rho}_n(1)}{1 + \widehat{\rho}_n(1)}. \tag{7.20}$$

The 95% confidence interval of d is centered around the estimate $\widehat{d}_n$.

7.1.2 Comments on Hurst Exponent

None of above methods is universal. Each one has its own applicability scope. This subject is widely discussed in literature [8, 24, 38, 42, 47]. Although the subject might be considered as a secondary one, it was addressed from the engineering point of view in [14].

One may say that Hurst exponent usage is *hit and miss* game [8]. One has to use extreme caution in real industrial cases. We cannot rely on any single estimator. It might be at least misleading. We have to be very cautious using any data preprocessing (filtering, detrending, removal of periodicity), especially in case of industrial data. It is very easy to loose information, bias the estimator or just bring wrong conclusions. From that perspective it is a good advice to verify obtained numbers with R/S plot. Especially in case, when crossover properties occur, as it simply ruins any estimator for single scaling number.

7.1.3 Multiple Scaling and Multi-persistence

Complex process dynamics frequently brings about multiple scaling exponents in the same range of scales. The ability to calculate them accurately is crucial and enables to build appropriate models for root-cause analysis, simulation and prediction. There are many methodologies to evaluate multiple scaling exponents. The simplest one is just through observation of slopes on the classical *log-log* R/S plot. This methodology is followed by different algorithms, like high-order correlation function [2] or wavelet leaders [37]. The approach has found several applications in meteorology, seismology, biology, medicine, epidemiology, telecommunication, networking. Unfortunately, similarly to mono-scaling fractal analysis, this methodology penetrates control engineering to a very limited extent [15, 52]. Observation of industrial controller data shows that crossover and multiple scales appear frequently. Analysis should not only be limited to the multiple Hurst exponents, but also crossover point position might carry on information [43].

Rescaled range R/S *log-log* plots need to be cross checked to investigate the existence of crossover behavior in any data. The analysis shows frequently that there are more than one slope in the R/S plot divided by the crossover. The data are characterized with different scaling exponents $H^{(i)}$ in different memory ranges. The evaluation of the real data shows three frequent scenarios of the multi-scaling behavior:

- **scen_1** with double scaling and single crossover: short term Hurst exponent is larger than the long term one $H^{(1)} > H^{(2)}$ (see Fig. 7.2),
- **scen_2** with double scaling and single crossover: short term Hurst exponent is smaller than the long term one $H^{(1)} < H^{(2)}$ (see Fig. 7.3),
- **scen_3** frequently observed in real data with three memory scales and two crossover points: $H^{(1)} > H^{(2)}$ and $H^{(2)} < H^{(3)}$ (see Fig. 7.4).

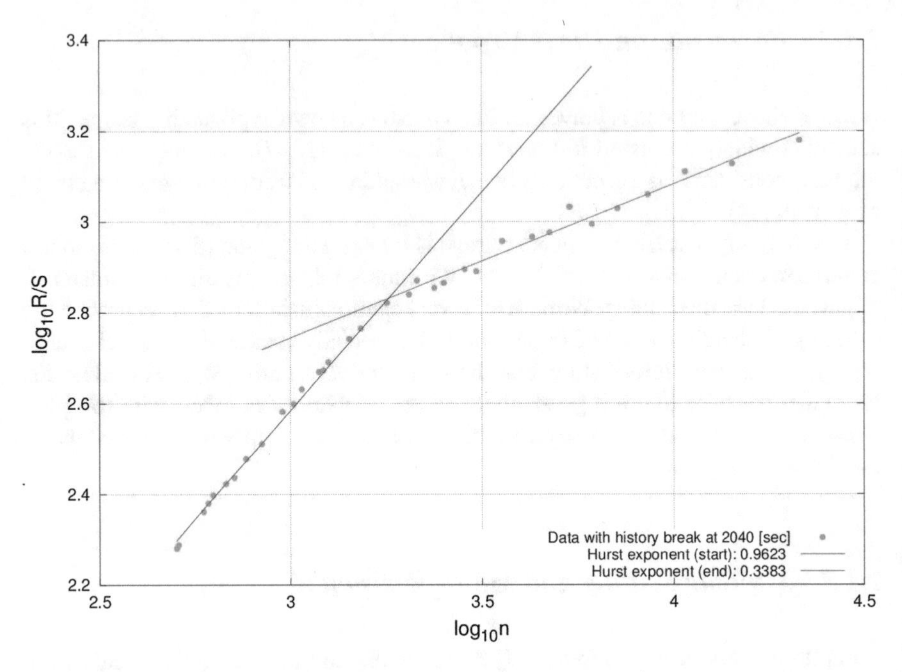

Fig. 7.2 Various control error R/S plot behavior—two scales—**scen_1**

It is visible that single scaling analysis without verification for possible multi-scaling properties leads to biased and irrelevant results [43]. Conflicting Hurst exponent estimates associated with different scales and estimators used by the methods generate erroneous results. Once crossover hypothesis is valid, we should understand the meaning of that fact, identify appropriate measures, find their applicability and limitations. Multiple scales followed by the crossover points reveals dichotomy of scaling behavior that is not predicted by existing simulation models, however empirically found data [53]. There might be different reasons for the multi scaling, i.e. for existence of the crossovers and multiple Hurst exponents.

One hypothesis is that it is the result of additive effects of nonlinearities and disturbances originating from different sources characterized by various delays (analogy with traffic flow models used in network simulations). It has been shown [49] that the multiplex of various flow sources generates more appropriate simulated traffic flows with clear evidence of the multiple scales and crossovers. Thus crossover points may appear more likely in the main process lines gathering income from side processes and installations then in auxiliary or isolated objects. This hypothesis also confirms well known practice that control loops situated at the end of installation process flow are hardly to be tuned as they cumulate errors and disturbances generated at former process stages.

The second possible crossover explanation is that specific control error fluctuations may be caused by disturbances impacting the process in different time scales, i.e. with different delays. The first Hurst exponent associated with short time scale

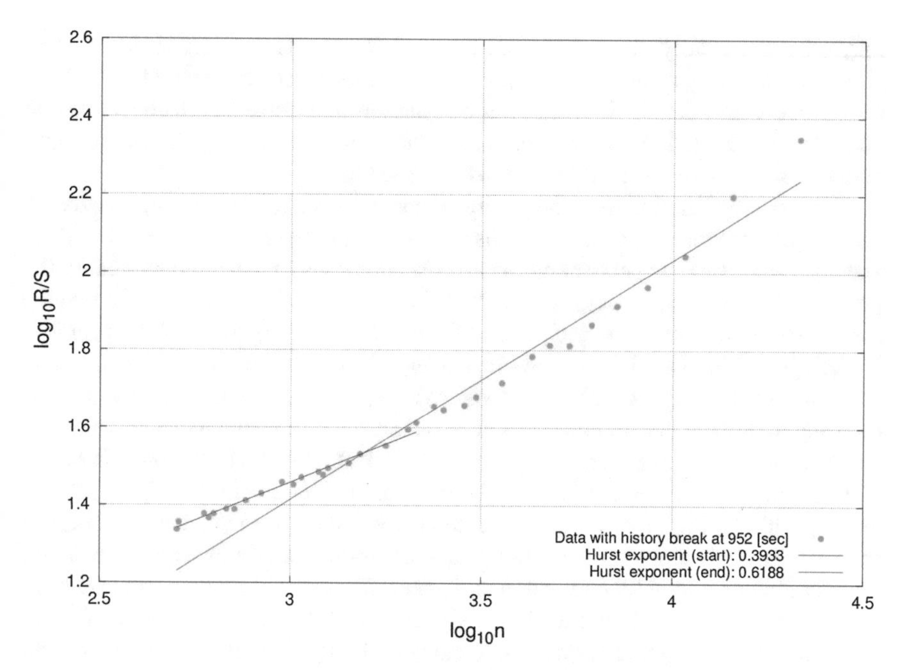

Fig. 7.3 Various control error R/S plot behavior—two scales—**scen_2**

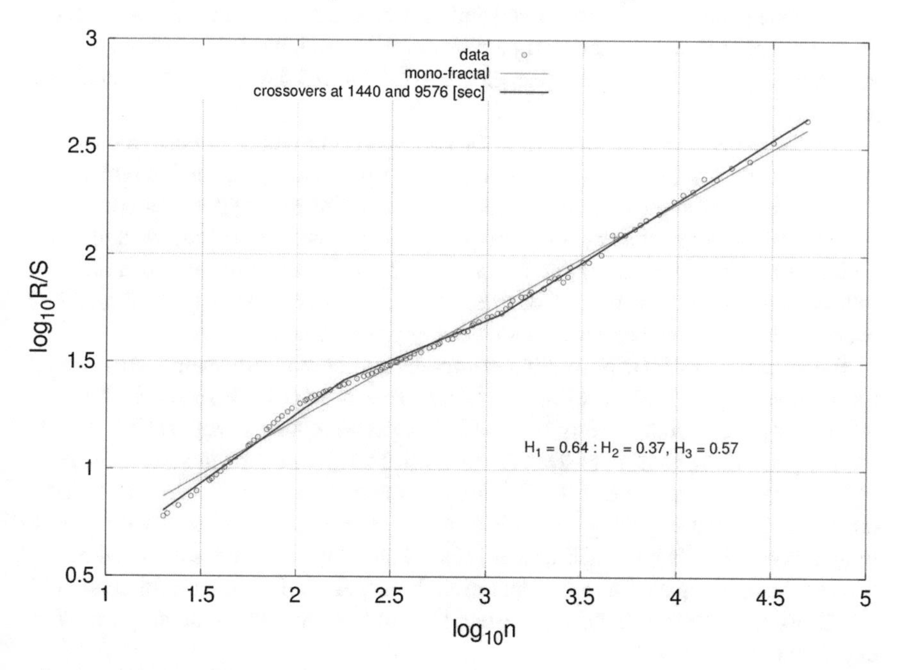

Fig. 7.4 Various control error R/S plot behavior—three scales—**scen_3**

may originate locally due to the local loop misfit (poor tuning, unsuitable control structure,impacting disturbances). Longer time scales (consecutive Hurst exponents) are probably associated with long term fluctuations and disturbances generated and added with larger delays. Most often it happens in the distributed processes or especially in evidence of strong internal cross coupled links.

The third explanation uses the nature of control system itself, i.e. the feedback. Let assume that there appears some disturbance. First, the controller starts to react and after some time, dependent on the process time constants and tuning quality the problem is solved and process is stabilized. Thus we may distinguish two time scale ranges: short time controller activity and fluctuations and long time stable operation.

Analysis of industrial data shows that **scen_1** is the most frequent. The data also show that R/S *log-log* plots with two crossovers is noticed more often as well. It is interesting to find out that all of **scen_1** plots have $H^{(1)} > 0.5$, which means persistent process. On the other hand, the long term exponent $H^{(2)} \approx 0.5$ tends towards Brownian motion with independent stochastic process. This observation can be explained exactly with the third hypothesis presented above. Short term persistent properties reflected by $H^{(1)}$ are associated with general controller quality of setpoint tracking, noises effect or internally coupled disturbances. These are actually narrow-banded and relatively week periodicities in the signal (below the period). Smaller $H^{(2)}$ informs about stable operation, just above the period [40]. The results were confirmed in simulations for PID and GPC [19] algorithms.

Opposite situation is also observed and may be explained as follows. Short memory exponent is in most cases anti-persistent $H^{(1)} < 0.5$, while the second one, the long term exponent has larger value, closer to $H^{(2)} \approx 0.5$ the independent stochastic process.

We can notice persistent (**scen_1**) or anti-persistent (**scen_2**) behavior in short memory scale, which decays with time (expressed by crossover) and is replaced by the process closer to the Gaussian properties. According to [78] persistent properties reflect sluggish controller performance, while anti-persistent ones indicate aggressive tuning often characterized by oscillatory operation. Observations confirm that control loops tuning often escapes from clearly visible process oscillations diminishing overshoots towards rather damped and slower operation.

Finally, **scen_3** with triple scaling exponents may be similarly explained as well. It follows the shape sketched in Fig. 7.4. Short memory scale is persistent with $H^{(1)} > 0.5$, then mid exponent converges towards independent process with $H^{(2)} \approx 0.5$ and the longest one is back persistent $H^{(3)} > 0.5$. First two exponents are interpreted in the same way as in **scen_1**. The longest memory scale exponent $H^{(3)}$ carries information about any long term disturbances affecting the process with a every long delays. They bring much longer effect than tuning time constant $crossover^{(1)}$ leaving the range space for stable operation. It may also reflect long term trend in the signal originating externally. They affect the process after much longer period than expressed in $crossover^{(2)}$.

It is very interesting to confront observations with explanations for the similar phenomenon found in other areas of research. Peng et al. in [67] commented that larger values of the Hurst exponent are more frequently associated with heartbeat time series originating from the severe heart diseases, while the smaller values come from healthy subjects. To explain it, authors suggest that it may be connected with the feedback mechanism existing and more efficient in healthy human. Following that hypothesis the large, persistent scaling factors may be a subject to the poorly working feedback, i.e. inappropriate or badly tuned control loop.

The phenomenon of crossover is investigated next. As it has been mention previously, its value is probably associated with control horizon/memory time (*settling time*), representing time when the controller settles down. As the settling time derives from two reasons: process dynamics and control quality, the crossover time should be closely connected with the dominant time constant. On the other hand, improvement in controller tuning should minimize it towards technological process limitation. In some cases it is affected by the characteristics of the setpoint signal variability.

Formally, subject and occurrence of crossover is not fully understood, requires further attention and, in case of control loop assessment, much larger amounts of regulation loop data from various sources. One possible explanation has been presented by Li and Mills [51]. They used the analogy between river flows and Internet packet delay dynamics. They suggested that crossover is the result of events of empty queue and queue overflow. When such situation happens, the model is broken and the nature of long term process is disrupted. In case of the control loops, it is somehow compliant with prosed explanation. Controller settling changes operation mode (tracking versus filtering). In case of double (or multiple) crossover, it might be caused nonlinearities, such as saturations, deadband or other ones embedded by instrumentation characteristics or other technological constraints.

7.1.4 Multi-fractal Behavior

Multifractals form an extension [15]. They were first used in the analysis of turbulent flows [57] and soon the notion started to be popular. Multifractality show the same irregularity at all scales. Multifractal spectrum consists of different singularities and fractal dimensions. The concept of spectrum is similar to the fractal dimension in monofractal theory. Understanding of multifractality gives insight into dynamics of the analyzed signal.

The notion of multifractality has been introduced as a measure to model turbulence. It has offered a model with attractive stochastic properties, which can reproduce various aspects of time series, like fat tails, long-term dependencies and multi-scaling. Multifractal analysis consists of decomposition and characterization of information at all scales. There are different algorithms to evaluate multifractal spectrum, like box-counting [71], Discrete Wavelet Multifractal Model [36], wavelet transform modulus maxima [63], wavelet leaders [75] and Multi-Fractal Detrended Fluctuation Analysis (MF-DFA).

MF-DFA algorithm shows less bias, it is less likely to give a false positive result and it is designed for data of a finite length [65, 74]. It makes the algorithm well fitted to the task of the analysis of control error time series. The method starts with disaggregation of time series e_i of length N, $i = 1, \ldots, N$.

1. Transform original data into mean-reduced cumulative sums $E_j = \sum_{i=1}^{j} (e_i - \overline{e})$, $j = 1, \ldots, N$, being aggregated time series with zero mean.
2. Divide time series E_j in non-overlapping segments of length s, starting from the beginning. As s does not have to divide evenly into N, we make another set of segments starting from the end of data coming to the beginning, so no piece of data is missed. As the result we obtain $2N_s$ boxes covering the whole dataset. Next, we fit polynomial p_k of order m to data in each of the segments $k = 1, \ldots, 2N_s$ using least squares.
3. Calculate mean square error $F^2(k, s)$ for estimate of each segment k of length s.

$$F^2(k, s) = \frac{1}{s} \sum_{i=1}^{s} (E\left[(k-1)s + i\right] - p_k\left[i\right])^2 \tag{7.21}$$

   for $k = 1, \ldots, N/s$ and

$$F^2(k, s) = \frac{1}{s} \sum_{i=1}^{s} (E\left[N - (k - N/s)s + i\right] - p_k\left[i\right])^2 \tag{7.22}$$

   for $k = N/s + 1, \ldots, 2N/s$.
4. For parameter q find the variance $F_q(s)$ of order q for positive and negative values of q for each size s

$$F_q(s) = \left(\frac{1}{2N/s} \sum_{k=1}^{2N/s} \left[F^2(k, s)\right]^{q/2} \right)^{1/q}. \tag{7.23}$$

   For $q = 0$ use

$$F_0(s) = \exp\left\{ \frac{1}{4N/s} \sum_{k=1}^{2N/s} \ln\left(F^2(k, s)\right) \right\}. \tag{7.24}$$

5. Repeat steps 2, 3, ..., 4 for different s evaluating new sets of variances $F_q(s)$.
6. Plot $F_q(s)$ for each q in double logarithmic sc ale scale and estimate linear fit with least squares. If slope $h(q)$ varies with q, multifractality is suspected. Single slope shows monofractal scaling.
7. Calculate multifractal exponent $\tau(q)$ as

$$\tau(q) = qh(q) - 1 \tag{7.25}$$

   and finally use Legendre transform to evaluate multifractal spectrum

$$\alpha(q) = \frac{d\tau(q)}{dq} \text{ and } f\left(\alpha\left(q\right)\right) = \alpha\left(q\right)q - \tau\left(q\right). \tag{7.26}$$

The algorithm enables to evaluate multifractal spectrum, using the graph $\log(F_q)$ versus $\log(s)$ to identify the crossover. The MFDFA1 implementation proposed in [35] is used further as the calculation methodology. Multifractality itself is interesting as it brings to the overall assessment new aspects of cascade effect existing in the complex multi-loop installations and human impacts, as for instance manual mode of operation, output biasing, loop re-tuning, etc.

Analysis of multifractality origins allows to perform further diagnosis [55, 76]. The literature depicts two main sources [4, 27, 40]. One is the time-correlation of small and large fluctuations. The second depicts the broadness and fat tails of the distribution of the underlying process. In control engineering time correlations (both short and long-time) are associated with the impeding disturbances and counter-flows of information in complex processes. The broadness of the control error variable reflects its dynamic properties (controller tuning).

These hypotheses may be validated with data manipulations. Correlation is checked through data shuffling [40], as it removes correlations from data and any remaining scaling is caused by probability density function broadness. The effect of distribution shape is checked by truncating the tails [28], i.e. removal of the extreme positive and negative numbers. Following above discussion multifractality and its origins analysis may help to identify other aspects of the loop performance.

Multifractal analysis extends classical approaches that are using statistical (Gaussian and non-Gaussian) or fractal measures. It has shown that multifractal properties itself and the analysis of the multifractality sources gives further insight into the subject [15]. The control issues that can be assessed include existence of cascade effects that may exist in complex systems and surely affects control, unknown cross-dependencies associated with un-decoupled disturbances (inappropriate use of feed-forward), poor control (bad tuning), superposition of contribution of many sources or human impact and possibly oscillating behavior. It delivers new degrees of freedom to the existing approach.

Review of industrial control error time trends depicts multifractal properties, being more explicit in short scale. Spectrum for longer scales is narrower and multifractal properties become less important. There should be dense discretization satisfying sufficient number of samples to perform fractal analysis in short scale. Correlations impact multifractal properties depending on the scale. There are rather minor source of multifractality in short scale, but they are more important with larger scales.

Short scale multifractal analysis enables to address controller dynamical performance issues. From the perspective of the tuning it is more important. Analysis of multifractal spectrum allows to incorporate into the picture the stable properties of the control error histogram with robust measures. As distribution fitting may fail (in the situation of strange shapes), the analysis of multifractal spectrum can still deliver information on control properties. Additionally complex cascaded behavior of the system, human impact or superposition of various impacts and oscillations (non-concave spectra) may be additionally assessed.

Simulation originated data cannot be definitely described as multifractal. Classical simulations used in the control engineering research do not reflect efficiently fractal and multifractal phenomena. Process complexity, cross-dependencies with various varying delays and human factors are not reflected in simulations. Simple re-enactment of the histogram properties through fat-tail disturbance is completely not enough. Although such simulation helps to analyze statistical or persistence properties, it fails with multifractality.

7.2 Entropy Measures

Any data-driven control performance assessment methodology requires appropriate and robust quality measures. Ideally, it is assumed that the shape of the probability density function of the control error signal $\varepsilon(t)$ should be as narrow as possible. A narrow PDF generally depicts that the uncertainty of the related variable, i.e. control error is small. Apart from the , which also corresponds to a small entropy value [89]. Entropy has been used as the index used for controller tuning. Thus, it is natural to investigate the opportunities that are offered by the entropy analysis.

Any random variable uncertainty might be measured by the information entropy. Its probability density function is denoted by $F(x)$. The Shannon entropy is usually utilized in that task. There are different definitions of the continuous-type entropy, like:

- the differential entropy, H^{DE} defined as:

$$H^{DE} = -\int_{-\infty}^{\infty} F(x)(x) \ln F(x)\, dx\ , \ x \in R. \tag{7.27}$$

- the rational entropy, H^{RE} defined as:

$$H^{RE} = -\int_{-\infty}^{\infty} F(x) \log\left(\frac{F(x)}{1+F(x)}\right) dx\ , \ x \in R. \tag{7.28}$$

The first application of the minimum entropy notion has been adopted as the Minimum Information Entropy (MIE) benchmark and the design of minimum information entropy index [61, 88]. The minimum MIE bound for both Gaussian (in Gaussian system, variance and entropy is equivalent) and nonlinear non-Gaussian cases is identically and defined by

$$H_{min} = \ln\left(\sqrt{2\pi \exp(1)}\sigma\right) \tag{7.29}$$

where exp(1) is Euler's number and the upper bound by

$$H_{min|upper} = \ln\left(\sqrt{2\pi \exp(1)}\sigma_{MV}\right) \tag{7.30}$$

Thus the CPA index which incorporates the MIE is denoted by the following formula:

$$\eta^H = \frac{\exp(H_{min})}{\exp(H_{act})}. \tag{7.31}$$

Knowledge of the process delay enables to apply the tight minimum information entropy upper bound $H_{min|upper}$ and allows to formulate the control measure as

$$\eta^H = \frac{\exp\left(H_{min|upper}\right)}{\exp(H_{act})}. \tag{7.32}$$

The Shannon entropy has some shortcuts, once used as the performance benchmark. It may receive negative or even negative infinite values for the continuous random variable. This feature can form potential limitations in its application as the benchmark measure. It has been proposed to use rational entropy instead [7, 90, 91]. This formulation of the entropy holds similar features of the Shanon entropy, simultaneously being consistent. For the random variable $x \in R$ and its probability density function denoted by $F(x)$, the rational entropy is defined as (7.28). Using the rational entropy definition two novel performance measures has been proposed [90]

$$\eta^{H1} = 1 - \frac{\arctan\left(\left|H_{act}^{RE} - H_{min}^{RE}\right|\right)}{\arctan\left(\max\left\{\left|H_{act}^{RE} - H_{min}^{RE}\right|\right\}\right)}. \tag{7.33}$$

$$\eta^{H2} = \frac{H_{max}^{RE} - H_{act}^{RE}}{H_{max}^{RE} - H_{min}^{RE}}, \tag{7.34}$$

where H_{max}^{RE} is the maximum rational entropy in some interval $[a, b]$. Simulations presented by the authors show that benchmarking indexes based on the rational entropy are effective and feasible covering uncertainties in control quality assessment.

References

1. Abry, P., Veitch, D.: Wavelet analysis of long-range-dependent traffic. IEEE Trans. Inf. Theory **44**(1), 2–15 (1998)
2. Barabási, A.L., Szépfalusy, P., Vicsek, T.: Multifractal spectra of multi-affine functions. Phys. A: Stat. Mech. Appl. **178**(1), 17–28 (1991)
3. Bardet, J.M., Bertrand, P.: Identification of the multiscale fractional Brownian motion with biomechanical applications. J. Time Ser. Anal. **28**(1), 1–52 (2007)
4. Barunik, J., Aste, T., Di Matteo, T., Liu, R.: Understanding the source of multifractality in financial markets. Phys. A: Stat. Mech. Appl. **391**(17), 4234–4251 (2012)
5. Beran, J.: Statistics for Long-Memory Processes. CRC Press, Boca Raton (1994)
6. Borak, S., Misiorek, A., Weron, R.: Models for heavy-tailed asset returns. In: Cizek, P., Härdle, K.W., Weron, R. (eds.) Statistical Tools for Finance and Insurance, 2nd edn, pp. 21–56. Springer, New York (2011)

7. Chen, L.J., Zhang, J., Zhang, L.: Entropy information based assessment of cascade control loops. In: Proceedings of the International MultiConference of Engineers and Computer Scientists, IMECS (2015)
8. Clegg, R.G.: A practical guide to measuring the Hurst parameter. Int. J. Simul.: Syst. Sci. Technol. **7**(2), 3–14 (2006)
9. Das, L., Srinivasan, B., Rengaswamy, R.: Data driven approach for performance assessment of linear and nonlinear kalman filters. In: 2014 American Control Conference, pp. 4127–4132 (2014)
10. Das, L., Srinivasan, B., Rengaswamy, R.: Multivariate control loop performance assessment with Hurst exponent and mahalanobis distance. IEEE Trans. Control Syst. Technol. **24**(3), 1067–1074 (2016a)
11. Das, L., Srinivasan, B., Rengaswamy, R.: A novel framework for integrating data mining with control loop performance assessment. AIChE J. **62**(1), 146–165 (2016b)
12. Domański, P.D.: Non-gaussian properties of the real industrial control error in SISO loops. In: Proceedings of the 19th International Conference on System Theory, Control and Computing, pp. 877–882 (2015)
13. Domański, P.D.: Fractal measures in control performance assessment. In: Proceedings of IEEE International Conference on Methods and Models in Automation and Robotics MMAR, Miedzyzdroje, Poland, pp. 448–453 (2016a)
14. Domański, P.D.: Non-Gaussian and persistence measures for control loop quality assessment. Chaos: Interdiscip. J. Nonlinear Sci. **26**(4), 043,105 (2016b)
15. Domański, P.D.: Multifractal properties of process control variables. Int. J. Bifurc. Chaos **27**(6), 1750,094 (2017)
16. Domański, P.D.: Control quality assessment using fractal persistence measures. ISA Trans. **90**, 226–234 (2019). https://doi.org/10.1016/j.isatra.2019.01.008
17. Domański, P.D., Gintrowski, M.: Alternative approaches to the prediction of electricity prices. Int. J. Energy Sect. Manag. **11**(1), 3–27 (2017)
18. Domański, P.D., Ławryńczuk, M.: Assessment of predictive control performance using fractal measures. Nonlinear Dyn. **89**, 773–790 (2017a)
19. Domański, P.D., Ławryńczuk, M.: Assessment of the GPC control quality using non-Gaussian statistical measures. Int. J. Appl. Math. Comput. Sci. **27**(2), 291–307 (2017b)
20. Domański, P.D., Ławryńczuk, M.: Control quality assessment of nonlinear model predictive control using fractal and entropy measures. In: Preprints of the First International Nonlinear Dynamics Conference NODYCON 2019, Rome, Italy (2019)
21. Domański, P.D., Golonka, S., Jankowski, R., Kalbarczyk, P., Moszowski, B.: Control rehabilitation impact on production efficiency of ammonia synthesis installation. Ind. Eng. Chem. Res. **55**(39), 10,366–10,376 (2016)
22. Feder, J.: Fractals. Springer Science & Business Media (2013)
23. Field, A.J., Harder, U., Harrison, P.G.: Measurement and modelling of self-similar traffic in computer networks. IEE Proc. Commun. **151**(4), 355–363 (2004)
24. Franzke, C.L.E., Graves, T., Watkins, N.W., Gramacy, R.B., Hughes, C.: Robustness of estimators of long-range dependence and self-similarity under non-Gaussianity. Philos. Trans. Royal Soc. A **370**(1962), 1250–1267 (2012)
25. Gao, J., Hu, J., Tung, W.: Entropy measures for biological signal analyses. Nonlinear Dyn. **68**(3), 431–444 (2012)
26. Geweke, J., Porter-Hudak, S.: The estimation and application of long memory time series models. J. Time Ser. Anal. **4**, 221–238 (1983)
27. Grahovac, D., Leonenko, N.N.: Detecting multifractal stochastic processes under heavy-tailed effects. Chaos Solitons Fractals **65**, 78–89 (2014)
28. Green, E., Hanan, W., Heffernan, D.: The origins of multifractality in financial time series and the effect of extreme events. Eur. Phys. J. B **87**(6), 1–9 (2014)
29. Grigolini, P., Palatella, L., Raffaelli, G.: Asymmetric anomalous diffusion: an efficient way to detect memory in time series. Fractals **9**(4), 439–449 (2001)

30. Habib, A., Sorensen, J.P.R., Bloomfield, J.P., Muchan, K., Newell, A.J., Butler, A.P.: Temporal scaling phenomena in groundwater-floodplain systems using robust detrended fluctuation analysis. J. Hydrol. **549**, 715–730 (2017)
31. Haslett, J., Raftery, A.E.: Space-time modelling with long-memory dependence: assessing ireland's0 wind power resource. J. Appl. Stat. **38**, 1–50 (1989)
32. Hassler, U.: Regression of spectral estimators with fractionally integrated time series. J. Time Ser. Anal. **14**(4), 369–380 (1993)
33. Higuchi, T.: Approach to an irregular time series on the basis of the fractal theory. Phys. D: Nonlinear Phenom. **31**(2), 277–283 (1988)
34. Hurst, H.E., Black, R.P., Simaika, Y.M.: Long-Term Storage: An Experimental Study. Constable and Co Limited, London (1965)
35. Ihlen, E.A.F.: Introduction to multifractal detrended fluctuation analysis in Matlab. Front. Physiol. **3**(141), 1–18 (2012)
36. Jaffard, S.: Multifractal formalism for functions. SIAM J. Math. Anal. **28**, 944–970 (1997)
37. Jaffard, S., Lashermes, B., Abry, P.: Wavelet leaders in multifractal analysis. In: Qian, T., Vai, M.I., Xu, Y. (Eds.), pp. 201–246. Wavelet analysis and applications, Birkhäuser Verlag (2007)
38. Jeong, H.D.J., McNiclke, D., Pawlikowski, K.: Hurst parameter estimation techniques: a critical review. In: 38th Annual ORSNZ Conference (2001)
39. Jia, Y., Zhou, J., Li, D.: Performance assessment of cascade control loops with non-Gaussian disturbances. In: 2018 Chinese Automation Congress, pp. 2451–2456 (2018)
40. Kantelhardt, J.W., Zschiegner, S.A., Stanley, H.E.: Multifractal detrended fluctuation analysis of nonstationary time series. Physica A **316**, 87–114 (2002)
41. Kaplan, L.M., Jay Kuo, C.C.: Fractal estimaton from noisy data via discrete fractional gaussian noise (DFGN) and the Haar basis. IEEE Trans. Signal Proccess. **41**(12) (1993)
42. Karagiannis, T., Faloutsos, M., Riedi, R.H.: Long-range dependence: now you see it, now you don't! In: IEEE Global Telecommunications Conference GLOBECOM '02 3 (2002)
43. Kavasseri, R.G., Nagarajan, R.: Evidence of crossover phenomena in wind-speed data. IEEE Trans. Circuits Syst. I: Regul. Papers **51**(11), 2255–2262 (2004)
44. Kavasseri, R.G., Nagarajan, R.: A multifractal description of wind speed records. Chaos Solitons Fractals **24**(1), 165–173 (2005)
45. Kenkel, N.C., Walker, D.J.: Fractals in the biological sciences. COENOSES **11**, 77–100 (1996)
46. Kettani, H., Gubner, J.A.: A novel approach to the estimation of the long-range dependence parameter. IEEE Trans. Circuits Syst. **53**(6), 463–467 (2006)
47. Kirichenko, L., Radivilova, T., Deineko, Z.: Comparative analysis for estimating of the hurst exponent for stationary and nonstationary time series. J. Int. Inf. Technol. Knowl. **5**, 371–387 (2011)
48. Koutsoyiannis, D.: Climate change, the hurst phenomenon, and hydrological statistics. Hydrol. Sci. J. **48**(1), 3–24 (2003)
49. Li, L., Li, Z., Zhang, Y., Chen, Y.: A mixed-fractal traffic flow model whose hurst exponent appears crossover. In: 2012 Fifth International Joint Conference on Computational Sciences and Optimization, pp. 443–447. IEEE (2012)
50. Li, M., Zhao, W., Cattani, C.: Delay bound: fractal traffic passes through network servers. Math. Problems Eng. (2013)
51. Li, Q., Mills, D.L.: Investigating the scaling behavior, crossover and anti-persistence of internet packet delay dynamics. In: Proceedings of the Seamless Interconnection for Universal Services Global Telecommunications Conference GLOBECOM'99 3 (1999)
52. Liu, K., Chen, Y.Q., Domański, P.D., Zhang, X.: A novel method for control performance assessment with fractional order signal processing and its application to semiconductor manufacturing. Algorithms **11**(7), 90 (2018)
53. Liu, K., Chen, Y.Q., Domański, P.D.: Control performance assessment of the disturbance with fractional order dynamics. In: Preprints of the First International Nonlinear Dynamics Conference NODYCON 2019, Rome, Italy (2019)
54. Lopes, R., Betrouni, N.: Fractal and multifractal analysis: a review. Med. Image Anal. **13**(4), 634–649 (2009)

55. L'vov, V.S., Pomyalov, A., Procaccia, I.: Outliers, extreme events, and multiscaling. Phys. Rev. E **63**(056), 118 (2001)
56. Majumder, B., Das, S., Pan, I., Saha, S., Das, S., Gupta, A.: Estimation, analysis and smoothing of self-similar network induced delays in feedback control of nuclear reactors. In: International Conference on Process Automation, Control and Computing (2011)
57. Mandelbrot, B.B.: Possible refinements of the lognormal hypothesis concerning the distribution of energy dissipation in intermitent turbulence. In: Rosenblatt, M., Van Atta, C. (Eds.) Statistical Models and Turbulence, Springer, New York (1972)
58. Mandelbrot, B.B.: Les objets fractals: forme, hasard, et dimension. Flammarion (1975)
59. Mandelbrot, B.B., Hudson, R.L.: The Misbehavior of Markets: A Fractal View of Financial Turbulence. Basic Books, New York (2004)
60. Mandelbrot, B.B., Fisher, A., Calvet, L.: A multifractal model of asset returns. Technical Report 1164, Cowles Foundation Discussion Paper (1997)
61. Meng, Q.W., Fang, F., Liu, J.Z.: Minimum-information-entropy-based control performance assessment. Entropy **15**(3), 943–959 (2013)
62. Movahed, M.S., Jafari, G.R., Ghasemi, F., Rahvar, S., Tabar, M.R.R.: Multifractal detrended fluctuation analysis of sunspot time series. J. Stat. Mech.: Theory Exp. 2006(02), 02,003 (2006)
63. Muzy, J.F., Bacry, E., Arneodo, A.: Wavelets and multifractal formalism for singular signals: application to turbulence data. Phys. Rev. Lett. **67**(25), 3515–3518 (1991)
64. Oh, G., Eom, C., Havlin, S., Jung, W.S., Wang, F., Stanley, H.E., Kim, S.: A multifractal analysis of asian foreign exchange markets. Eur. Phys. J. B **85**(6), 1–6 (2012)
65. Oświecimka, P., Kwapień, J., Drozdz, S., Rak, R.: Investigating multifractality of stock market fluctuations using wavelet and detrending fluctuation methods. Acta Phys. Polonica B **36**(8), 2447 (2005)
66. Peng, C.K., Buldyrev, S.V., Havlin, S., Simons, M., Stanley, H.E., Goldberger, A.L.: Mosaic organization of DNA nucleotides. Phys. Rev. E **49**(2), 1685–1689 (1994)
67. Peng, C.K., Havlin, S., Stanley, H.E., Goldberger, A.L.: Quantification of scaling exponents and crossover phenomena in nonstationary heartbeat time series. Chaos (Woodbury, NY) **5**(1), 82–87 (1995)
68. Perkiömäki, J.S., Mäkikallio, T.H., Huikuri, H.V.: Fractal and complexity measures of heart rate variability. Clin. Exp. Hypertens. **27**(2–3), 149–158 (2005)
69. Peters, E.E.: Chaos and Order in the Capital Markets: A New View of Cycles, Prices, and Market Volatility, 2nd edn. John Wiley & Sons Inc (1996)
70. Pillay, N., Govender, P.: A data driven approach to performance assessment of PID controllers for setpoint tracking. Procedia Eng. **69**, 1130–1137 (2014)
71. Russel, D., Hanson, J., Ott, E.: Dimension of strange attractors. Phys. Rev. Lett. **45**(14), 1175–1178 (1980)
72. Schmitt, D.T., Ivanov, P.C.: Fractal scale-invariant and nonlinear properties of cardiac dynamics remain stable with advanced age: a new mechanistic picture of cardiac control in healthy elderly. Am. J. Physiol. Regul. Integr. Comp. Physiol. **293**(5), R1923–R1937 (2007)
73. Schröder, M.: Fractals, Chaos, Power Laws: Minutes from an Infinite Paradise. W. H. Freeman and Company, New York, NY (1991)
74. Schumann, A.Y., Kantelhardt, J.W.: Multifractal moving average analysis and test of multifractal model with tuned correlations. Phys. A: Stat. Mech. Appl. **390**(14), 2637–2654 (2011)
75. Serrano, E., Figliola, A.: Wavelet leaders: a new method to estimate the multifractal singularity spectra. Phys. A: Stat. Mech. Appl. **388**(14), 2793–2805 (2009)
76. Sornette, D.: Dragon-kings, black swans and the prediction of crises. Int. J. Terraspace Sci. Eng. **2**(1), 1–18 (2009)
77. Spinner, T.: Performance assessment of multivariate control systems. Ph.D. thesis, Texas Tech University (2014)
78. Spinner, T., Srinivasan, B., Rengaswamy, R.: Data-based automated diagnosis and iterative retuning of proportional-integral (PI) controllers. Control Eng. Pract. **29**, 23–41 (2014)
79. Srinivasan, B., Spinner, T., Rengaswamy, R.: Control loop performance assessment using detrended fluctuation analysis (DFA). Automatica **48**(7), 1359–1363 (2012)

80. Taleb, N.N.: Fooled by Randomness: The Hidden Role of Chance in Life and in the Markets. Penguin Books, New York (2001)
81. Taqqu, M.S., Teverovsky, V.: On estimating the intensity of long-range dependence in finite and infinite variance time series. In: A Practical Guide To Heavy Tails: Statistical Techniques and Applications, pp. 177–217 (1996)
82. Taqqu, M.S., Teverovsky, V., Willinger, W.: Estimators for long-range dependence: an empirical study. Fractals **03**(04), 785–798 (1995)
83. Teverovsky, V., Taqqu, M.S.: Testing for long-range dependence in the presence of shifting means or a slowly declining trend using a variance type estimator. J. Time Ser. Anal. **18**, 279–304 (1997)
84. Tosatti, E., Pietrelli, L.: Fractals in physics. In: Proceedings of the Sixth Trieste International Symposium on Fractals in Physics, Italy, North Holland (1985)
85. Turcotte, D.L.: Fractals in Geology and Geophysics. In: Fractals in Geophysics, pp. 171–196. Springer (1989)
86. Weron, R.: Estimating long range dependence: finite sample properties and confidence intervals. Physica A **312**, 285–299 (2002)
87. Wu, Y., Li, J.: Hurst parameter estimation method based on haar wavelet and maximum likelihood estimation. J. Huazhong Normal Univ. **52**(6), 763–775 (2013)
88. You, H., Zhou, J., Zhu, H., Li, D.: Performance assessment based on minimum entropy of feedback control loops. In: 2017 6th Data Driven Control and Learning Systems (DDCLS), pp. 593–598 (2017)
89. Yue, H., Wang, H.: Minimum entropy control of closed-loop tracking errors for dynamic stochastic systems. IEEE Trans. Autom. Control **48**(1), 118–122 (2003)
90. Zhang, J., Jiang, M., Chen, J.: Minimum entropy-based performance assessment of feedback control loops subjected to non-Gaussian disturbances. J. Process Control **24**(11), 1660–1670 (2015a)
91. Zhang, J., Zhang, L., Chen, J., Xu, J., Li, K.: Performance assessment of cascade control loops with non-Gaussian disturbances using entropy information. Chem. Eng. Res. Design **104**, 68–80 (2015b)
92. Zhang, Q., Wang, Y., Lee, F., Chen, Q., Sun, Z.: Improved Renyi entropy benchmark for performance assessment of common cascade control system. IEEE Access **7**, 6796–6803 (2019a)
93. Zhang, Q., Wang, Y., Lee, F., Zhang, W., Chen, Q.: Performance assessment of cascade control system with non-Gaussian disturbance based on minimum entropy. Symmetry **11**(3), 379 (2019b)
94. Zhou, J., Jia, Y., Jiang, H., Fan, S.: Non-Gaussian systems control performance assessment based on rational entropy. Entropy **20**(5), 331 (2018)

Chapter 8
Business Key Performance Indicators—KPIs

The calamity of the information age is that the toxicity of data increases much faster than its benefits.
– Nassim Nicholas Taleb

Previous chapters described results of the scientific research in the area of CPA. On the other hand it is well known that the industry has its own rules. The plant owners are mostly (or even only) interested in the installation Total Cost of Operation (TCO). The process monitoring and supervisory decision making is driven by financial incentives. Nowadays the financial efficiency is not only dependent on the production throughput maximization or the cost cutting. New factors, like environmental protection, social responsibility, sustainable development, energy management are seriously considered in the plant TCO. All these impacts, being considered in the long term, financially contribute to the general plant numbers.

The control system quality is often perceived only as a very minor factor from the general operational perspective. There are several reports showing potential financial benefits of just the better control [3, 6, 7, 9]. Nonetheless, the awareness of control quality importance gets general understanding only to a very limited extend. There is still much to be done in the propagation of this knowledge. The wide acceptance is often limited by simplified guide to plant numbers using only Excel files. These aspects will be further described in the next Chap. 9 describing the methods that aim at proving financial efficacy for the control rehabilitation projects.

This part focuses at the presentation of business-like control performance approaches and measures that are understood and used in industry. These measures are called Key Performance Indicators (KPIs). A KPI is defined as a "*computable performance assessment, as derived from a combination of metrics*" [15]. One has to be aware that the comprehensive coverage of all such approaches is almost impossible. Each corporation, each site and even each installation frequently uses custom

P. D. Domański, *Control Performance Assessment: Theoretical Analyses and Industrial Practice*, Studies in Systems, Decision and Control 245,
https://doi.org/10.1007/978-3-030-23593-2_8

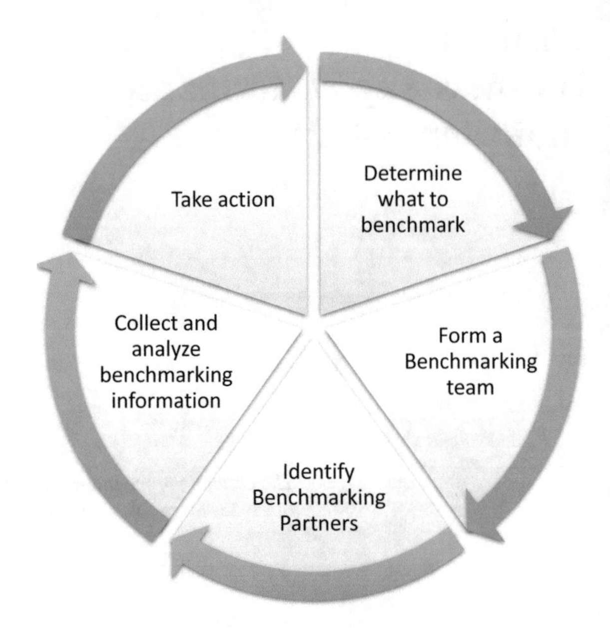

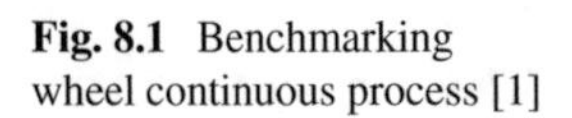
Fig. 8.1 Benchmarking wheel continuous process [1]

approaches originating from the local practices, habits, understandings, knowledge, cultural and social background further impacted by procedures and regulations issued by superior corporate authority.

Comparison with the others and the measures of analogy between similar units has been the first straightforward method used. Once we have two identical or similar process, driven by two control strategies it is easy to indicate the better one. Existence of the larger number of such plants enables to develop benchmarking comparison measures reflecting process performance and following that control quality. This historical approach of comparison by analogy has soon developed into the whole research and practical area of benchmarking. Feigenbaum in 1951 [5] described it as the process of continuously measuring and comparing one's business processes against comparable processes in leading organizations to obtain information that will help the organization identify and implement improvements. Benchmarking as a measure tool has been first introduced for business processes by Rank Xerox [2] in the late *70s*. Nowadays, it is wide and mature science with many proven theoretical and practical methodologies and algorithms in vast application domain.

The benchmarking is the ongoing continuous process that follows the PDCA (plan, do, check, act) cycle [1] (Fig. 8.1).

Application of the benchmarking idea into control design and maintenance process cycle enables to achieve and sustain the significant results. From one side, which in the industrial life is crucial, it has integrated the CPA task with other business management processes. Control engineers and technology staff started to use common language and terminology as the managers. The usage of benchmarking enabled structuralization and proceduralization of the CPA task making it transparent, exchangeable, repeatable and measurable.

One of the outcomes from the adoption of benchmarking philosophy to the task of control system quality assessment is the need to define appropriate measures. Previous paragraphs described the technical ones. Review of the literature [10, 12] and the industrial practice shows a list of various measures, that are commonly called KPIs. They are relatively simple in calculation and are easily understandable by the plant personnel at all hierarchy levels. The KPIs can be evaluated through a visual inspection or with the calculations that are mostly implemented and visualized with MS Excel sheets and charts:

- presentation of the time trends for control loop variables,
- percentage of the time the loop operates (is kept by the operators) in MAN (manual) or AUTO (automatic) mode (sometimes called a *service factor* [8]),
- percentage of time the process output (CV—controlled variable) violates upper/lower constraint,
- percentage of time the controller output (MV—manipulated variable) violates upper/lower constraint or the change rate limitation,
- percentage of time the controller output (MV—manipulated variable) saturates at full close/open position,
- mean value and standard deviation (the simplest statistics) of the CV, MV or the control error variable,
- maximum and/or minimum values of the loop variables (CV, MV or the control error),
- the loop oscillation tendency and period,
- cumulative travel time per some period (day, week, month) for the actuator, mostly the valve,
- number of direction changes in the actuator (valve) travel per the selected time period,
- noise level in the loop and the respective process variables,
- number of alarms associated with the loop and the respective process variables,
- number of the operator interventions into the loop, i.e. MAN/AUTO switch, bias modifications, setpoint changes, loop retuning, loop bypassing and the overlapping actions, etc.,
- any combination of the various measures and KPIs used in the operational practice (sometimes called a *overall loop performance index* [14]).

Industrial measures are often associated with some business-like diagrams, which should visualize the results of the assessment in a simplified form. The classical way to identify abnormal operation is to check whether the trend of any variable exceeds some selected minimum and maximum limitations using Quality Control Charts (QCC). This chart is often called a Shewhart Control Chart. The Upper Control Limit (UCL) and Lower Control Limit (LCL) define the normal operating zone. Process variable standard deviation σ is often used to evaluate the values of the constraints around the normal operating point x^{norm}:

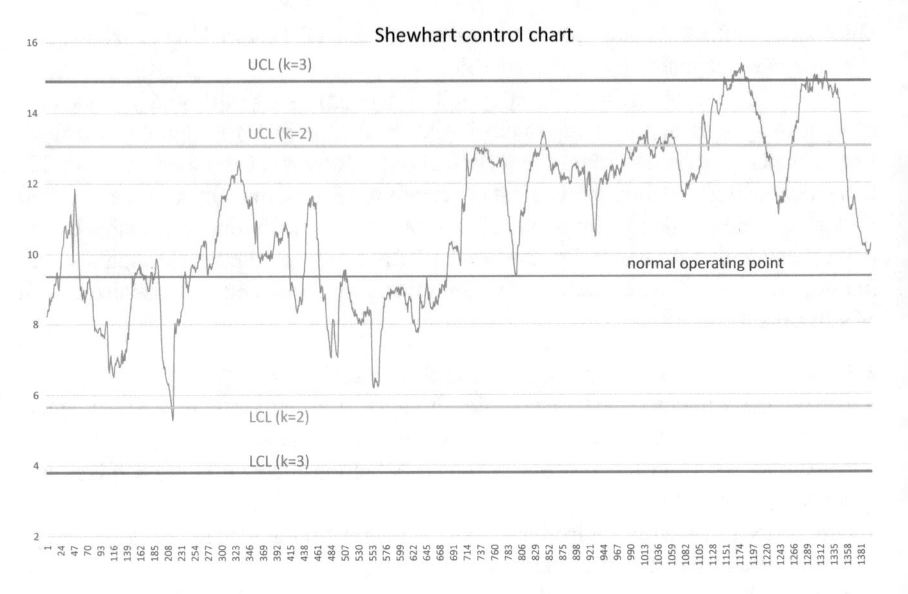

Fig. 8.2 Exemplary Shewhart Control Chart

$$UCL = x^{norm} + k \cdot \sigma \tag{8.1}$$

$$LCL = x^{norm} - k \cdot \sigma \tag{8.2}$$

where multiplier k defines the width of the normal operation range. Assuming normal properties of the variable parameter $k = 2$ defines that approximately 5% of the variable measurements produces abnormal signals. the selection $k = 3$ widens the normal operation regimes causing only approximately 1% of the occurrences being in abnormal range (see example in Fig. 8.2). An application of the Shewhart charts may be found in [13].

Another way of representing loop quality is to use the so called radar plots (sometimes called a web, star or spider chart [11]). Radar Chart enables to compare multiple quantitative variables and enables some kind of a multi-criteria loop assessment. This allows to identify what loop properties addressed by some measure scores high or low the control quality and to detect at least some of the behavior causes. An example from some CPA project is sketched in Fig. 8.3 showing comparison between six various control quality measures.

Further considerations of the industrial multi-criteria data considerations and results representation may be found in [4]. As the comprehensive approach is required and there is no single universal measures it is expected that this issue should be addressed in further research.

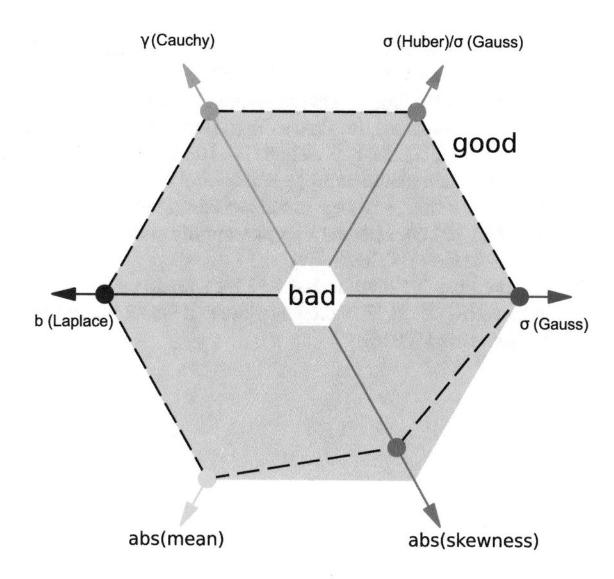

Fig. 8.3 Exemplary radar chart for the CPA task

References

1. Bhutta, K.S., Huq, F.: Benchmarking best practices: an integrated approach. Benchmarking: Int. J. **6**(3), 254–268 (1999)
2. Cross, R., Iqbal, A.: The Rank Xerox experience: benchmarking ten years on. In: IFIP Advances in Information and Communication Technology, Benchmarking—Theory and Practice, pp. 3–10 (1995)
3. Domański, P.D., Golonka, S., Jankowski, R., Kalbarczyk, P., Moszowski, B.: Control rehabilitation impact on production efficiency of ammonia synthesis installation. Ind. Eng. Chem. Res. **55**(39), 10,366–10,376 (2016)
4. Dziuba, K., Góra, R., Domański, P.D., Ławryńczuk, M.: Multicriteria control quality assessment for ammonia production process (in Polish). In: Zalewska, A. (Ed.) 1st Scientific and Technical Conference Innovations in the Chemical Industry, pp. 80–90. Warszawa, Poland (2018)
5. Feigenbaum, A.V.: Quality control: principles, practice and administration: an industrial management tool for improving product quality and design and for reducing operating costs and losses. McGraw-Hill, New York, USA (1951)
6. Gabor, J., Pakulski, D., Domański, P.D., Świrski, K.: Closed loop NOx control and optimization using neural networks. In: IFAC Symposium on Power Plants and Power Systems Control, Brussels, Belgium, pp. 188–196 (2000)
7. Jelali, M., Thormann, M., Wolff, A., Müller, T., Loredo, L.R., Sanfilippo, F., Zangari, G., Foerster, P.: Enhancement of product quality and production system reliability by continuous performance assessment of automation systems (AUTOCHECK). Technical Report Final report to Contract No RFS-CR03045, EUR 23205, Office for Official Publications of the European Communities, Luxemburg (2008)
8. Kinney, T.: erformance monitor raises service factor of mpc. In: Proceedings of the ISA, Houston, USA (2003)
9. Marlin, T.E., Perkins, J.D., Barton, G.W., Brisk, M.L.: Benefits from process control: results of a joint industry-university study. J. Process Control **1**(2), 68–83 (1991)
10. Mitchell, W., Shook, D.: Finding the needle in the haystack—an innovative means of visualizing control performance problems (2004)

11. Ordys, A., Uduehi, D., Johnson, M.A.: Process Control Performance Assessment—From Theory to Implementation. Springer, London (2007)
12. Smuts, J.F., Hussey, A.: Requirements for successfully implementing and sustaining advanced control applications. In: Proceedings of the 54th ISA POWID Symposium, pp. 89–105 (2011)
13. Wu, P., Guo, L., Duan, Y., Zhou, W., He, G.: Control loop performance monitoring based on weighted permutation entropy and control charts. Can. J. Chem. Eng. (2018)
14. Xia, C., Howell, J.: Loop status monitoring and fault localisation. J. Process Control **13**(7), 679–691 (2003), selected Papers from the sixth IFAC Symposium on Bridging Engineering with Science - DYCOPS - 6
15. Zimmerman, T.: Metrics and KPI for robotic cybersecurity performance analysis. Technical Report NISTIR 8177, National Institute of Standards and Technology (NIST), U.S. Department of Commerce (2016)

Chapter 9
Control Improvement Benefit Estimation

The curious task of economics is to demonstrate to men how little they really know about what they imagine they can design.

– Friedrich A. Hayek

In the midst of chaos, there is also opportunity.

– Sun Tzu

Industrial processes are mostly non-stationary time-varying systems impeded by many internal and external disturbances and/or correlations. Their complexity generates a lot of challenges for the control system implementation, design, tuning and maintenance. Dynamic goals are addressed by the base control with single point or cascade PID loops. There are many reports showing that base control tuning brings significant financial benefits [9, 13]. Further improvement is often obtained with supervisory applications of APC and PO [10, 11].

There arises a need for methodologies to compare control rehabilitation cost against expected economic benefits. Such decisions should be based on the financial basis. Estimation techniques allowing calculation of the benefits resulting from the control system improvement have been proposed [4, 13, 15]. The cost element of the decision is simple. It may be easily derived from past projects or just requested from control system vendor. The benefit part should be evaluated specifically for each case. The methodology is based on the mitigation of process variability, leading towards quantitative results [1]. Frequently, one may assume upper or lower limitation for the variable. Reduction of its variability through better control enables to shift it closer to the constraint and thus to generate benefit. As the variable is explicitly linked with the performance, the benefits may be calculated. The method assumes that the shape

P. D. Domański, *Control Performance Assessment: Theoretical Analyses and Industrial Practice*, Studies in Systems, Decision and Control 245,
https://doi.org/10.1007/978-3-030-23593-2_9

of the variable histogram is Gaussian and standard deviation is used as the variability measure.

The task to predict possible improvements associated with upgrade of a control system exists in literature for a long time [14]. From the early days it was mostly associated with the implementation of APC. There are three well established approaches called: *same limit*, *same percentage* and *final percentage* rules [3, 4]. The method is based on the evaluation of normal distribution for some variable informing about economic performance. Thus the method assumes Gaussian properties of the process behavior (Fig. 9.1).

This section presents summary and extension of the results that can be found in [7, 8]. Improvement potential is evaluated on the basis of the well-established algorithm presented below:

1. Evaluate histogram of the selected variable or the performance index.
2. Fit normal distribution to the obtained histogram which is described by two parameters: mean value and standard deviation σ.
3. It is assumed that mean value (M_{improv} for the improved system and M_{now} for the original one) is kept within the same defined distance from potential upper (or lower limitation). The idea is to shift the mean value towards the respective constraint. For the confidence level of 95% it is equal to $a = 1.65$. Such a value is used in the calculations. The mean value for the improved operation is estimated. Standard deviation σ_0 relates to the original system and σ_1 to the improved one.

$$M_{improv} = M_{now} \cdot a \cdot (\sigma_0 - \sigma_1) \tag{9.1}$$

4. Finally percentage improvement is calculated on the basis of the following equation:

$$\Delta M = 100 \cdot \frac{M_{improv} - M_{now}}{M_{improv}} \tag{9.2}$$

Let assume that we fit PDF to the histogram with parameters (x_0, σ_0). We need to keep *the same limit* at point of $x^* = x_0 + k \cdot \sigma_0$, where k determines the point shift. We assume that better control improves variability by factor c, i.e. new $\sigma_1 = c \cdot \sigma_0$. Thus we may maintain the same limit with density function shifted by benefit (9.3).

$$M = k \cdot \sigma_0 - c \cdot \sigma_0 \sqrt{2(\frac{k}{2} - \ln c)} \tag{9.3}$$

Let us select $k = 2.0,\ c = 0.70,\ u_0 = 1.0,\ \gamma_0 = 0.65$ (solid black line). Limit point is $x = 2.0$ with function value *0.106*. We obtain improved variability with $\gamma_1 = 0.455$ (blue dashed line). Thus we may shift the function by benefit factor $M = 0.312$ towards the *same limit* (green dotted line). Figure 9.2 shows example graphical visualization. This approach is very popular despite some deficiencies.

As it has been already shown industrial data frequently does not have Gaussian properties. This fact caused the need to extend *the same limit* algorithm from the classical form to other, fat-tailed probability density functions.

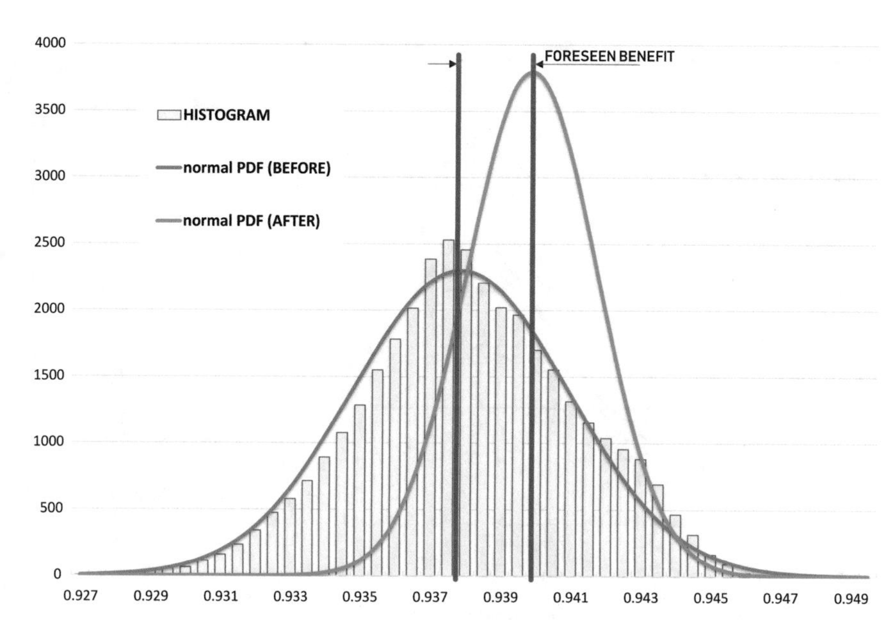

Fig. 9.1 Same limit rule graphical representation

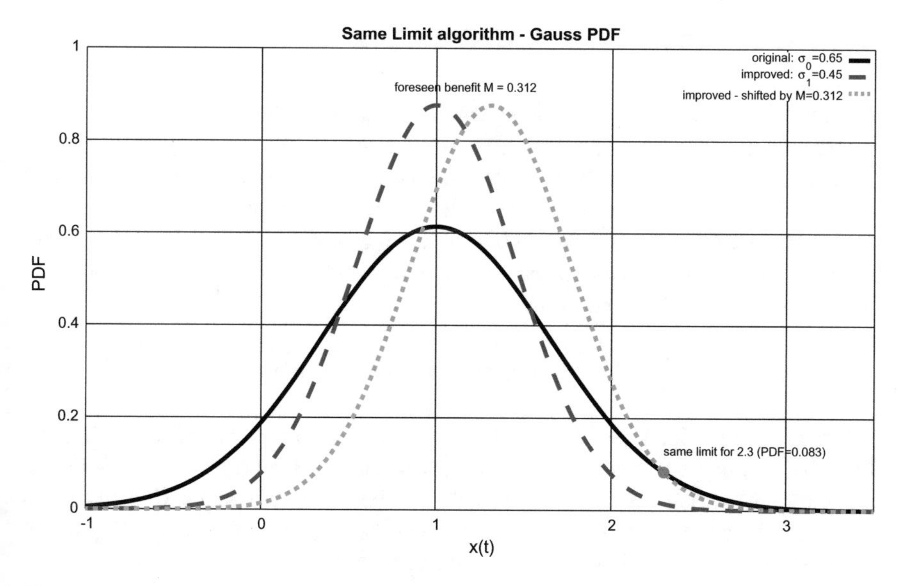

Fig. 9.2 Gauss *same limit* rule graphical example

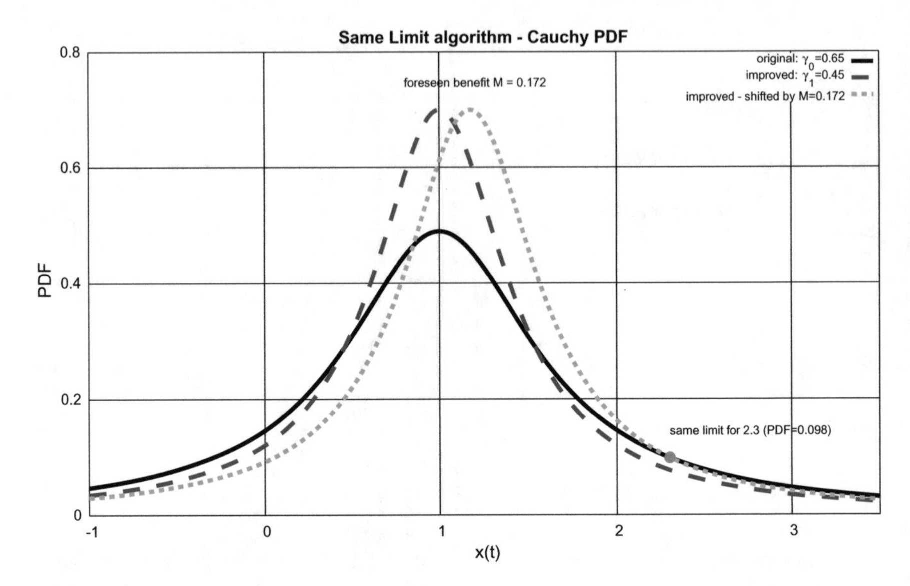

Fig. 9.3 Cauchy *same limit* rule graphical example

9.1 Algorithm for Cauchy Distribution

Cauchy version of the methodology extends classical Gaussian approach using Cauchy distribution [7]. The algorithm is similar and control system improvement is calculated using the notion of PDF broadness (σ). Cauchy distribution is presented in (4.14). Assume we have fitted PDF to the histogram with parameters (x_0, γ_0). We need to maintain *the same limit* at point of $x_0 = x_0 + k \cdot \gamma_0$. We assume that better control decreases the variability by factor c with new $\gamma_1 = c \cdot \gamma_0$. We keep the same limit with a PDF function shifted by:

$$M = k \cdot \gamma_0 - \gamma_0 \sqrt{c(1-c)(1+k^2)}. \tag{9.4}$$

Let us assume that $k = 2.0,\ c = 0.07,\ u_0 = 1.0,\ \gamma_0 = 0.65$ (solid black line). Resulting limiting value is at point $x = 2.0$ with function value of *0.127*. We obtain new improved variability of $\gamma_1 = 0.38$ (blue dashed line). We may shift distribution by benefit factor $M = 0.172$ towards the *same limit* (green dotted line). Figure 9.3 presents graphical visualization of the Cauchy example.

9.2 Algorithm for Laplace Distribution

Laplace version of the methodology shows the extension of the classical Gaussian approach with the Laplace double exponential distribution [7]. The basic function is sketched in (4.15). We identify histogram fitting distribution with parameters (x_0, b_0).

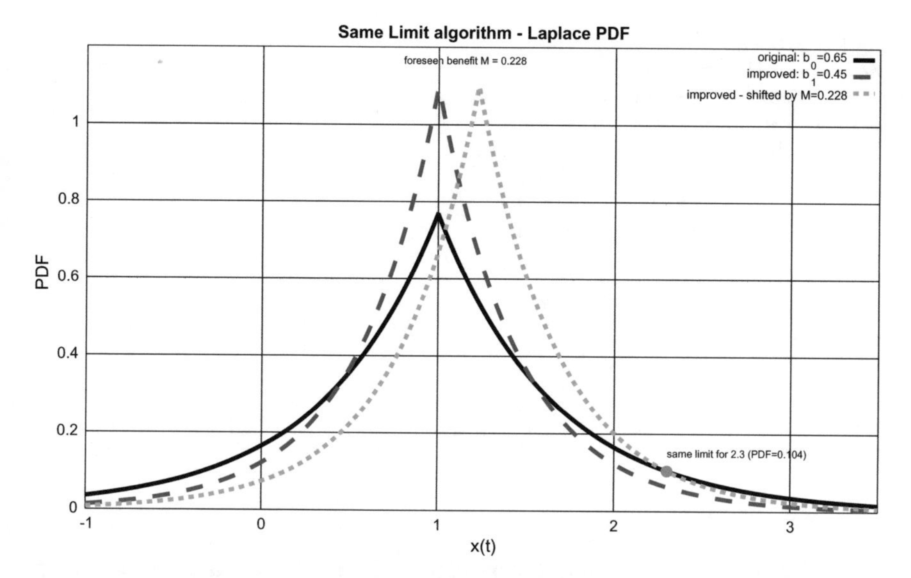

Fig. 9.4 Laplace *same limit* rule graphical example

One wants to keep *the same limit* at point of $x_0 = x_0 + k \cdot b_0$. Additionally we estimate that improvement in control diminishes variability by factor c, i.e. new $b_1 = c \cdot b_0$. Thus we may maintain the same limit with density function shifted by benefit:

$$M = k \cdot b_0 - (c - b_0) \cdot (k - \ln c). \tag{9.5}$$

Let us assume that $k = 2.0,\ c = 0.70,\ u_0 = 1.0,\ b_0 = 0.65$ (solid black line). Resulting limiting value is at point $x = 2.0$ with function value of *0.14*. We obtain new improved variability of $b_1 = 0.38$ (blue dashed line). Thus we may shift distribution by benefit factor $M = 0.228$ towards the *same limit* (green dotted line). Figure 9.4 presents graphical visualization of the Laplace example.

9.3 Algorithm for Stable Distribution

Control improvement benefit estimation using α-stable density function is not as straightforward as for normal PDF, when only standard deviation is responsible for variability [8]. There are three coefficients responsible for the shape. Scaling γ is responsible for the function broadness, stability α for tails fatness and skewness β impacting tail limits and asymmetry. We need to take into account combinations of these parameters in solving *the same limit* task. It is infeasible analytically. Furthermore the parameters address different aspects of the control loop tuning. Distribution dispersion γ is a measure of control quality. It plays the role of a robust CPA measure [6].

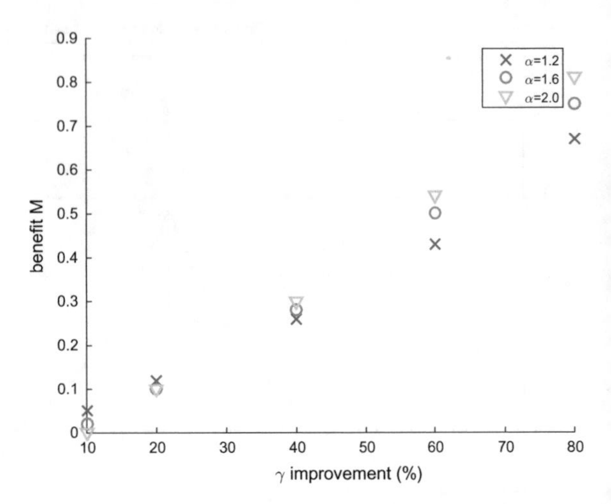

Fig. 9.5 Predicted improvement M versus PDF scale γ assuming unchanged skewness $\beta = 0.1$ for different stability α

Stability factor (α) also impacts control performance. It reflects persistence properties of the variable, i.e. control error. Skewness informs about control quality as well. Symmetry is not always desirable. In some cases we may allow variations in the direction opposite to the constraint reducing the ones towards the limitation, like in steam spray control. Analyses of the α-stable factors impacting *the rule of the same limit* are evaluated through simulations.

A numerical approach was used to calculate the benefit obtained after control system tuning, as the function describing the PDF is complex. The method relies on finding the zero of the difference of two distribution functions in the reference point, i.e. the function with "*after tuning*" parameters and the original one "*before tuning*" being shifted. Any simple optimization algorithm may be used to obtain the result. The developed method is efficient and can be used with any PDF.

Simulation experiments were performed according to the predefined procedure, designed to verify assumed hypothesis. The results for the situation, when skewing factor remains unchanged during loop improvement ($\beta = 0.1$) is sketched in Fig. 9.5. Situation, when tuning causes symmetrization of the control error histogram, i.e. skewness factor after loop improvement becomes zero ($\beta = 0.0$) is presented in Fig. 9.6.

Diagrams below present respective shapes of the probability density functions change during the experiments. Selected plots show extreme cases only. Minor improvement (10%) in scale and situation of unchanged persistent behavior ($\alpha = 1.2 \rightarrow 1.2$) is sketched in Fig. 9.7, while Fig. 9.8 reflects shifting the system towards uncorrelated properties with $\alpha = 1.2 \rightarrow 2.0$. Figures 9.9 and 9.10 depict respective plots for large scale improvement of 80 %. Respective presentation of all the results in tabular form is presented in Tables 9.1 and 9.2.

The 10% change in γ, when $\alpha = 1.2 \rightarrow 2.0$ results in no improvement for both skewness scenarios. In all the other scenarios the δ shift is accompanied with M benefit. There are two other arising observations. Points form curves may be fitted

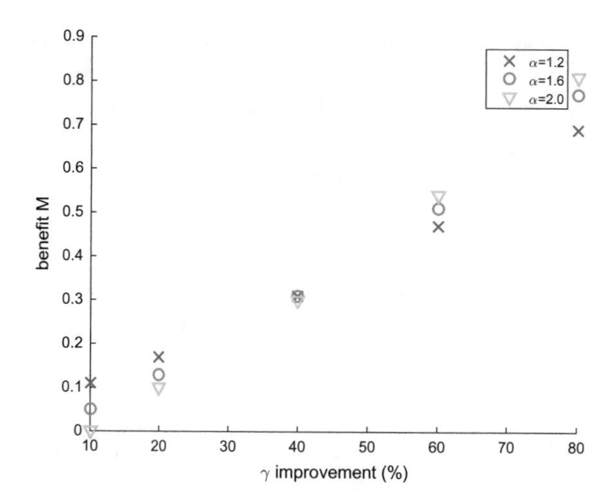

Fig. 9.6 Predicted improvement M versus PDF scale γ assuming symmetrization with skewness $\beta = 0.0$ for different stability α

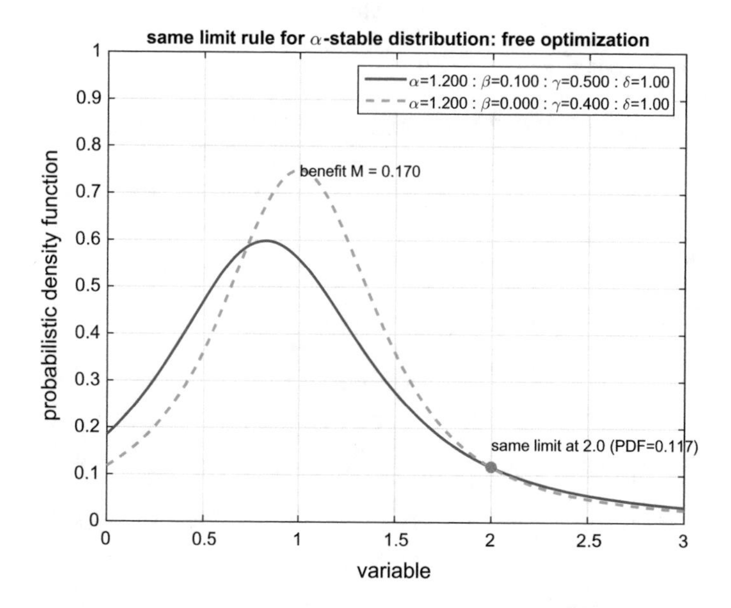

Fig. 9.7 10% change in γ, $\alpha = 1.2 \rightarrow 1.2$

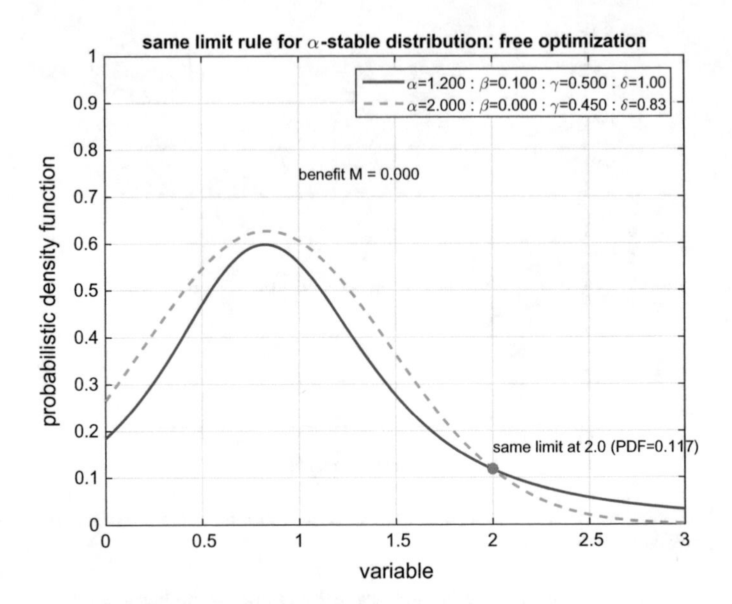

Fig. 9.8 10% change in γ, $\alpha = 1.2 \to 2.0$

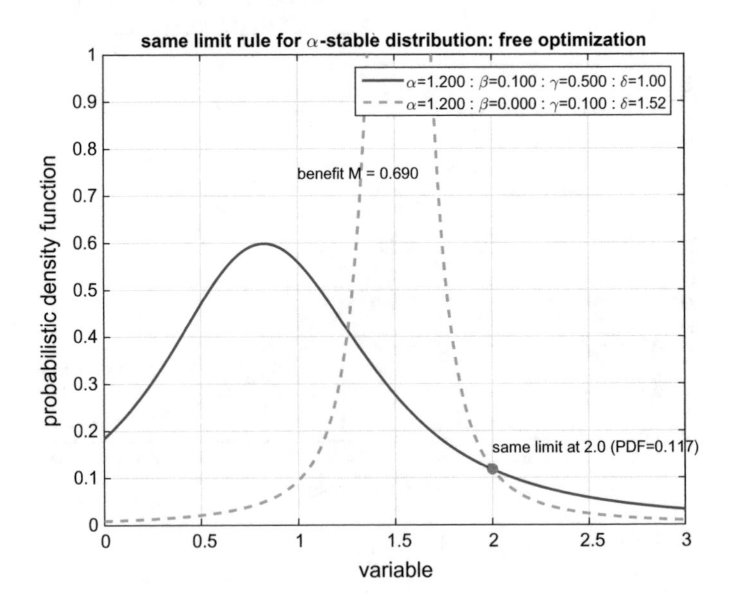

Fig. 9.9 80% change in γ, $\alpha = 1.2 \to 1.2$

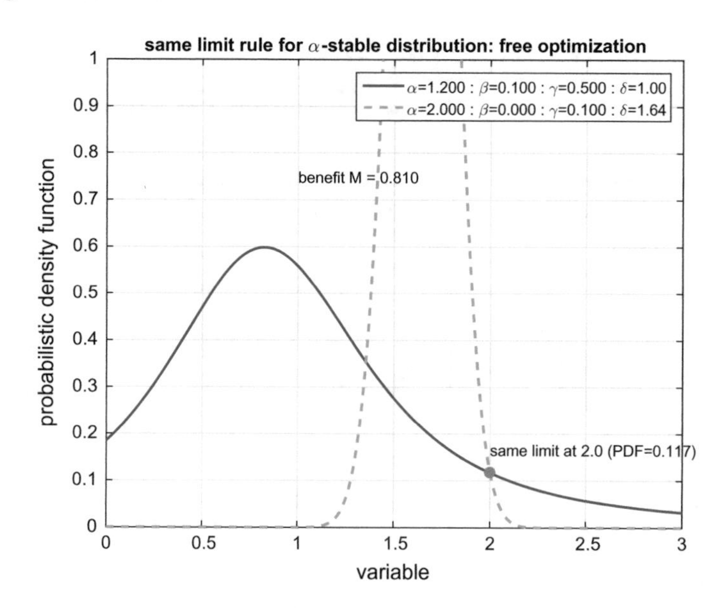

Fig. 9.10 80% change in γ, $\alpha = 1.2 \rightarrow 2.0$

Table 9.1 The *same limit* algorithm values of M and δ shifts in symmetric case $\beta = 0.0$

α	$\gamma = 0.45$		$\gamma = 0.40$		$\gamma = 0.30$		$\gamma = 0.20$		$\gamma = 0.10$	
	M	δ	M	δ	M	δ	M	δ	M	δ
2.0	0.00	0.83	0.10	0.93	0.30	1.13	0.54	1.37	0.81	1.64
1.6	0.05	0.88	0.13	0.96	0.31	1.14	0.51	1.34	0.77	1.60
1.2	0.11	0.94	0.17	1.00	0.31	1.14	0.47	1.30	0.69	1.52

Table 9.2 The *same limit* algorithm values of M and δ shifts in asymmetric case $\beta = 0.1$

α	$\gamma = 0.45$		$\gamma = 0.40$		$\gamma = 0.30$		$\gamma = 0.20$		$\gamma = 0.10$	
	M	δ	M	δ	M	δ	M	δ	M	δ
2.0	0.00	0.83	0.10	0.93	0.30	1.13	0.54	1.37	0.81	1.64
1.6	0.02	0.89	0.10	0.97	0.28	1.14	0.50	1.34	0.75	1.59
1.2	0.05	1.04	0.12	1.08	0.26	1.19	0.43	1.33	0.67	1.53

with the second order polynomial and the curves cross. Larger benefit is obtained with persistent properties for small improvement in PDF dispersion γ, while in case of significant narrowing of the histogram, the bigger profits appear when stability factor shifts towards uncorrelated value of $\alpha = 2$.

An interesting result is obtained once free optimization algorithm is run to check improvement potential. Figure 9.11 shows the results. We may see that the modifications are done in all PDF factors, although not relatively large.

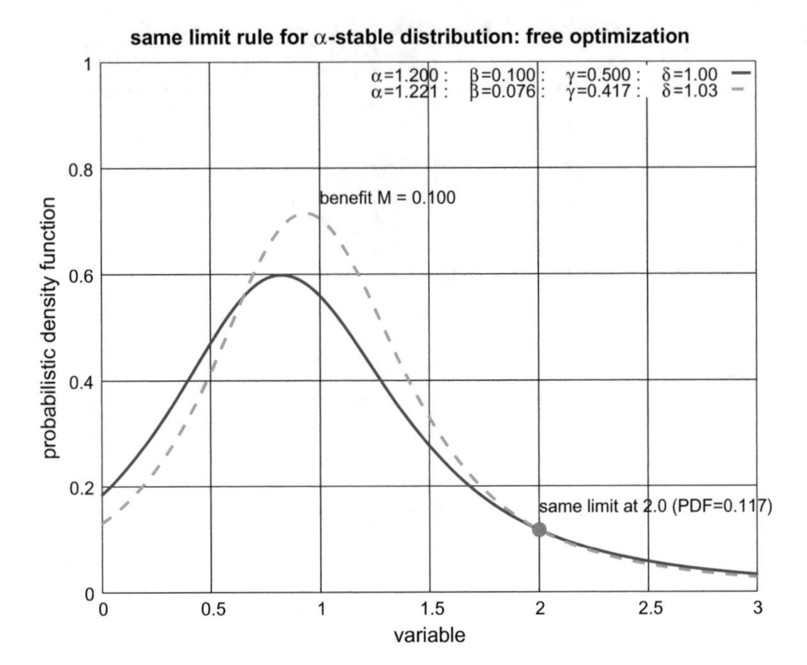

Fig. 9.11 Stable distribution—free optimization

9.4 Comparison and Other Approaches

Benefit estimation approaches may be compared. The solutions and comparison is performed for three proposed distributions, i.e. Gauss, Cauchy and Laplace. In all the cases, despite the same improvement scale and similar broadness, different benefit estimation has been obtained (Table 9.3).

Gaussian method is the most optimistic, i.e. it predicts the highest benefits using the same assumptions. In contrary the use of Cauchy PDF is the most conservative. One may call it the realistic solution according to the engineering practice. Laplace PDF approach situates itself in between. The differences originate from the heaviness of the distribution tails. One may notice potential high risk in overestimation of the benefits using Gaussian approach, while fat tail exhibits in the histogram.

Table 9.3 Comparison of benefit estimates

Density function	PDF constraint	Value at constraint	Benefit value—M
Gauss	$0.70 \cdot \sigma$	0.083	0.312
Cauchy	$0.70 \cdot \gamma$	0.098	0.172
Laplace	$0.70 \cdot b$	0.104	0.228

Proposed methodology is general and does not consider any assumptions on the variable properties, like non-stationarity or non-linearity. Results confirm complexity of the fat tail distributions. More degrees of freedom in histogram shaping enable more possibilities. The biggest impact i put on improvements, which directly influence control error histogram shape, i.e. dispersion factor γ. Earlier works confirm the hypothesis that this coefficient is a good measure of loop dynamic performance and its enhancement may be obtained with the controller settings. The other two interesting coefficients of the α-stable distribution enable to exploit further degrees of freedom.

Stability parameter α is associated with the loop persistence properties. Their reasons are very often connected to non-Gaussian noises, process complexity reflected in embedded correlations or human interventions into the loop operation. It is not the result of direct controller settings. The improvement results from process modifications (technology/construction). Asymmetric performance is the result of human interventions or process nonlinearities and can be associated with installation equipment.

Concluding, we see that application of the fat tail α-stable distribution into the process of potential benefit estimation out of control improvement delivers more degrees of freedom extending standard the same limit rule, which reflects controller tuning, with much more comprehensive perspective covering process technology and installation equipment.

The approach has been validated in several real industrial projects [7–9]. The effect of the minimized variance due to the better control has been further investigated towards economic performance indexes, like a quadratic function, a linear function with constraints and the so-called *clifftent* performance function [4]. The authors have called it the profit index and have applied it to the estimation of benefits of the APC application, which is often associated with such embedded indexes. The proposed profit index assesses *a priori* the economic control system performance for the improved system relative to the maximum achievable quality of the process.

Completely different approach has been proposed in [2]. The authors have proposed the method originating in economics using the option to expand, considering four scenarios: safe scenario—invest only in APC, value-added and risky scenario—invest in APC and RTO and gamble scenario—reject APC. It should be considered as the purely economical approach, with the technical measures being considered in the background, although the proposal might be interesting for further investigation.

Apart from the above classical methodology industry have developed several other, simplified customized approaches. One of them is called the Data Reduction Method [5, 12] and uses histogram. It consists of the following steps:

(1) Collect historical raw data over some predefined time interval.
(2) Normalize the data to a production rate.
(3) Evaluate and plot the histogram.
(4) Subtract the values at the median and the 15% points.
(5) Difference indicates potential economic benefit.

It is interesting to notice that the method uses median as the robust mean value estimator, and thus assumes non-Gaussian behavior in data. Another approach is strictly heuristic and is called the Best Operator Method [5]. It is based on the comparison of the historical operational data, like:

- operating data between the historical best versus average results,
- normal operating results versus when operators are running,
- normal versus those obtained during continuous monitoring demanded by a process optimization or alteration,
- comparison between two similar plants except for one having automation and the other manual control.

The method is very conservative as it does not take into consideration improvement potential, which has not been first penetrated.

References

1. Ali, M.K.: Assessing economic benefits of advanced control. In: 5th Asian Control Conference, Process Control in the Chemical Industries, Chemical Engineering Department, pp. 146–159. King Saud University, Riyadh, Kingdom of Saudi Arabia (2002)
2. Asawachatroj, A., Banjerdpongchai, D., Busaratragoon, P.: Economic assessment of APC and RTO using option to expand. Eng. J. **20**(5), 115–134 (2016). http://engj.org/index.php/ej/article/view/947
3. Bauer, M., Craig, I.K.: Economic assessment of advanced process control a survey and framework. J. Process Control **18**(1), 2–18 (2008)
4. Bauer, M., Craig, I.K., Tolsma, E., de Beer, H.: A profit index for assessing the benefits of process control. Ind. Eng. Chem. Res. **46**(17), 5614–5623 (2007)
5. Dolenc, J.: Estimating benefits from process automation. https://www.emersonautomationexperts.com/presentations/EstimatingBenefitsFromProcessAutomation_JohnDolenc.pdf Emerson Global Users Exchange. Grapevine, TX (2007)
6. Domański, P.D.: Non-Gaussian and persistence measures for control loop quality assessment. Chaos: An Interdiscip. J. Nonlinear Sci. **26**(4), 043,105 (2016)
7. Domański, P.D.: Non-Gaussian assessment of the benefits from improved control. Preprints of the IFAC World Congress 2017, pp. 5092–5097. Toulouse, France (2017)
8. Domański, P.D., Marusak, P.M.: Estimation of control improvement benefit with α-stable distribution. In: Kacprzyk, J., Mitkowski, W., Oprzedkiewicz, K., Skruch, P. (Eds.) Trends in Advanced Intelligent Control, Optimization and Automation, vol 577, pp. 128–137. Springer International Publishing AG (2017)
9. Domański, P.D., Golonka, S., Jankowski, R., Kalbarczyk, P., Moszowski, B.: Control rehabilitation impact on production efficiency of ammonia synthesis installation. Ind. Eng. Chem. Res. **55**(39), 10,366–10,376 (2016)
10. Gabor, J., Pakulski, D., Domański, P.D., Świrski, K.: Closed loop NOx control and optimization using neural networks. In: IFAC Symposium on Power Plants and Power Systems Control, Brussels, Belgium, pp. 188–196 (2000)
11. Laing, D., Uduehi, D., Ordys, A.: Financial benefits of advanced control. Benchmarking and optimization of a crude oil production platform. Proc. Am. Control Conf. **6**, 4330–4331 (2001)
12. Latour, P.R., Sharpe, J.H., Delaney, M.C.: Estimating benefits from advanced control. ISA Trans. **25**(4), 13–21 (1986)

13. Marlin, T.E., Perkins, J.D., Barton, G.W., Brisk, M.L.: Benefits from process control: results of a joint industry-university study. J. Process Control **1**(2), 68–83 (1991)
14. Tolfo, F.: A methodology to assess the economic returns of advanced control projects. In: American Control Conference 1983, pp. 1141–1146. IEEE (1983)
15. Wei, D., Craig, I.: Development of performance functions for economic performance assessment of process control systems. In: 9th IEEE AFRICON, pp. 1–6 (2009)

Part II
Validation with Simulation Examples

Several different methods and approaches to the Control Performance Assessment have been presented and discussed in the previous chapters. That presentation covered the wide spectrum of measures that have been proposed for industrial applications. The approaches vary from simple and well-known step response indexes, through well established benchmarking methods up to novel advanced entropy and persistence measures.

This part presents the simulations of the control quality assessment methods and algorithms described in Part I. There were presented methods and measures defined in various domains with different approaches. Some of these methods are very popular and widely used in industry, some of them are considered as the standard benchmarks, while the other ones are scarcely known.

The visualization of the considered CPA approaches starts with the introduction to the adopted simulation setup. Chapter 10 describes three main aspects of the simulations:

- considered assessment methods and measures,
- validated control algorithms,
- and assessed exemplary processes.

Selected CPA methods consist of the compromise between available and industry validated approaches. The validation is performed for two the most popular control algorithms in process industry, i.e. PID algorithm and model predictive control in two configurations: GPC and DMC. Additionally, the standard single loop PID controller is expanded (if required) into feedforward-feedback configuration or the cascaded control. Such three-element control constitutes one of the main base control philosophies used in process industry.

The presented algorithms are applied and extensively validated for various process, starting from SISO single element linear benchmarks (Chap. 11), being followed in Chap. 12 by the multi-loop PID cascaded and feedforward-feedback two element configurations. PID control validations is concluded in Chap. 13 with two industrial simulated process of drum-level control and linear binary distillation column.

Finally, Model Predictive Control is addressed in Chap. 14 with SISO and MIMO examples using GPC and DMC predictive algorithm.

Chapter 10
Simulation Setup

The enemy of a good plan is the dream of a perfect plan.

– Carl von Clausewitz

The idea of this part is to compare and visualize representative methods with the simulation benchmarks. Three areas of the industrial control are covered and reviewed in the chapter:

- assessment measures and methodology,
- considered control algorithms,
- and assessed processes and control philosophies.

10.1 Assessment Measures

In fact it is almost impossible to compare and benchmark all available CPA methods. Some selection has to be done. One of the criteria was industrial popularity and applicability. The measures reflect different properties and behavior of the control quality. Comparison is dedicated towards these measures which aim at the loop dynamic properties. Thus in general those indexes which reflect steady state error, like all location statistical factors will be excluded. The same happens with the indexes describing the loop asymmetric operation, like all skewness factors. Though the selection is as always subjective, selected *22* indexes cover step response measures and data-driven methods (Table 10.1). Step response indicators are considered as the industry common best practices, while data-driven indicators do not require additional process knowledge, only normal operation time series of the control error variable.

The indexes mentioned above are evaluated and compared within several simulation scenarios. These scenarios are used to evaluate properties of the considered

P. D. Domański, *Control Performance Assessment: Theoretical Analyses and Industrial Practice*, Studies in Systems, Decision and Control 245,
https://doi.org/10.1007/978-3-030-23593-2_10

Table 10.1 Selected representative CPA indexes

No	Method domain	Index	Acronym
1	Step response	Overshoot	κ
2		Settling time	t_s
3		Decay ratio	DR
4		Idle Index	II
5		Area Index	AI
6		Output Index	OI
7		R-Index	RI
8	Time-domain methods	Mean Square Errors	MSE
9		Integral Absolute Error	IAE
10		Quadratic Error	QE
11	Statistical (Gauss)	Standard deviation	σ
12	Cauchy PDF factors	Scale (dispersion) factor	γ^C
13	α-stable PDF factors	Stability factor (characteristic exponent)	α
14		Scale (dispersion) factor	γ^L
15	Laplace PDF factors	Scale	b
16	Robust statistics	M-estimator of standard deviation	σ^{rob}
17	Alternative	Multiple Hurst exponents	H^0, H^1, H^2, H^3
18		Differential entropy	H^{DE}
19		Rational entropy	H^{RE}

measures, their scope of applicability, strengths and weaknesses. These simulation also show how each measure may be applied and how it behaves.

The observations allows to make some kind of the generalizations and enables to finally propose a general and comprehensive assessment methodology that might be proposed for industrial use.

10.2 Control Algorithms

Above summarized processes and control configurations desire specific control algorithms being described in the following section. They are presented below.

10.2.1 PID Feedback Controller

The considered simulation environment can use different control algorithms. The PID control algorithm in the standard parallel form is applied as the benchmark algorithm.

$$G_{PID}(s) = k_p \left(1 + \frac{1}{T_i s} + T_d s\right). \tag{10.1}$$

Actually, there are several reasons behind such a choice. First of all, the PID algorithm constitutes the majority (95%) of the industrial control loops in process industry [2, 20]. Secondly, the algorithm itself is not included itself in any of the CPA strategies. Selection of the applied algorithm does not affects the performance assessment process. The algorithm is really simple and its control performance depends on three tuning parameters. From that perspective the visualization of the results is much easier and does not over complicates the presentation form. Finally, it is not practical to use any APC control strategy for the simple linear SISO control loop. The predictive control will be added into the consideration into the further analysis, when nonlinear multivariate case is considered.

10.2.2 Cascaded Control

Cascaded control is used when the process may be decomposed into two separate modes with different time horizons. In such a case the appropriate control philosophy consists of two feedback controllers with the upstream $R_1(s)$ (also called the master, outer or top) and the downstream $R_2(s)$ controller (also called the slave, inner or bottom) taking care for the inner faster loop. The considered case studies are based on the structure sketched in Fig. 10.1.

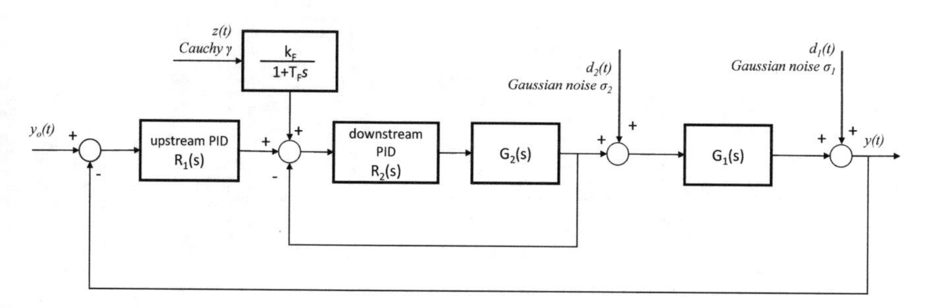

Fig. 10.1 Cascaded two element control scheme

10.2.3 Disturbance Decoupling Feedforward

In several cases the feedback control loop is disturbed by external impacts (see Fig. 10.2). Once one cannot measure these disturbances, nothing much can be done. However, in the opposite case measured disturbance can be decoupled by the specialized feedforward unit. There are several approaches to its design. The canonical one uses inverted process model and is defined by the following formula.

$$R_{FF}(s) = -\frac{G_D(s)}{G(s)}. \quad (10.2)$$

Such an approach enables full disturbance decoupling only in an ideal case. In case of delays or plant unstable zeros the modifications to the approach have to be done. Nonetheless the construction of the feedforward unit requires good knowledge of the plant and disturbance transfer functions. Furthermore, theoretical structure of the feedforward (10.2) is hardly achievable in industrial implementations, when PLC/DCS embedded block-ware limits input of any wished transfer function. Thus, simplified structures of the feedforward disturbance decoupling is used in practice.

The feedforward module must be constructed out of the blocks, which are available in the control system. One may find few possible versions of the feedforward. The most popular two versions of the industrial disturbance decoupling feedforward design are presented in Fig. 10.3. In general they are very similar to each other and the selection is mostly the matter of a habit or a corporate template.

The feedforward module version A described by the resulting formula (10.3) has been selected as a disturbance decoupling feedforward template used in the further analyzes.

$$R_{FF}^{A}(s) = K\left(\frac{1}{T_1 s + 1} - \frac{1}{T_2 s + 1}\right), \; T_1 < T_2. \quad (10.3)$$

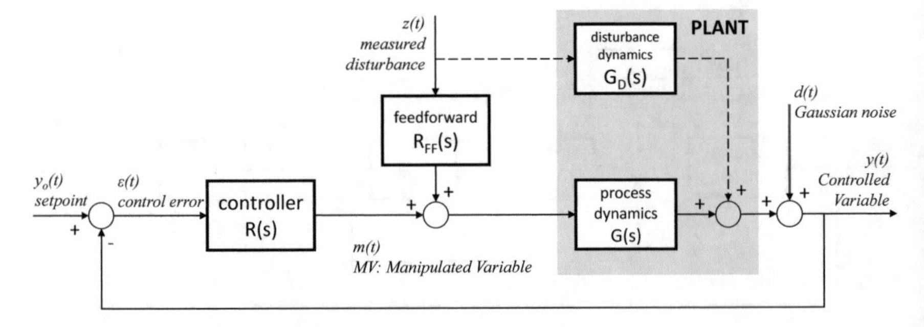

Fig. 10.2 Two element control scheme with feedback-feedforward configuration

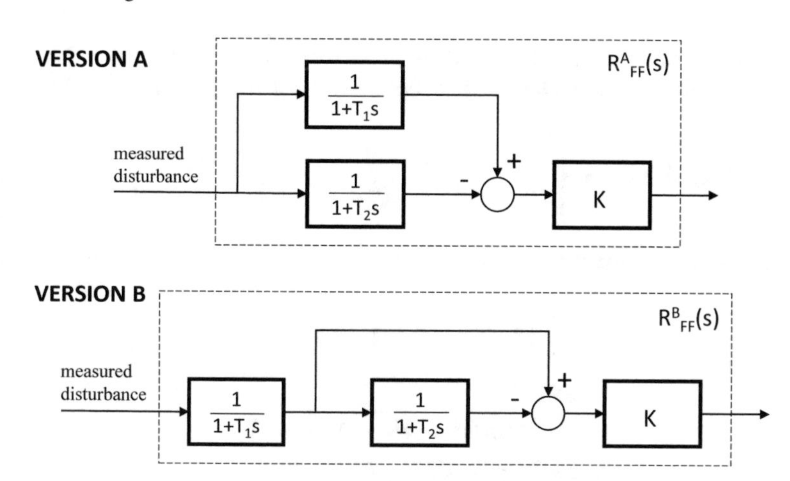

Fig. 10.3 Two versions of the practical feedforward design—in both cases $T_1 < T_2$

10.2.4 MPC—Model Predictive Control

Model Predictive Control (MPC) [5, 14] constitutes one of the main APC approaches that is used in process industry. Once PID loops are not enough the MPC is then selected as the improvement alternative. MPC history may be traced back to the works of Kalman in the early 1960s [12] with the linear quadratic regulator (LQR) designed to minimize an unconstrained quadratic objective function of states and inputs. Further development accelerated in late 1970s with the introduction of the Model Predictive Heuristic Control [17] (nowadays known under the name of Model Algorithmic Control—MAC) and Dynamic Matrix Control (DMC) [8]. The Generalized Predictive Control (GPC) has been proposed almost 10 years later [6, 7]. Continuous theoretical development of the MPC approach [18] is being successfully followed by thousands of the industrial application [9, 10, 13, 16].

MPC algorithms share several advantages, as for instance: control rule minimizing performance index is automatically evaluated, the algorithm incorporates constraints directly in its definition and the control philosophy is universal enabling further specific modifications and updates. The necessity to find out precise process model which is used in the control rule evaluation constitutes the main shortcut.

Let us consider Multiple-Input Multiple-Output (MIMO) processes with n_{u} MVs and n_{y} CVs. It means that the vector of MVs is $u = \left[u_1 \ldots u_{n_{\mathrm{u}}}\right]^{\mathrm{T}}$, The vector of CVs is $y = \left[y_1 \ldots y_{n_{\mathrm{y}}}\right]^{\mathrm{T}}$. At each sampling, $k = 0, 1, \ldots$, the MPC algorithm calculates on-line the increments of the future manipulated variables (a vector of length $n_{\mathrm{u}} N_{\mathrm{u}}$)

$$\Delta \boldsymbol{u}(k) = \begin{bmatrix} \Delta u(k|k) \\ \vdots \\ \Delta u(k + N_{\mathrm{u}} - 1|k) \end{bmatrix} \tag{10.4}$$

where N_u is the control horizon. The increments for the future sampling instant $k+p$ calculated at the current instant k are denoted by $\Delta u(k+p|k)$. They are defined as

$$\Delta u(k+p|k) = \begin{cases} u(k|k) - u(k-1) & \text{for } p = 0 \\ u(k+p|k) - u(k+p-1|k) & \text{for } p \geq 1 \end{cases} \tag{10.5}$$

The MPC optimization problem may be expressed in a vector-matrix formulation

$$\begin{aligned} &\min_{\Delta \boldsymbol{u}(k)} \left\{ \left\| \boldsymbol{y}^{\text{sp}}(k) - \hat{\boldsymbol{y}}(k) \right\|^2 + \|\Delta \boldsymbol{u}(k)\|^2_{\boldsymbol{\Lambda}} \right\}, \\ &\text{subject to} \\ &\boldsymbol{u}^{\min} \leq \boldsymbol{J} \Delta \boldsymbol{u}(k) + \boldsymbol{u}(k-1) \leq \boldsymbol{u}^{\max}, \\ &-\Delta \boldsymbol{u}^{\max} \leq \Delta \boldsymbol{u}(k) \leq \Delta \boldsymbol{u}^{\max}, \\ &\boldsymbol{y}^{\min} \leq \hat{\boldsymbol{y}}(k) \Delta \boldsymbol{u}(k) \leq \boldsymbol{y}^{\max}. \end{aligned} \tag{10.6}$$

where the norms are defined as $\|\boldsymbol{x}\|^2 = \boldsymbol{x}^{\text{T}}\boldsymbol{x}$ and $\|\boldsymbol{x}\|^2_{\boldsymbol{\Lambda}} = \boldsymbol{x}^{\text{T}}\boldsymbol{\Lambda}\boldsymbol{x}$, the set-point trajectory vector $\boldsymbol{y}^{\text{sp}}(k) = [y^{\text{sp}}(k+N_1|k) \ldots y^{\text{sp}}(k+N_2|k)]^{\text{T}}$, the predicted trajectory vector $\hat{\boldsymbol{y}}(k) = \left[\hat{y}(k+N_1|k) \ldots \hat{y}(k+N_2|k)\right]^{\text{T}}$ and the vectors which define the output constraints, i.e. $\boldsymbol{y}^{\min} = \left[y^{\min} \ldots y^{\min}\right]^{\text{T}}$, $\boldsymbol{y}^{\max} = [y^{\max} \ldots y^{\max}]^{\text{T}}$, are of length $N_2 - N_1 + 1$. The vectors which define the input constraints, i.e. $\boldsymbol{u}^{\min} = \left[u^{\min} \ldots u^{\min}\right]^{\text{T}}$, $\boldsymbol{u}^{\max} = [u^{\max} \ldots u^{\max}]^{\text{T}}$, $\Delta \boldsymbol{u}^{\max} = [\Delta u^{\max} \ldots \Delta u^{\max}]^{\text{T}}$ and the vector $\boldsymbol{u}(k-1) = [u(k-1) \ldots u(k-1)]^{\text{T}}$ are of length N_u, the matrices $\boldsymbol{\Lambda} = \text{diag}(\lambda, \ldots, \lambda)$ and

$$\boldsymbol{J} = \begin{bmatrix} 1 & 0 & 0 & \ldots & 0 \\ 1 & 1 & 0 & \ldots & 0 \\ \vdots & \vdots & \vdots & \ddots & \vdots \\ 1 & 1 & 1 & \ldots & 1 \end{bmatrix}, \tag{10.7}$$

are of dimensionality $N_u \times N_u$.

In all MPC algorithms a dynamic model of the controlled process is used to predict the future values of output variable, $\hat{y}(k+p|k)$, over the prediction horizon, i.e. for $p = N_1, \ldots, N_2$. The receding horizon predictive control principle formulated above is presented graphically in Fig. 10.4. The main difference between the DMC and GPC lies in the type of the model used.

10.2.4.1 DMC Algorithm

Discrete-time step-response model of the controlled process used for prediction calculation, i.e. finding the values of $\hat{y}_m(k+p|k)$, is the main DMC algorithm feature. Although it is limited to stable processes only, its main advantage is that the step-response process model may be easily obtained in practice. Since the step-response

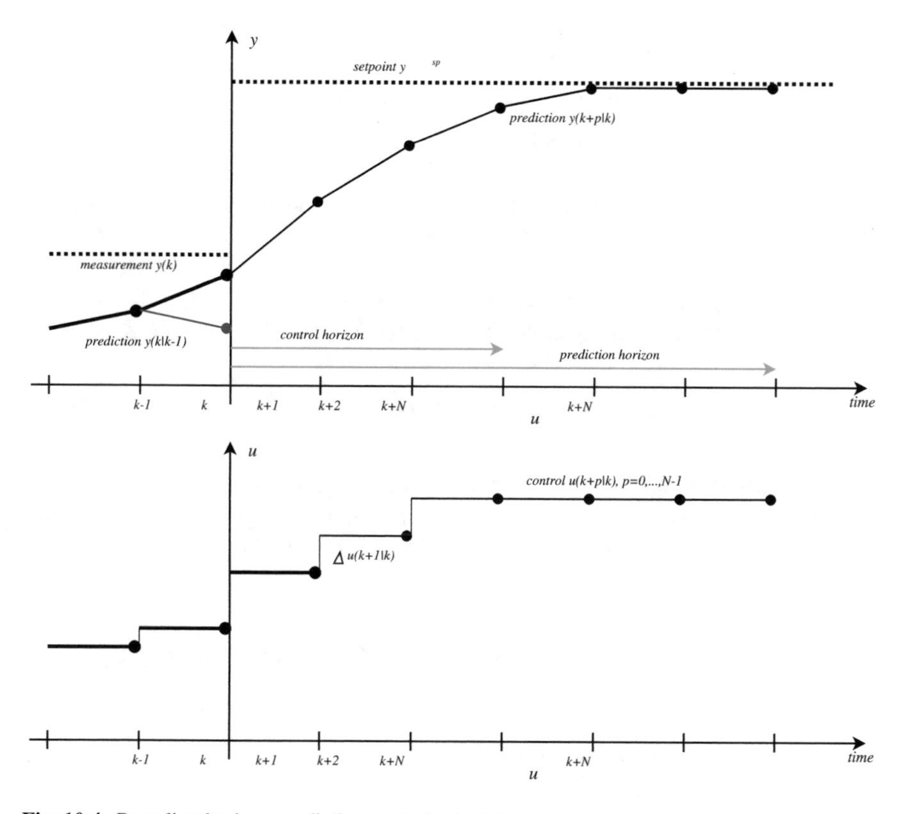

Fig. 10.4 Receding horizon predictive control principle

model is linear in terms of the manipulated variables, minimization of the general MPC cost function (10.6) leads to computationally simple quadratic optimization task. When there are no constraints imposed on the process variables, the solution may be evaluated analytically. The unconstrained MPC optimization may be projected onto the admissible set determined by the constraints [18].

10.2.4.2 Generalized Predictive Control

In the GPC algorithm model of the process has form of discrete difference equation

$$\boldsymbol{A}(q^{-1})y(k) = \boldsymbol{B}(q^{-1})u(k-1) + \boldsymbol{C}(q^{-1})\frac{\epsilon(k)}{\Delta}, \tag{10.8}$$

where polynomials in q^{-1} shift operator are

$$A(q^{-1}) = 1 + a_1 q^{-1} + \cdots + a_{n_A} q^{-n_A}, \tag{10.9}$$

$$B(q^{-1}) = b_1 q^{-1} + \cdots + b_{n_B} q^{-n_B}, \tag{10.10}$$

$$C(q^{-1}) = 1 + c_1 q^{-1} + \cdots + c_{n_C} q^{-n_C}. \tag{10.11}$$

$\epsilon(k)$ denotes vector of white noises with zero mean and $\triangle = 1 - q^{-1}$ backward-difference operator ($1/\triangle$ means integration). Model (10.8) may be called Auto-Regressive Integrated Moving Average with eXogenous Input (ARIMAX) or Controlled Auto-Regressive Integrated Moving Average (CARIMA) [18]. Assuming that the process is affected by integrated white noise ($C(q^{-1}) = 1$) model (10.8) becomes

$$A(q^{-1})y(k) = B(q^{-1})u(k-1) + \frac{\epsilon(k)}{\triangle}. \tag{10.12}$$

It is necessary to point out that the future predictions of CV are linear functions of the calculated decision vector $\triangle \boldsymbol{u}(k)$ according to the GPC prediction and the free trajectory depends only on the past. Thus the general MPC optimization problem (10.6) gives GPC minimization task

$$\begin{aligned}
&\min_{\triangle \boldsymbol{u}(k)} \left\{ \left\| \boldsymbol{y}^{\mathrm{sp}}(k) - \boldsymbol{G}\triangle \boldsymbol{u}(k) - \boldsymbol{F}\boldsymbol{y}^{\mathrm{PG}}(k) - \boldsymbol{G}^{\mathrm{PG}}\triangle \boldsymbol{u}^{\mathrm{PG}}(k) \right\|^2 + \|\triangle \boldsymbol{u}(k)\|^2_{\boldsymbol{\Lambda}} \right\}, \\
&\text{subject to} \\
&\boldsymbol{u}^{\min} \le \boldsymbol{J}\triangle \boldsymbol{u}(k) + \boldsymbol{u}(k-1) \le \boldsymbol{u}^{\max}, \\
&-\triangle \boldsymbol{u}^{\max} \le \triangle \boldsymbol{u}(k) \le \triangle \boldsymbol{u}^{\max}, \\
&\boldsymbol{y}^{\min} \le \boldsymbol{G}\triangle \boldsymbol{u}(k) + \boldsymbol{F}\boldsymbol{y}^{\mathrm{PG}}(k) \\
&\qquad + \boldsymbol{G}^{\mathrm{PG}}\triangle \boldsymbol{u}^{\mathrm{PG}}(k)\triangle \boldsymbol{u}(k) \le \boldsymbol{y}^{\max}.
\end{aligned} \tag{10.13}$$

As prediction equation is linear in terms of vector $\triangle \boldsymbol{u}(k)$, obtained optimization problem (10.13) is of Quadratic Programming (QP) type, i.e. cost-function is quadratic and all constraints are linear.

10.3 Assessed Processes

The analysis starts with the univariate linear SISO loops. As the test process the PID control benchmarks proposed by Åström [3] are used:

- multiple equal poles transfer function (Fig. 10.5a):

$$G_1(s) = \frac{1}{(s+1)^n}, \quad n = 4, \tag{10.14}$$

- fourth order system (Fig. 10.5b):

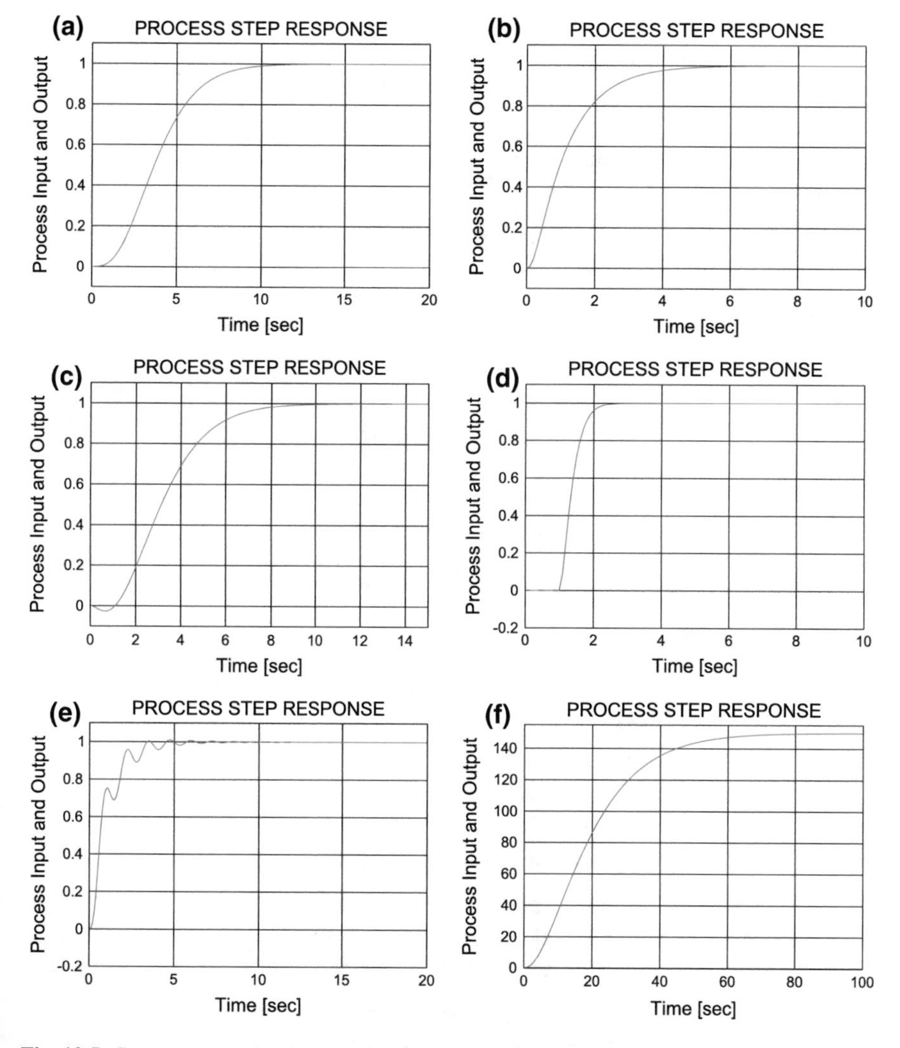

Fig. 10.5 Step responses for the considered linear transfer function

$$G_2(s) = \frac{1}{(s+1)(\alpha \cdot s+1)\left(\alpha^2 \cdot s+1\right)\left(\alpha^3 \cdot s+1\right)},\ \alpha = 0.2, \tag{10.15}$$

- nonminimumphase plant (Fig. 10.5c):

$$G_3(s) = \frac{(1-\alpha \cdot s)}{(s+1)^3},\ \alpha = 0.5, \tag{10.16}$$

- time-delay and double lag plant (Fig. 10.5d):

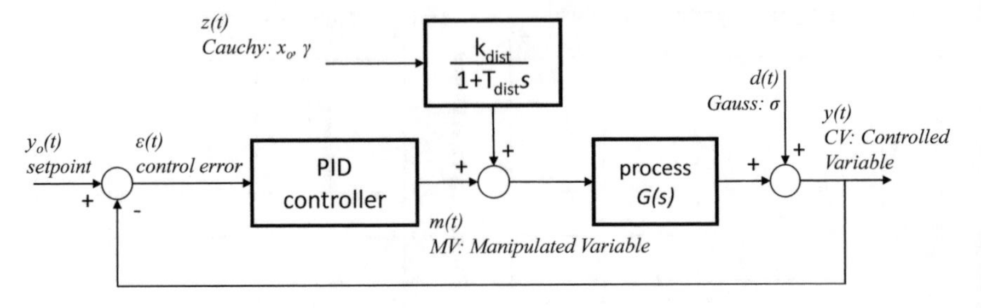

Fig. 10.6 SISO univariate simulation environment

$$G_4(s) = \frac{1}{(0.2s+1)^2} e^{-s}, \tag{10.17}$$

- oscillatory transfer function (Fig. 10.5e):

$$G_5(s) = \frac{1}{(s+1)\left(0.04s^2+0.04s+1\right)}, \tag{10.18}$$

- fast and slow modes plant (Fig. 10.5f):

$$G_6(s) = \frac{100}{(s+10)^2}\left(\frac{1}{(s+1)} + \frac{0.5}{(s+0.05)}\right), \tag{10.19}$$

The step response for each of the considered plant is sketched in Fig. 10.5.

The above transfer functions will be tested in the simulation environment sketched in Fig. 10.6. This environment enables to test the following two scenarios:

(1) undisturbed SISO loop with small σ measurement noise $d(t)$ as in Table 10.2,
(2) disturbed SISO loop (with relatively significant noise and disturbance). Simulations for each transfer function have specific $z(t)$ properties (see Table 10.2).

The linear SISO feedback configuration, called in industry the *single element* control is followed by more complicated and realistic configurations. Once there is a

Table 10.2 Parameters of the simulation disturbances

Transfer function	σ^{noDist}	σ^{Dist}	γ	k_{dist}	T_{dist}
G_1	0.05	0.5	0.6	1	0.1
G_2	0.05	1.0	0.2	2	0.25
G_3	0.05	1.0	0.2	1	0.1
G_4	0.05	1.0	0.2	2	4
G_5	0.05	1.0	0.2	1	2
G_6	0.05	1.0	0.2	1	0.5

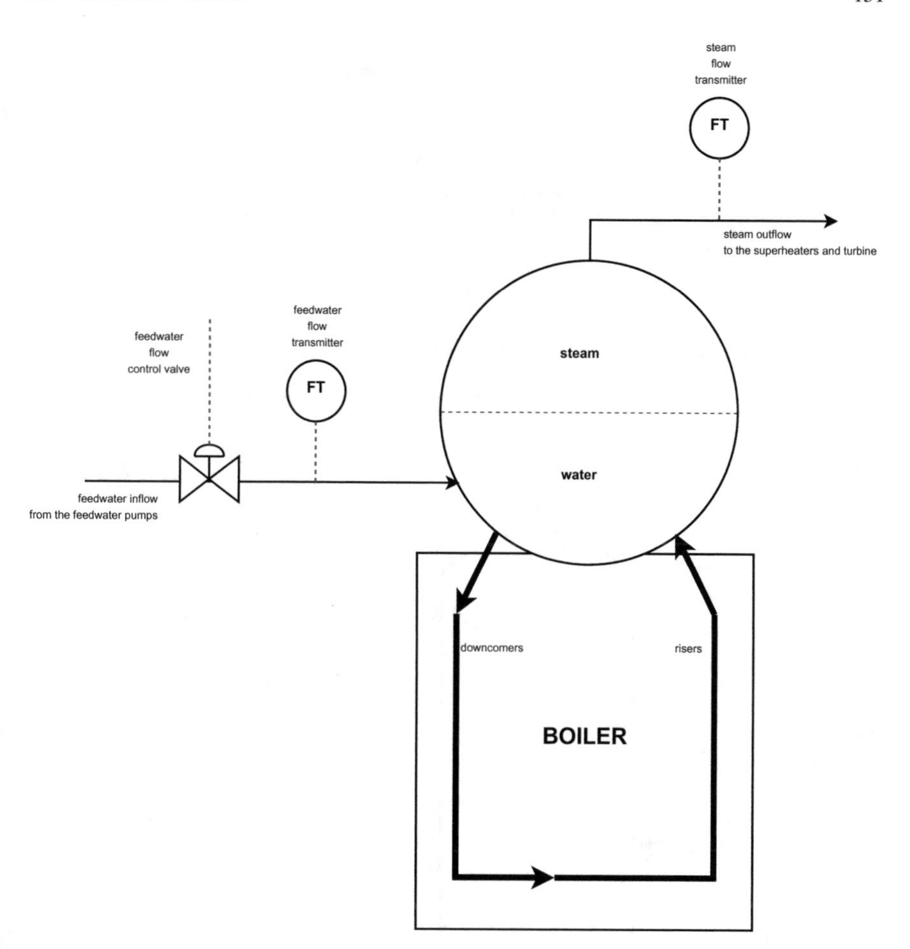

Fig. 10.7 Drum level diagram

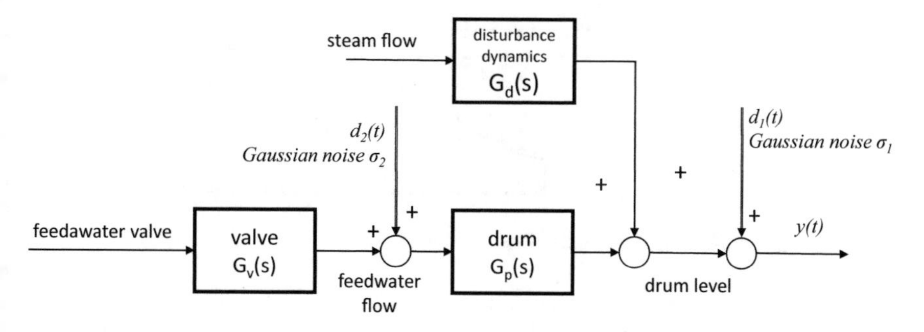

Fig. 10.8 Drum level linear model with transfer functions

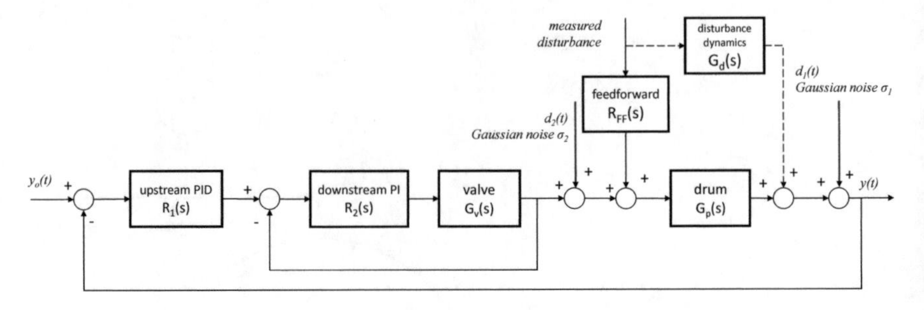

Fig. 10.9 Drum level three element simulation environment

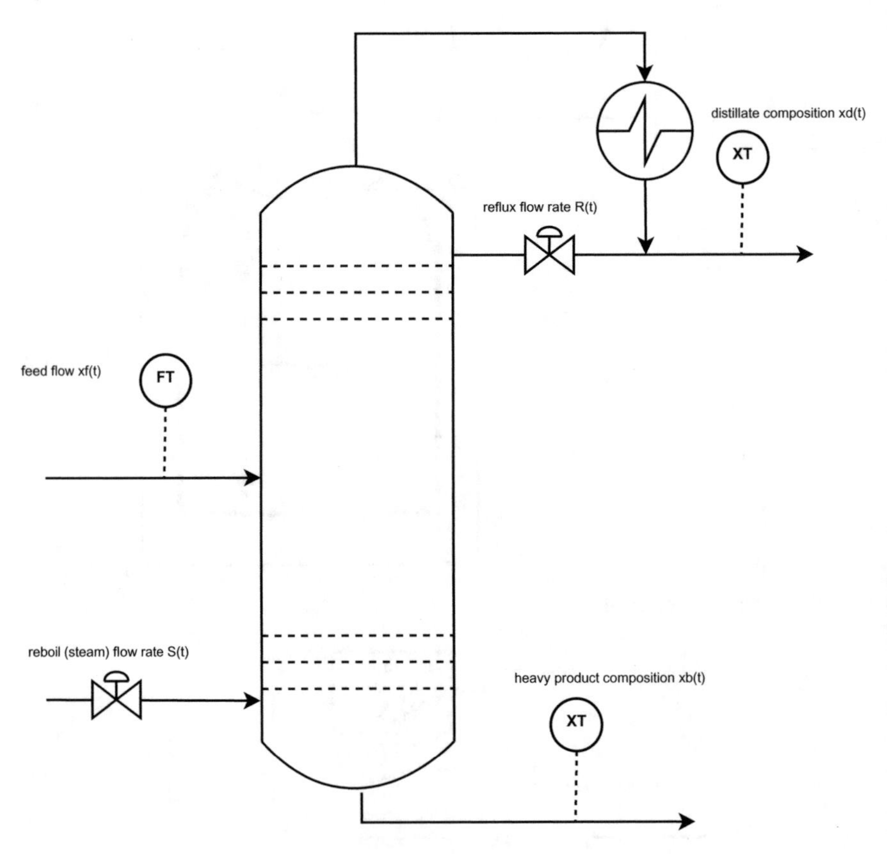

Fig. 10.10 Simplified diagram of a distillation column

possibility to measure and incorporate into the univariate control system additional variables the control philosophy expands towards two element configuration. We may distinguish two types of such structures: cascaded control and feedforward disturbance decoupling.

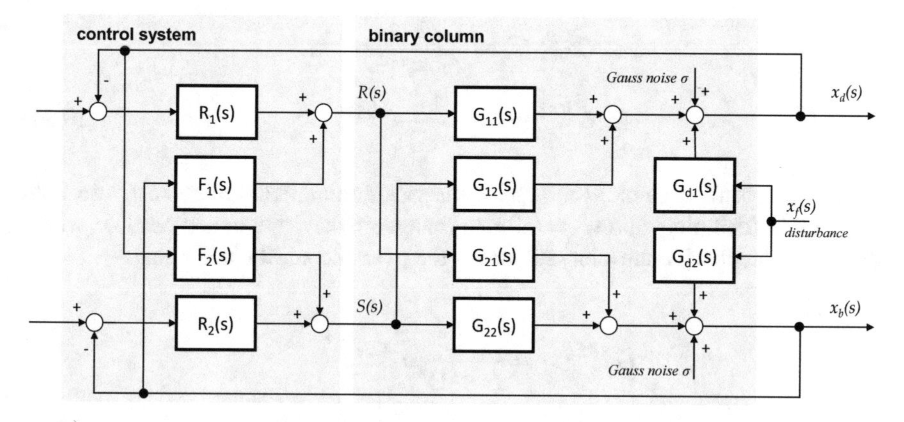

Fig. 10.11 Distillation binary column simulation environment

Cascaded control includes two feedback controllers with the upstream $R_1(s)$ and the downstream $R_2(s)$ controller taking care for the inner loop as in Fig. 10.1. There are considered two modified benchmark simulation scenarios originating from [11, 15] as simulation benchmarks:

1. Cascaded control case 1 ($R_1(s)$ is PID and $R_2(s)$ a PID controller):

$$G_1^{C1}(s) = \frac{1}{5s+1}e^{-2s}, \ \sigma_1 = 0.01, \tag{10.20}$$

$$G_2^{C1}(s) = \frac{1}{s+1}e^{-s}, \ \sigma_2 = 0.001, \tag{10.21}$$

2. Cascaded control case 2 (both $R_1(s)$ and $R_2(s)$ are PID controllers):

$$G_1^{C2}(s) = \frac{10(-0.5s+1)}{(3s+1)^3(s+1)^2}e^{-0.5s}, \ \sigma_1 = 0.02, \tag{10.22}$$

$$G_2^{C2}(s) = \frac{3}{(1.33s+1)}e^{-0.3s}, \ \sigma_2 = 0.006, \tag{10.23}$$

The other type of the *two element* control consists of SISO feedback controller supported with the disturbance decoupling feedforward as sketched in Fig. 10.2. Two feedforward simulation examples are reviewed [1]. The first example is the two tank process with the transfer functions depicted below:

$$G^{FF1}(s) = \frac{1}{(s+1)^2} e^{-0.2s},$$
$$G_D^{FF1}(s) = \frac{1}{s+1} e^{-0.1s}. \tag{10.24}$$

In the second example, the plant witnesses nonminimumphase response with oscillations (damping ration $\zeta = 0.23$) to changes in the manipulated variable, which delays the feedback controller and makes the overall control challenging:

$$G^{FF2}(s) = \frac{1}{(s+1)^2} e^{-0.2s},$$
$$G_D^{FF2}(s) = \frac{-2s+1}{4s^2+0.92s+1} e^{-1.5s}. \tag{10.25}$$

Two MIMO benchmarks originating from the industry are assessed:

Ex. 1 *Boiler drum level three element control*

A steam boiler is the indispensable power equipment in industrial process, with the task to supply steam of certain parameters (pressure and temperature) according to the varying flow setpoint demands. To achieve this goal, every parameter of boiler production process must be controlled strictly. Effective control over all load levels may significantly improve plant operational efficiency.

Drum is one of the main elements. Its goal is to separate superheated steam from the boiling water (see Fig. 10.7). Drum water level is the main parameter informing about quality and safety of its operation. The water level in a drum needs to be kept around the middle line, cause too low levels may make the water in drum transform into steam, and boiler tubes will be damaged by overheating. If the water level is too high, water is carried over into the superheater or the turbine and may cause damages resulting in extensive maintenance costs or forced outages. The dynamics of the drum water level is defined by three transfer functions [4] in a simplified form, as sketched in Fig. 10.8.

$$G_d(s) = \frac{-0.25(-s+1)}{s(s+1)(2s+1)},$$
$$G_p(s) = \frac{0.25(-s+1)}{s(2s+1)},$$
$$G_v(s) = \frac{1}{0.15s+1}. \tag{10.26}$$

Three element control philosophy is commonly used to maintain the drum level with cascaded feedback PID upstream and PI downstream control accompanied with the feedforward disturbance decoupling. The overall controlled process simulation environment is presented in Fig. 10.9.

Ex. 2 *Linear binary distillation column MIMO control*
A bench-scale distillation column for ethanol and water mixture is proposed as the second example [19]. The column separates a two component mixture of water and methanol. Its schematic diagram is presented in Fig. 10.10. A stream of a feed $x_f(t)$ enters the column in the middle. The light distillate product is drawn from the top and the heavy product is obtained from the bottom. The objective of the controller is to keep the composition of light product $x_d(t)$ and heavy product $x_b(t)$ at their desired values by manipulating the reflux flow rate $R(t)$ and steam flow rate $S(t)$. The feed flow rate is considered as the disturbance.
In case that the upstream process changes, the feed flow rate is considered as the disturbance. All variables are deviations from a typical operating point. The linearized model (10.27) in continuous-time domain consists of four model SISO transfer functions and two disturbance SISO models.

$$\begin{bmatrix} X_d(s) \\ X_b(s) \end{bmatrix} = \begin{bmatrix} G_{11}(s) & G_{12}(s) \\ G_{21}(s) & G_{22}(s) \end{bmatrix} \begin{bmatrix} R(s) \\ S(s) \end{bmatrix} + \begin{bmatrix} G_{d1}(s) \\ G_{d2}(s) \end{bmatrix} X_f(s)$$

$$\begin{bmatrix} X_d(s) \\ X_b(s) \end{bmatrix} = \begin{bmatrix} \frac{12.8e^{-s}}{16.7s+1} & \frac{-18.9e^{-3s}}{21s+1} \\ \frac{6.6e^{-7s}}{10.9s+1} & \frac{-19.4e^{-2s}}{14.4s+1} \end{bmatrix} \begin{bmatrix} R(s) \\ S(s) \end{bmatrix} + \begin{bmatrix} \frac{3.8e^{-8s}}{14.9s+1} \\ \frac{4.9e^{-3s}}{13.2s+1} \end{bmatrix} X_f(s) \tag{10.27}$$

Both composition measurements are disturbed with the zero-mean Gaussian noise with standard deviation $\sigma = 0.02$. The feed ratio signal is disturbed with a signal consisting of the varying step wave with added α-stable symmetrical, zero-mean noise and $\alpha = 1.8$ and $\gamma = 1.6$. The MIMO control system using PID feedback controllers $(R_1(s),\ R_2(s))$ and feedforward disturbance decoupling $(F_1(s),\ F_2(s))$ is used as the assessed control philosophy (Fig. 10.11). As the industrial interpretation is the goal of the presentation the feedforward used in the analysis will use the industrial form (10.3).

References

1. Adam, E.J., Marchetti, J.L.: Designing and tuning robust feedforward controllers. Comput. Chem. Eng. **28**(9), 1899–1911 (2004)
2. Åström, K.J., Hägglund, T.: New tuning methods for PID controllers. In: Proceedings 3rd European Control Conference, pp. 2456–2462 (1995)
3. Åström, K.J., Hägglund, T.: Benchmark systems for pid control. In: IFAC Digital Control: Past, Present and Future of PID Control, pp. 165–166 (2000)
4. Bequette, B.: Process Control: Modeling, Design, and Simulation, 1st edn. Prentice Hall Press, Upper Saddle River, NJ, USA (2002)
5. Camacho, E.F., Bordons, C.: Model Predictive Control. Springer, London, UK (1999)
6. Clarke, W., Mohtadi, C., Tuffs, P.S.: Generalized predictive control—I. the basic algorithm. Automatica **23**(2), 137–148 (1987a)
7. Clarke, W., Mohtadi, C., Tuffs, P.S.: Generalized predictive control—II. extensions and interpretations. Automatica **23**(2), 149–160 (1987b)

8. Cutler, R., Ramaker, B.: Dynamic matrix control—a computer control algorithm. In: Proceedings AIChE National Meeting, Houston, TX, US (1979)
9. Ferreau, H.J., Almér, S., Peyrl, H., Jerez, J.L., Domahidi, A.: Survey of industrial applications of embedded model predictive control. In: 2016 European Control Conference, pp. 601–601 (2016)
10. Forbes, M.G., Patwardhan, R.S., Hamadah, H., Gopaluni, R.B.: Model predictive control in industry: challenges and opportunities. IFAC-PapersOnLine **48**(8), 531–538, 9th IFAC Symposium on Advanced Control of Chemical Processes ADCHEM 2015 (2015)
11. Jeng, J.C., Lee, M.W.: Identification and controller tuning of cascade control systems based on closed-loop step responses. IFAC Proc. **45**(15), 414–419, 8th IFAC Symposium on Advanced Control of Chemical Processes (2012)
12. Kalman, R.E.: Contribution to the theory of optimal control. Boletin de la Sociedad Mathematica Mexicana **5**, 102–119 (1960)
13. Lee, J.H.: Model predictive control: review of the three decades of development. Int. J. Control Autom. Syst. **9**(3), 415 (2011)
14. Maciejowski, J.M.: Predictive Control With Constraints. Prentice Hall, Harlow (2002)
15. Mehta, U., Majhi, S.: On-line identification of cascade control systems based on half limit cycle data. ISA Trans. **50**(3), 473–478 (2011)
16. Qin, S.J., Badgwell, T.A.: A survey of industrial model predictive control technology. Control Eng. Pract. **11**, 733–764 (2003)
17. Richalet, J., Rault, A., Testud, J.L., Papon, J.: Model algorithmic control of industrial processes. IFAC Proc. **10**(16), 103–120, preprints of the 5th IFAC/IFIP International Conference on Digital Computer Applications to Process Control, The Hague, The Netherlands, 14–17 June, 1977
18. Tatjewski, P.: Advanced Control of Industrial Processes, Structures and Algorithms. Springer, London (2007)
19. Wood, R.K., Berry, M.W.: Terminal composition control of a binary distillation column. Chem. Eng. Sci. **28**(9), 1707–1717 (1973)
20. Yamamoto, S., Hashimoto, I.: Present status and future needs: the view from Japanese industry. In: Proceedings 4th International Conference on Chemical Process Control, pp. 1–28 (1991)

Chapter 11
PID Single Element Control

All life is problem solving.

– Karl Popper

This part presents the comparison of the performance of different CPA strategies used for the CPA task with the single element control linear case studies described in the previous chapter. The review is organized according to the structured procedure described below. The procedure starts with finding of the *so called* optimal PID controller. This evaluation is done for the undisturbed case. Standard Octave function, which minimizes weighted performance index consisting of ITAE criterion (weighting factor $\mu_1 = 1$), maximum overshoot ($\mu_2 = 10$) and sensitivity ($\mu_3 = 20$) [1] is used. The analysis procedure is repeated twice according to the scheme sketched in Fig. 10.6. First, the undisturbed and unconstrained loop is analyzed. It is followed by the constrained and disturbed case. All the simulations are conducted with the use of the same setpoint profile.

(1) Optimal value for each of the considered measures is evaluated.
(2) The optimal controller setting is visualized with the according step response, time trends for the SISO loop variables, control error histogram diagram and its extended range R/S plot.
(3) As the optimal PID setting is known the domain of the controller parameters k_P, T_i, T_d is scanned to evaluate the behavior of each of the measures and their ability to detect good control. Each simulation scenario consists of many loop simulations for different PID controller settings. Generally the properties of each CPA measure fill the 4D dimension space. As such the visualization would be unreadable, the presentation uses 2D surfaces presenting three cross-section of the parameters space, cut with the optimal values of each of the parameters. Selecting optimal value of the T_d the PI cross-section surface k_P, T_i

P. D. Domański, *Control Performance Assessment: Theoretical Analyses and Industrial Practice*, Studies in Systems, Decision and Control 245,
https://doi.org/10.1007/978-3-030-23593-2_11

is investigated, for constant optimal T_i value the PD surface k_P, T_d is plotted and for constant k_p the ID surface T_d, T_i is presented.

(4) The procedure concludes with the observations and summary of the results.

The effect of the loop disturbances is discussed to conclude the analysis.

11.1 Multiple Equal Poles Transfer Function

The analysis starts with the evaluation of the optimal PID controller and is followed by the analysis of the tuning surfaces for all considered CPA indexes.

11.1.1 Optimal Controller Performance

The optimal controller transfer function $R1_{PID}^{opt}(s)$ has been designed according to the predefined rules. The evaluated tuning parameters are presented below:

- gain: $k_p^{opt} = 1.05$
- integration: $T_i^{opt} = 3.00$
- derivation: $T_d^{opt} = 0.93$

The graphical representation of the loop setpoint step response is sketched in Fig. 11.1, while Fig. 11.2 shows the disturbance step response. We see that the obtained tuning delivers the response with almost invisible overshoot. The controller output is smooth without any oscillations.

Next, the loop simulation has been run. The time trends (sample of first 120 s) for the selected loop profiles for setpoint, CV and MV are sketched in Fig. 11.3. The figure shows comparison of both undisturbed and disturbed behavior.

Respective control error histogram is shown in Fig. 11.4. It is clearly visible that its shape is not Gaussian with the significant tails. The Histogram can be modeled the most precisely by the α-stable distribution. The analog histogram for the disturbed simulation is sketched in Fig. 11.6.

Finally, the extended range R/S plot has been evaluated (Fig. 11.5). We clearly see that its character is not monofractal. The multi-persistent properties are well reflected with the two crossovers ($cross^1$ and $cross^2$) and three memory ranges described by the respective Hurst exponents (H^1, H^2 and H^3). The analog plot for the disturbed simulation is sketched in Fig. 11.7.

All the evaluated indexes for both undisturbed (nonDist) and disturbed (Dist) loop simulations are summarized in Table 11.1. It is also interested to see what tuning, i.e. controller parameters have been pointed out by each method. Table 11.2 summarizes these results. The table consists only of these measures that can detect the solution, i.e. point out the single result as the extreme index value (minimum or maximum). The settings highlighted show the results which are close enough to the optimal solution $k_p = k_p^{opt} \pm 0.2$ and $T_i = T_i^{opt} \pm 0.5$, i.e. $k_p \in [0.85, 1.25]$ and $T_i \in [2.5, 3.5]$.

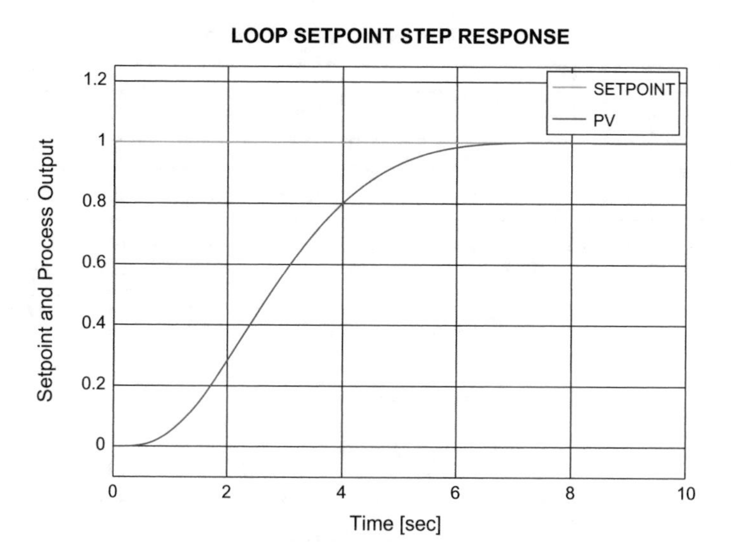

Fig. 11.1 Loop setpoint step response for $G_1(s)$

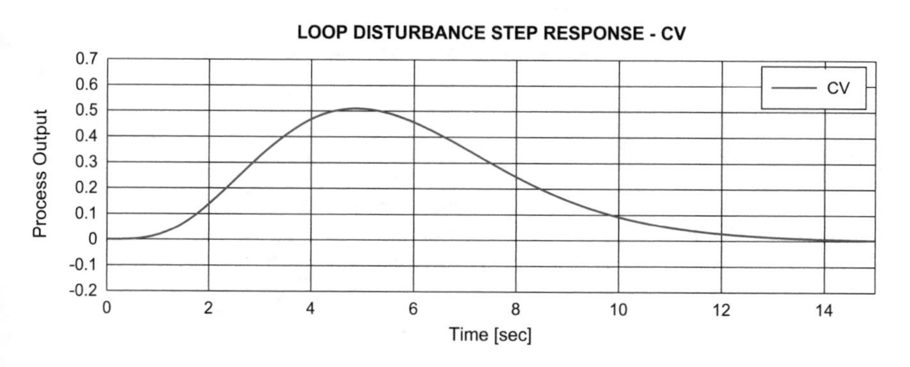

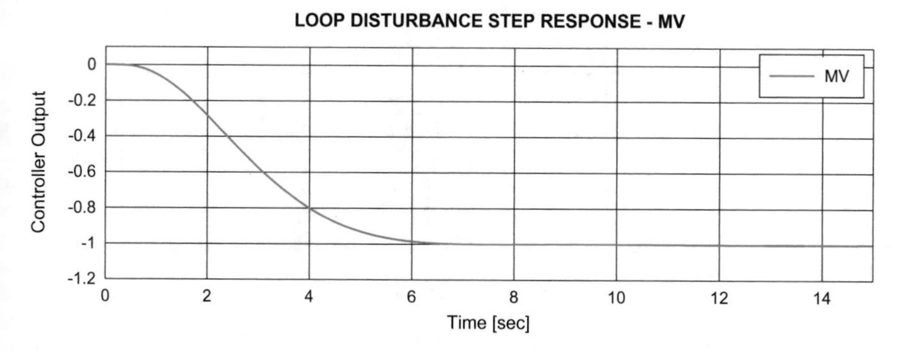

Fig. 11.2 Loop disturbance step response for $G_1(s)$

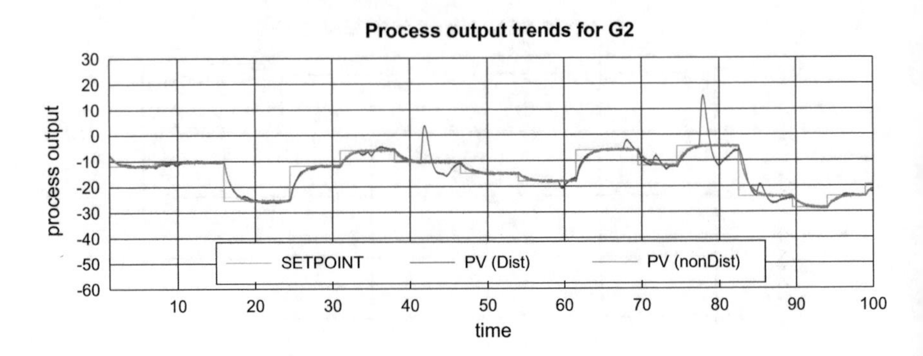

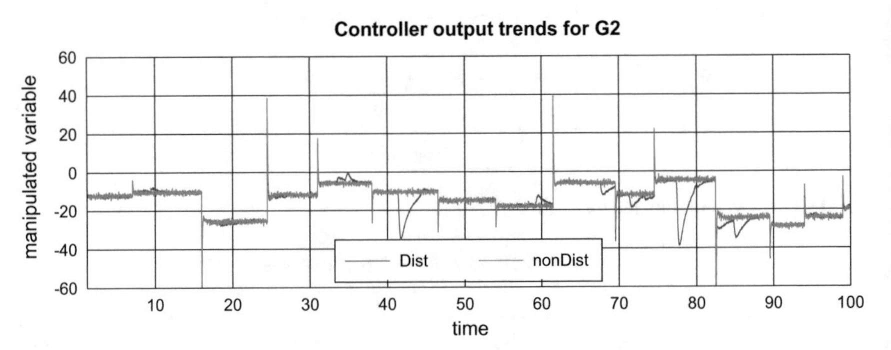

Fig. 11.3 Loop time trends for first 100 s of $G_1(s)$

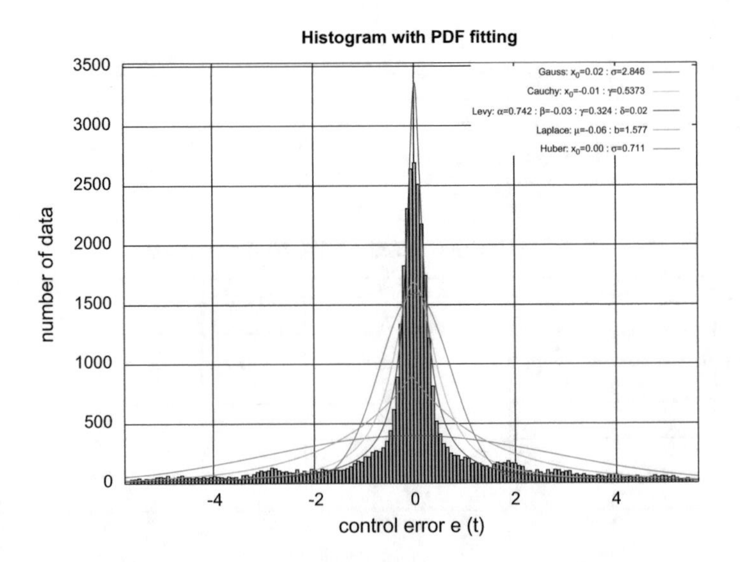

Fig. 11.4 Control error histogram with the fitted different PDFs for $G_1(s)$

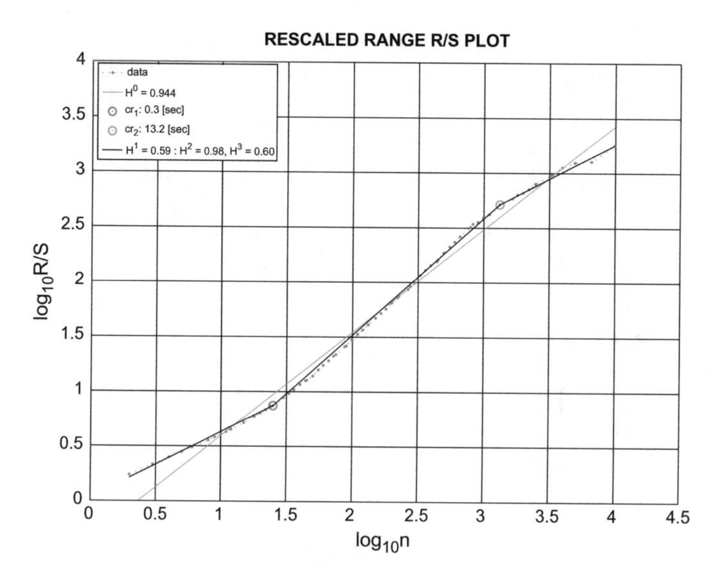

Fig. 11.5 Extended range R/S plot of the control error for $G_1(s)$

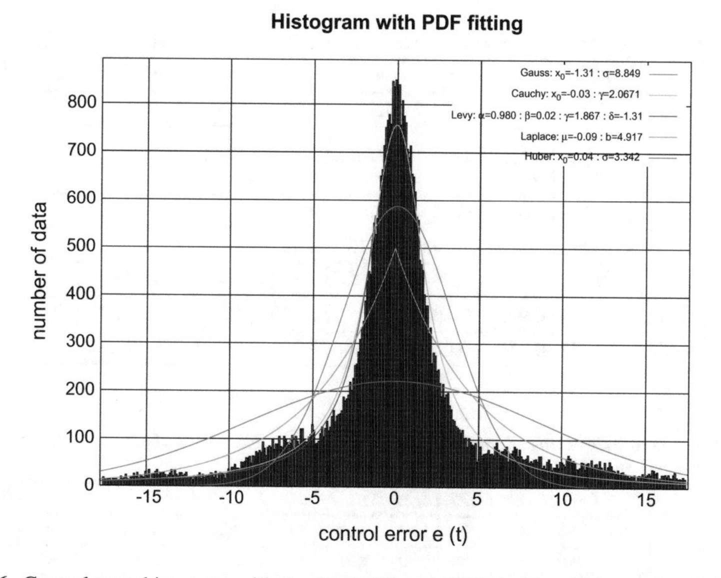

Fig. 11.6 Control error histogram with the fitted different PDFs in disturbed case for $G_1(s)$

11.1.2 Control Performance Detection Using Step Indexes

The analysis of the performance assessment potential for the considered quality indexes (KPIs) is presented in form of the control quality surfaces. The drawings are

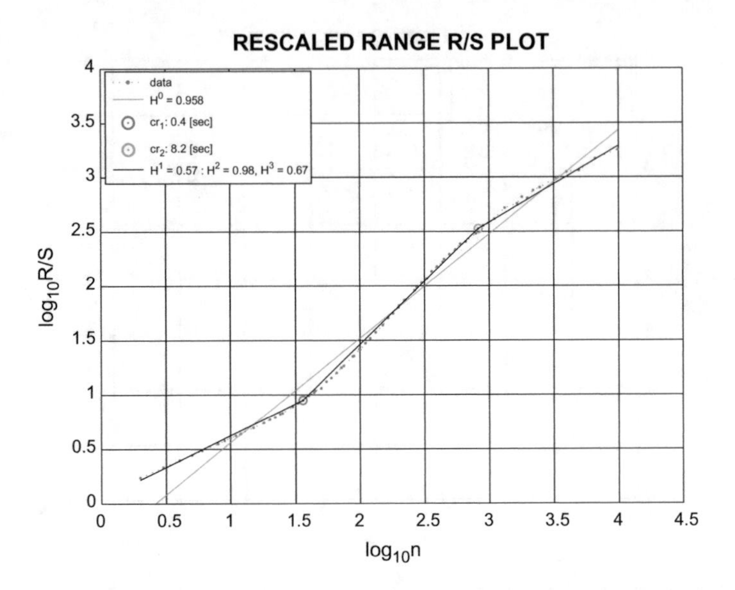

Fig. 11.7 Extended range R/S plot in the control error in disturbed case for $G_1(s)$

Table 11.1 Indexes for the optimal $G_1\,(s)$ loop controller (undisturbed and disturbed)

	T_s	$\kappa[\%]$	DR	II	AI	OI	RI	MSE	IAE	QE
nonDist	6.19	0.30	0.00	−0.22	0.99	0.003	0.67	8.08	1.54	34.8
Dist	–	–	–	–	–	–	–	45.26	3.51	86.0
	σ	γ^C	α	γ^L	b	σ^{rob}				
nonDist	2.84	0.33	0.67	0.23	1.54	0.540				
Dist	6.73	1.606	1.365	1.605	3.515	2.62				
	H^0	H^1	H^2	H^3	$cr^1[s]$	$cr^2\,[s]$	H^{RE}	H^{DE}		
nonDist	0.944	0.591	0.976	0.598	0.25	13.20	10.54	−15820		
Dist	0.958	0.565	0.980	0.669	0.36	8.25	34.80	−57050		

Table 11.2 Controller tuning pointed out by the selected measures for $G_1\,(s)$—highlighted are the solutions close to the optimal one

		T_s	$\kappa[\%]$	DR	MSE	IAE	QE	σ	γ^C	γ^L	b	σ^{rob}	H^{RE}	H^{DE}
nonDist	k_p	**1.05**	1.90	n/a	2.0	1.80	0.95	2.00	**1.25**	**1.25**	1.80	1.30	–	–
	T_i	**3.0**	10.0	n/a	4.80	3.50	10.0	4.80	**3.20**	**3.20**	3.80	3.20	–	–
Dist	k_p	n/a	n/a	n/a	1.85	1.85	0.95	1.85	2.00	1.40	1.85	1.85	1.50	0.45
	T_i	n/a	n/a	n/a	9.50	5.00	10.00	9.50	10.00	5.00	5.00	3.50	6.20	4.20

prepared according to the same common scheme. The plots are in form of the two dimensional surfaces presenting each KPI as the function of two PID parameters: k_p and T_i. The third parameter, $T_d = 0.93$ is kept constant at its optimal value, so the

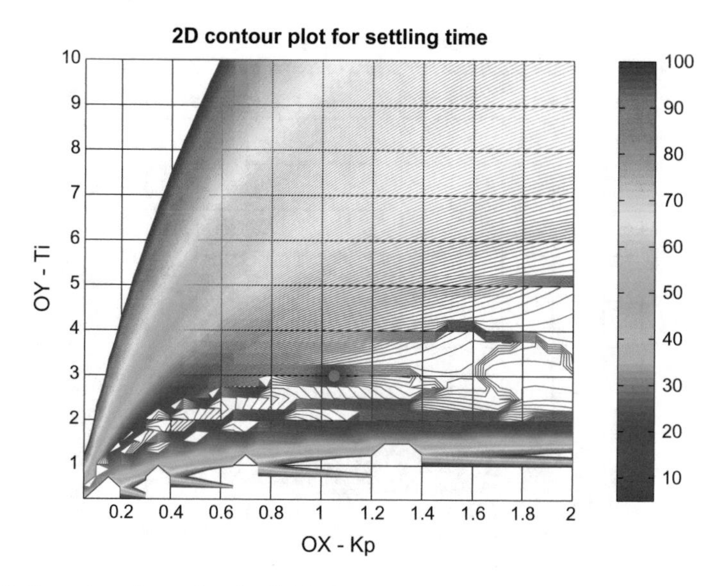

Fig. 11.8 Control quality surface for $G_1(s)$: T_s

surface is a 2D cross-section. Due to the lack of place other cross-sections are not presented as they do not bring additional value to the analysis.

The position of the found optimal controller is depicted with the magenta circle marker •. There are two variants of the diagram interpretation. The best index is at its minimum value, or there is some optimal value of the index. In the first case the red diamond marker ♦ depicts minimum index position. The black bold contour line for the selected the best value of the KPI is used in the second case.

Step based indexes, i.e. settling time, overshoot, II, AI, OI and RI are evaluated without the disturbance.

Review of the attached drawings for step response indexes enables to bring up the following observations:

- Classical indexes of the settling time (Fig. 11.8) and the overshoot (Fig. 11.9) work fine. It is clearly seen that the calculated optimal controller $R1_{PID}^{opt}(s)$ is very close to the minimum index one. On the other hand, the overshoot treated independently does not enable to find single solution.
- Optimal controller is relatively close to the contour line depicting ideal value of the Idle Index (Fig. 11.10).
- Ideal literature value of the Area Index $0.3 < AI < 0.7$ is relatively far away from the optimal controller (Fig. 11.11) and reflects rather sluggish control. Similar situation is visible for the Output Index (Fig. 11.12), when optimal controller $OI = 0.003$ is quite far away from the literature ideal $OI < 0.35$.
- R-Index surface is not smooth and very scattered (Fig. 11.13). It shows that detection ability with RI is very limited and does not work for the considered loop.

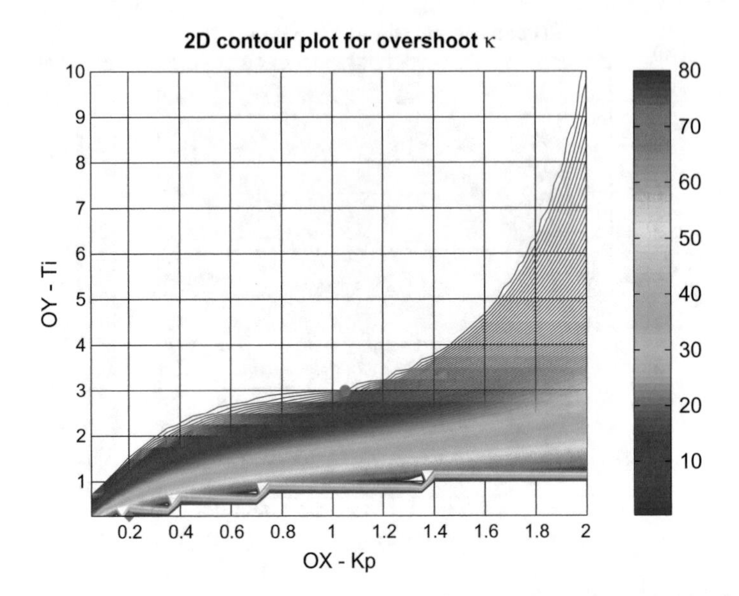

Fig. 11.9 Control quality surface for $G_1(s)$: κ

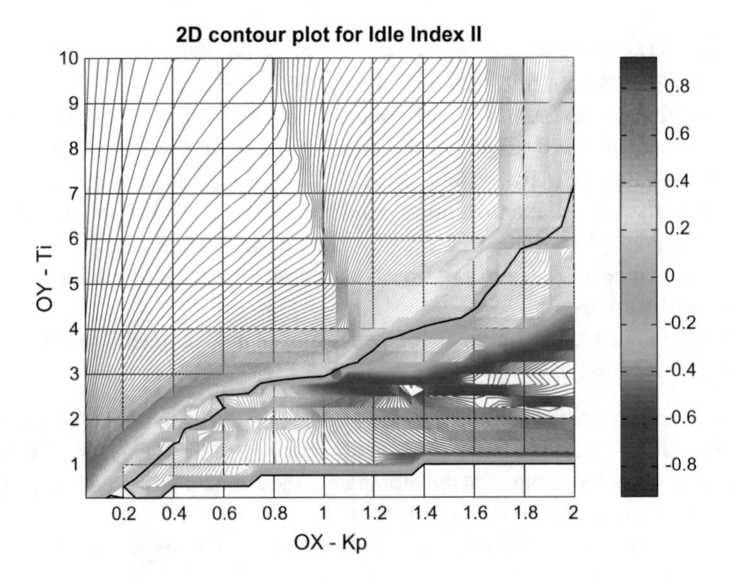

Fig. 11.10 Control quality surface for $G_1(s)$: II (the best value $II = 0$)

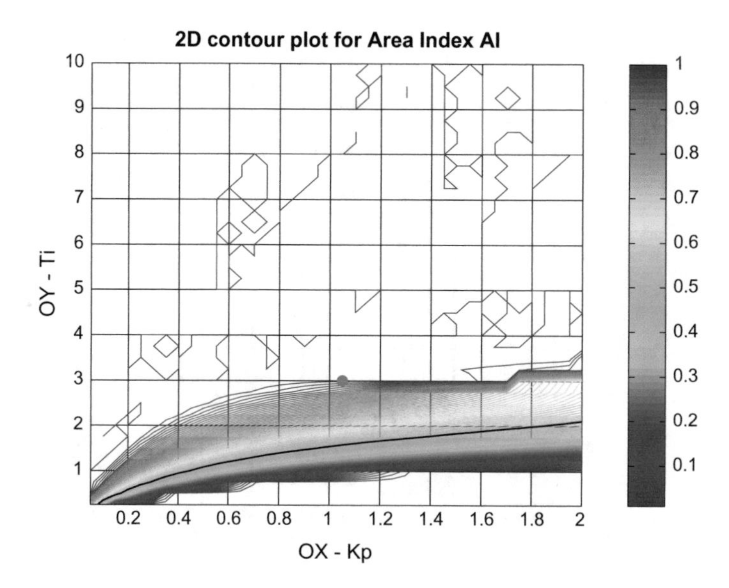

Fig. 11.11 Control quality surface for $G_1(s)$: AI (the best value $AI = 0.5$)

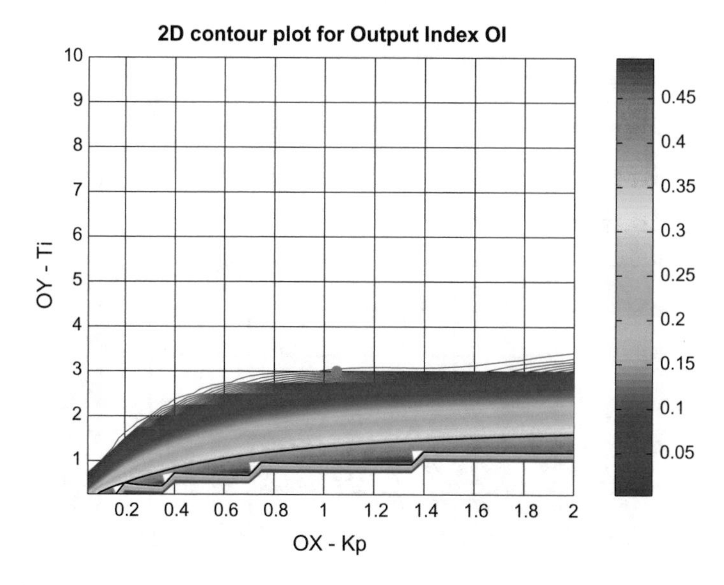

Fig. 11.12 Control quality surface for $G_1(s)$: OI (the best value $OI = 0.35$)

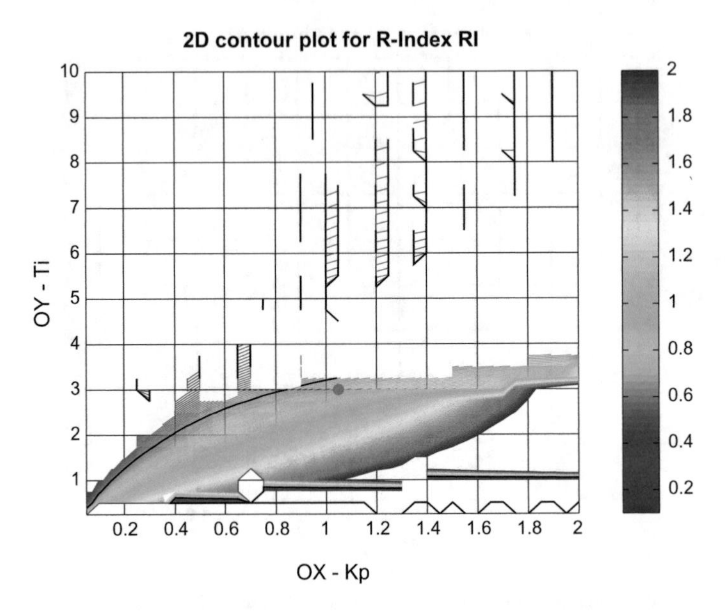

Fig. 11.13 Control quality surface for $G_1(s)$: RI (the best value $RI = 0.6$)

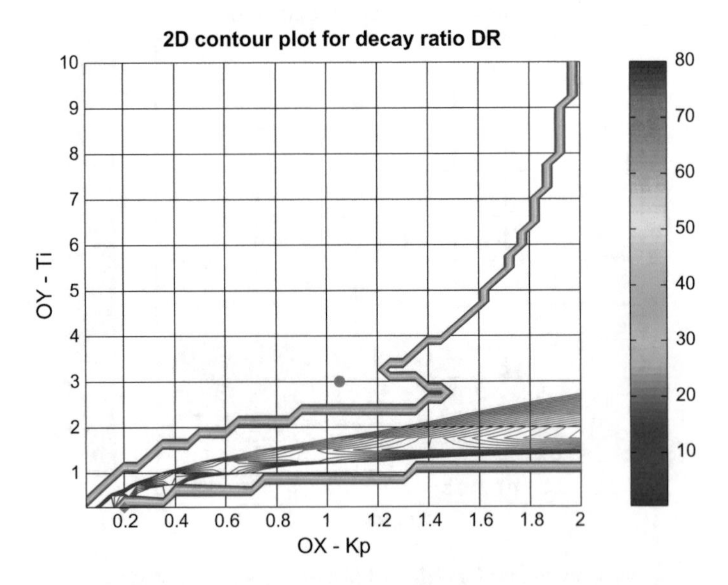

Fig. 11.14 Control quality surface for undisturbed $G_1(s)$: DR

- Decay Ratio surface (Fig. 11.14) is not evident and the detection ability with DR is very limited and does not work for the considered loop.

11.1.3 Control Performance Detection for Undisturbed Loop

The simulation of the loop using predefined setpoint profile is realized in two forms: undisturbed and disturbed one. Thus for such experiment the integral, statistical, fractal and entropy indexes are evaluated and analyzed. Diagrams sketched below present surfaces for the undisturbed simulation mode.

Review of the above undisturbed surfaces allows to bring about the following observations:

- Gaussian standard deviation σ (Fig. 11.18) is very close to the MSE (Fig. 11.15), while Laplace scale factor b (Fig. 11.23) is very close to the IAE (Fig. 11.16). The results identified with them differ significantly. Square error and σ tend to the sluggish and non-overshoot control highly disliking aggressiveness in tuning. On the other hand absolute error Laplace scale factor b is more realistic and closer to the expectations. Additionally, Quadratic Error QR (Fig. 11.17) is close to the MSE.
- Usage of the robust standard deviation (Fig. 11.19) seems to be more effective with the best found controller close to the optimal one. Cauchy scale factor γ^C (Fig. 11.20) performs similarly. Detection with the scale factor of the stable distribution (Fig. 11.21) is even more accurate (it points out exactly the same point) and seems to be the best KPI.
- Interpretation of the stable distribution stability factor α (Fig. 11.22) is not clear. The shape of the surface is close to the scale γ^L. However, the solution tends

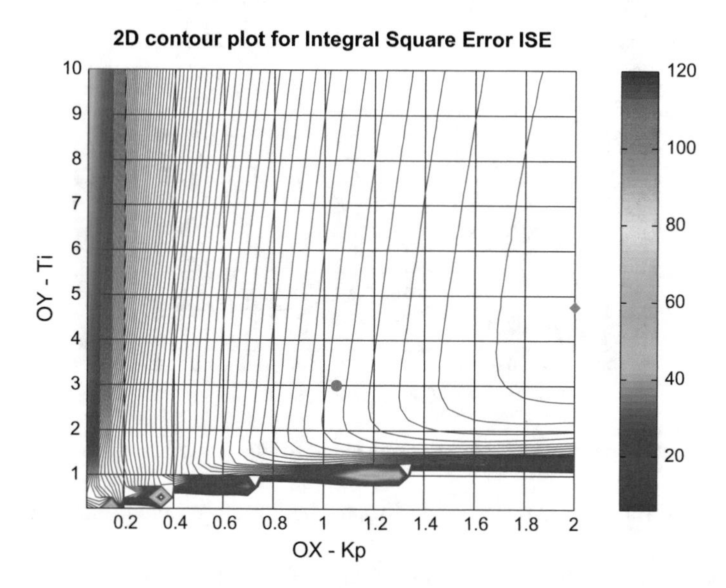

Fig. 11.15 Control quality surface for undisturbed $G_1(s)$: MSE

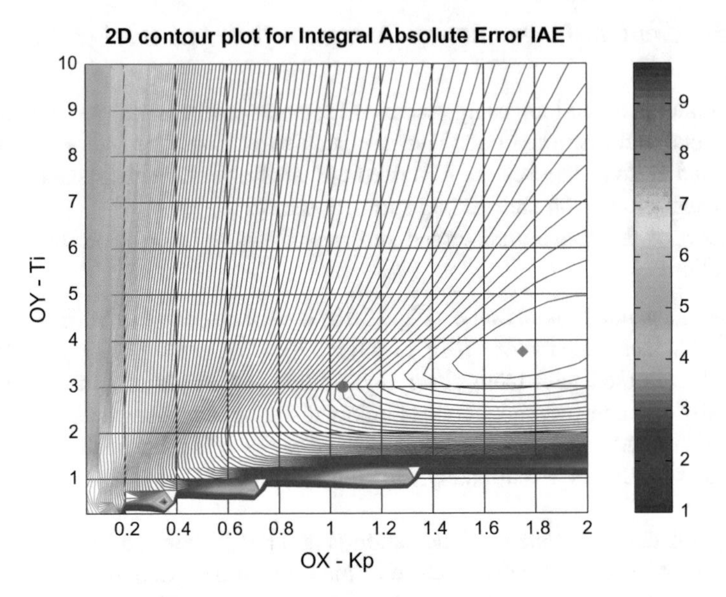

Fig. 11.16 Control quality surface for undisturbed $G_1(s)$: IAE

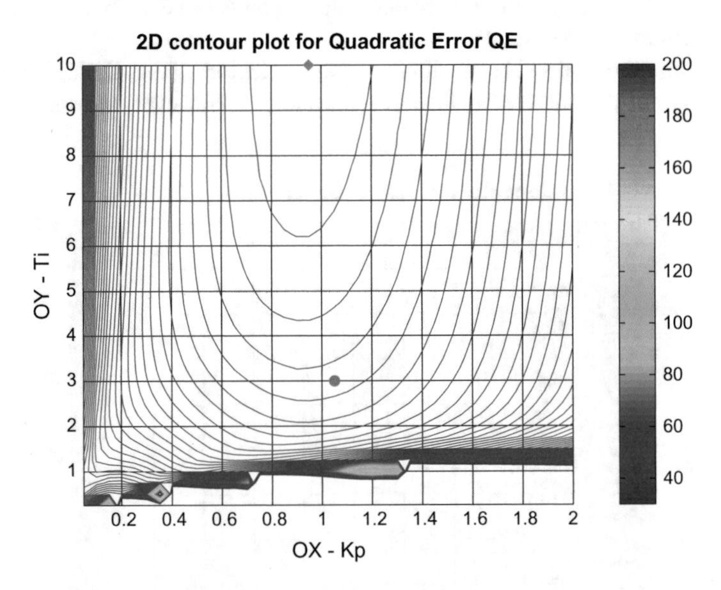

Fig. 11.17 Control quality surface for undisturbed $G_1(s)$: QE

towards lower values, which according to the literature reflect rather aggressive tuning.

- All the Hurst exponents share similar shape of the surfaces, however on different levels. The single Hurst exponent H^0 averages the overall behavior. It is smooth and consistent with other shapes.

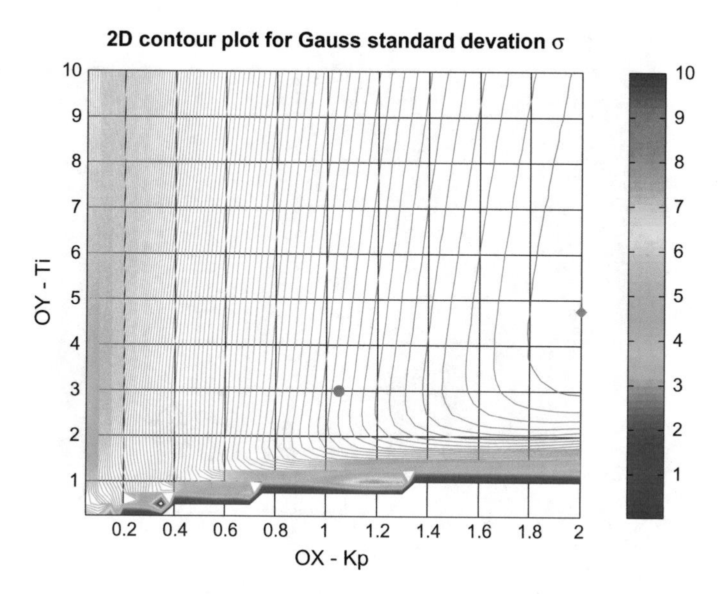

Fig. 11.18 Control quality surface for undisturbed $G_1(s)$: Gauss σ

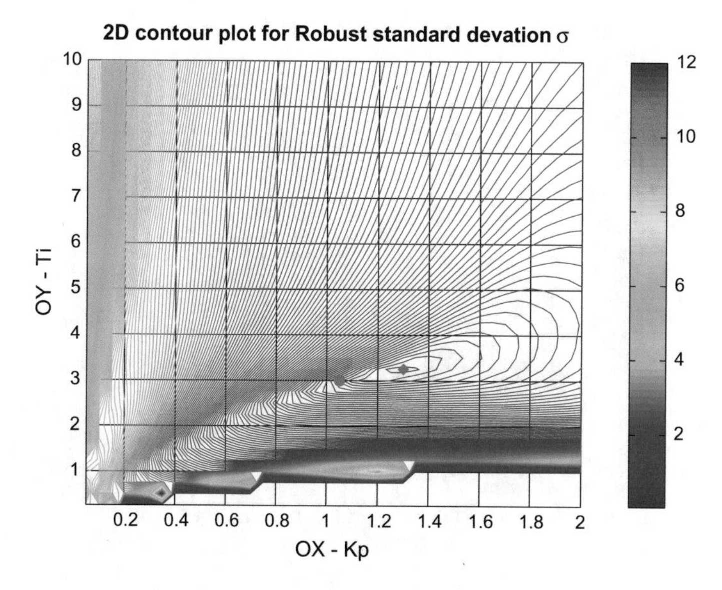

Fig. 11.19 Control quality surface for undisturbed $G_1(s)$: σ^{rob}

- The short memory exponent H^1 (Fig. 11.24) should be the most informative according to the literature [4, 5]. The value for the optimal control is not as expected, i.e. $H \approx 0.5$, but is relatively close. Its values vary in a relatively narrow range of $0.55 \div 0.65$. Thus, the controller, according to the literature, is slightly

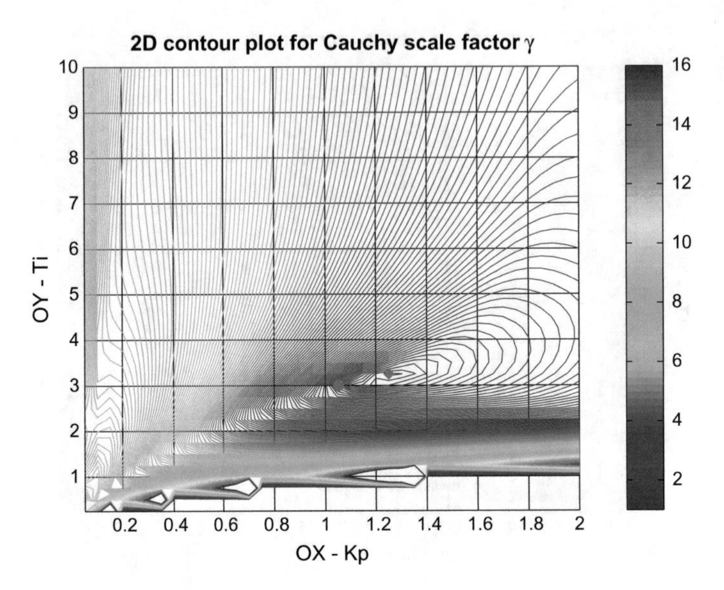

Fig. 11.20 Control quality surface for undisturbed $G_1(s)$: Cauchy γ^C

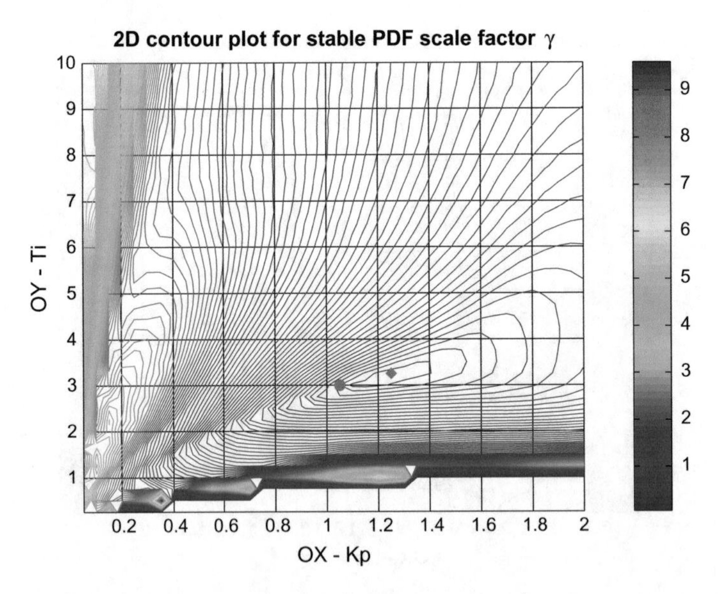

Fig. 11.21 Control quality surface for undisturbed $G_1(s)$: stable γ^L

sluggish. Longer memory Hurst exponents are clearly less informative. H^2 varies in a very narrow range of ≈ 0.95, while the longest one has no practical meaning (Figs. 11.25, 11.26 and 11.27).

- Both entropies (Figs. 11.28 and 11.29) share very similar shape. However, the detection (minimum value) clearly tends towards very sluggish control.

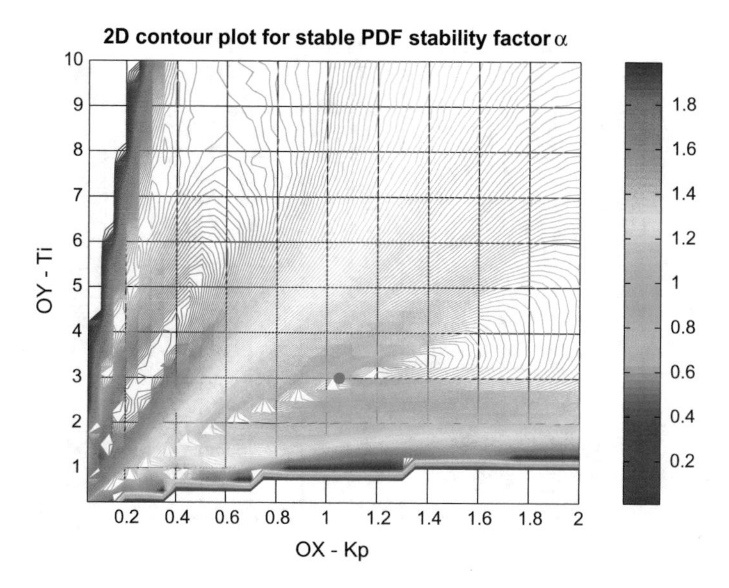

Fig. 11.22 Control quality surface for undisturbed $G_1(s)$: stable α (the best value $\alpha = 2.0$)

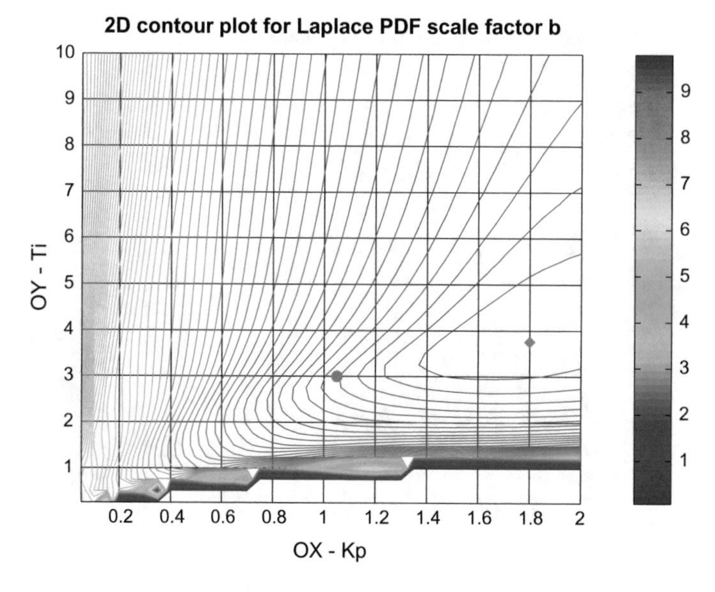

Fig. 11.23 Control quality surface for undisturbed $G_1(s)$: Laplace scaling b

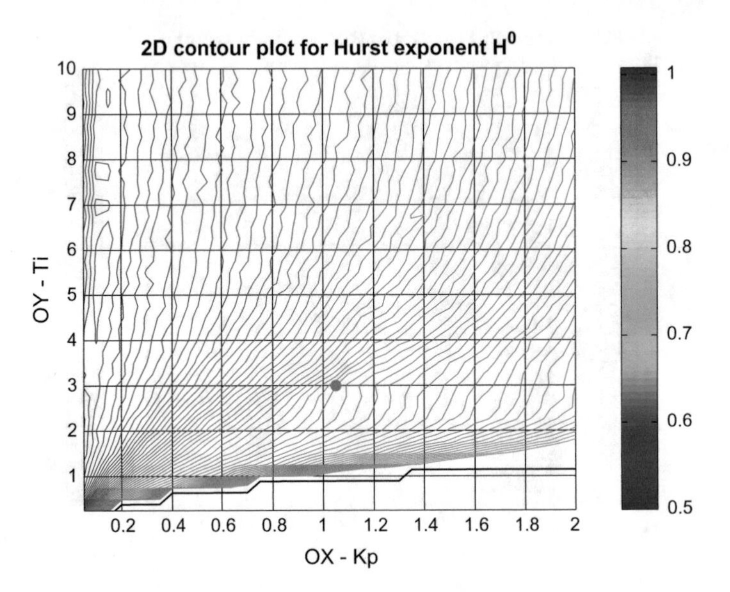

Fig. 11.24 Control quality surface for undisturbed $G_1(s)$: H^0 (the best value $H^0 = 0.5$)

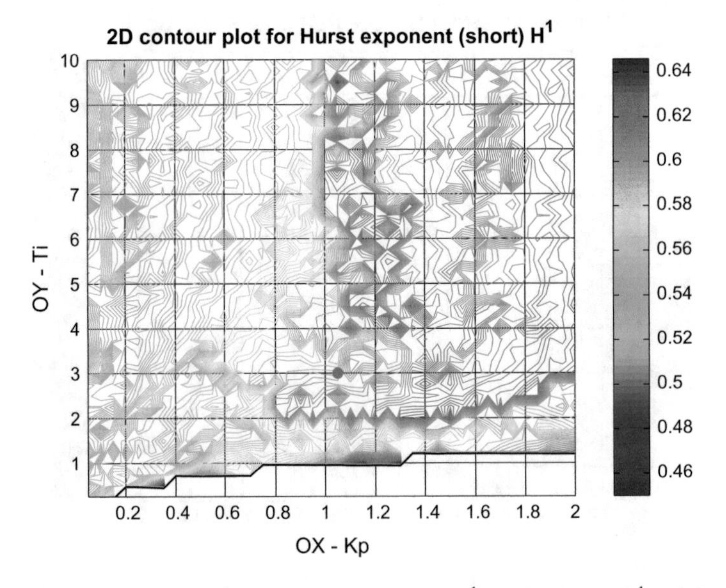

Fig. 11.25 Control quality surface for undisturbed $G_1(s)$: H^1 (the best value $H^1 = 0.5$)

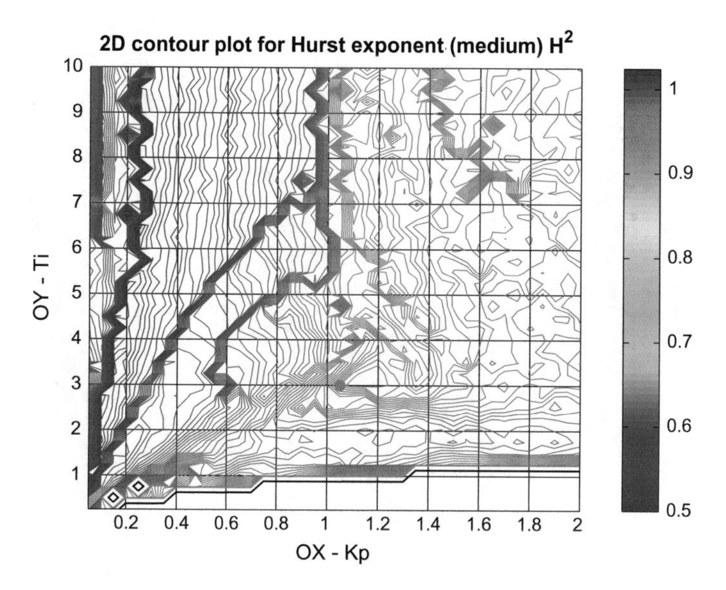

Fig. 11.26 Control quality surface for undisturbed $G_1(s)$: H^2 (the best value $H^2 = 0.5$)

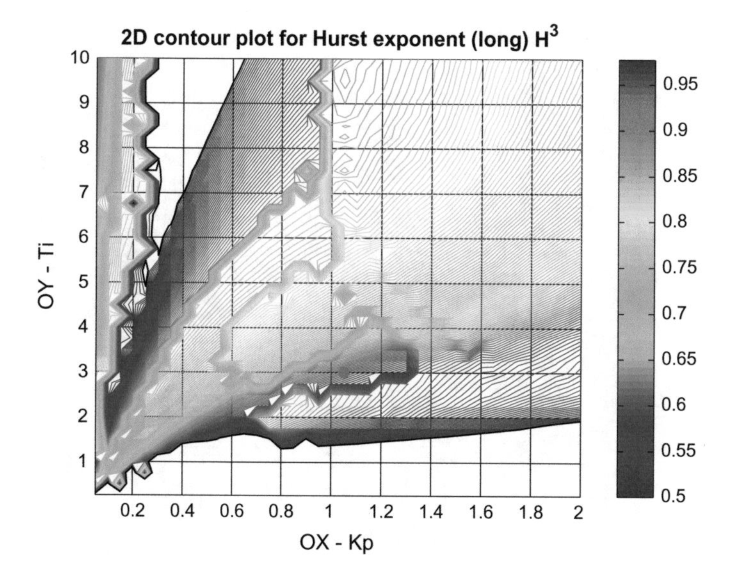

Fig. 11.27 Control quality surface for undisturbed $G_1(s)$: H^3 (the best value $H^3 = 0.5$)

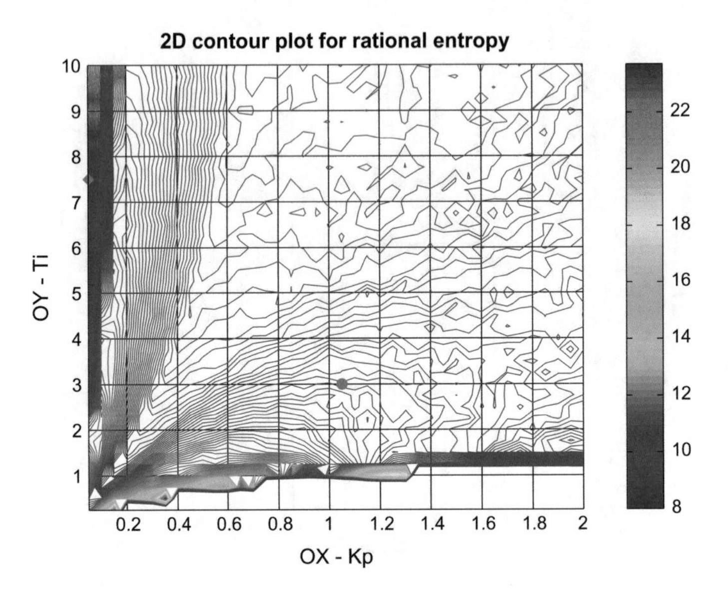

Fig. 11.28 Control quality surface for undisturbed $G_1(\mathrm{s})$: H^{RE}

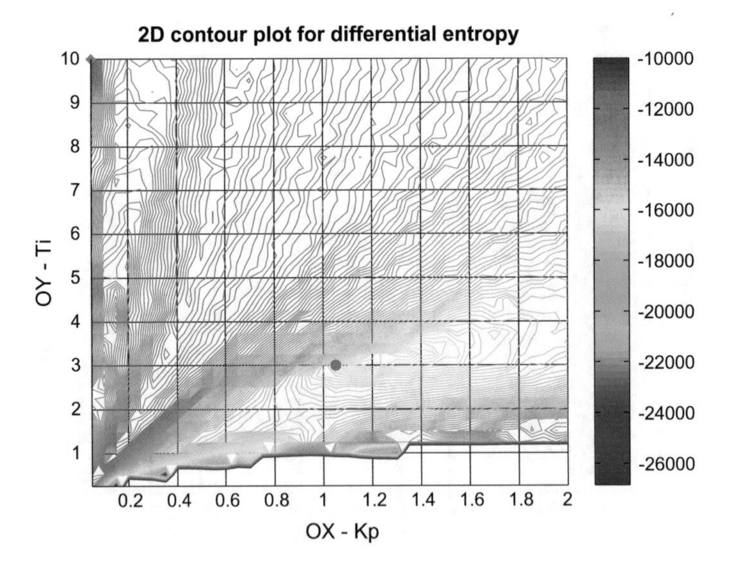

Fig. 11.29 Control quality surface for undisturbed $G_1(\mathrm{s})$: H^{DE}

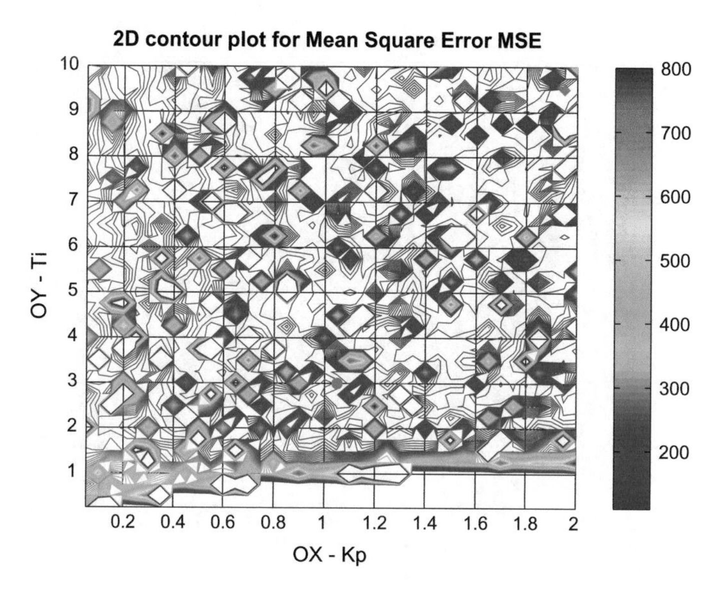

Fig. 11.30 Control quality surface for disturbed $G_1(s)$: MSE

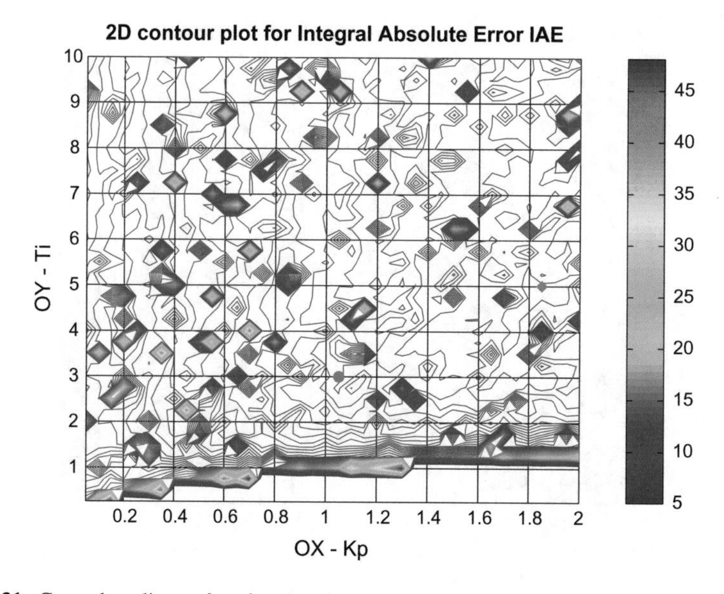

Fig. 11.31 Control quality surface for disturbed $G_1(s)$: IAE

11.1.4 Control Performance Detection for Disturbed Loop

Next, the control loop is disturbed and the same set of the analyzes is performed. The way of the drawings preparation and surfaces analysis is exactly the same.

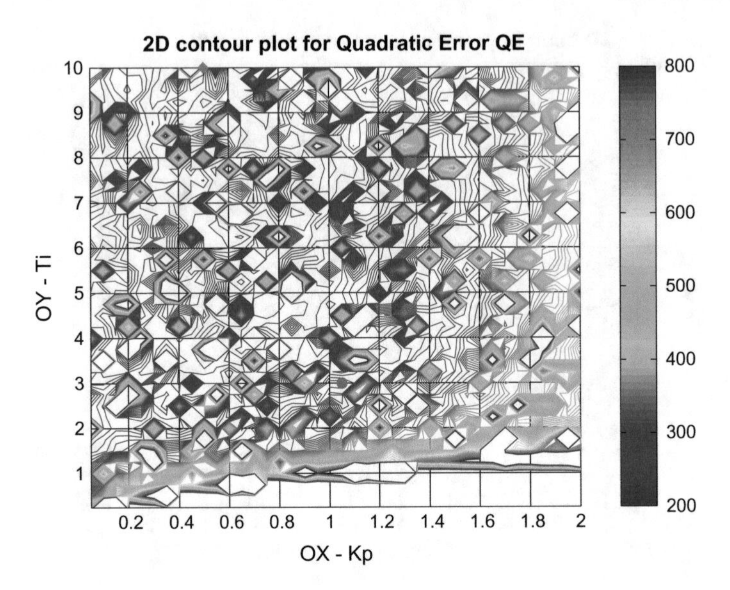

Fig. 11.32 Control quality surface for disturbed $G_1(s)$: QE

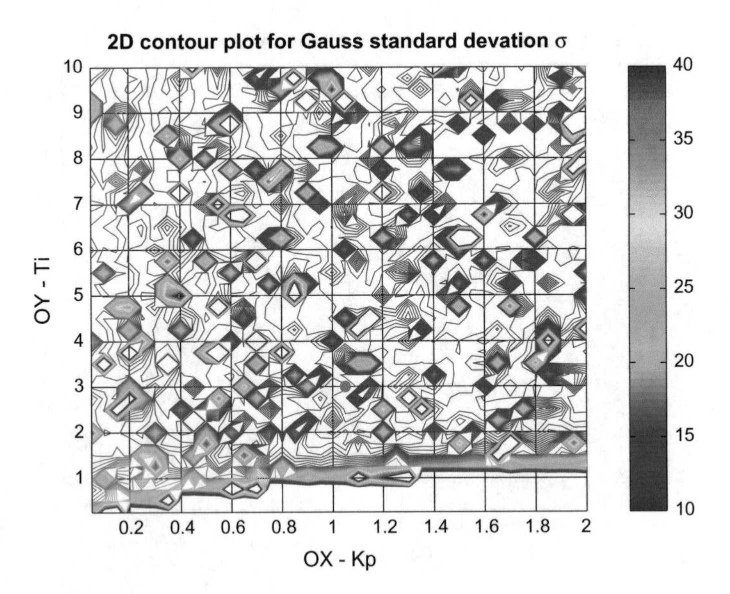

Fig. 11.33 Control quality surface for disturbed $G_1(s)$: Gauss σ

Review of the attached drawings enables to bring up the following observations:

- The main difference is clear. The surfaces are not smooth and they consist of various patters of the *hills and valleys*. Although the general shapes are close to the undisturbed case we see scattered local minima and maxima added to the

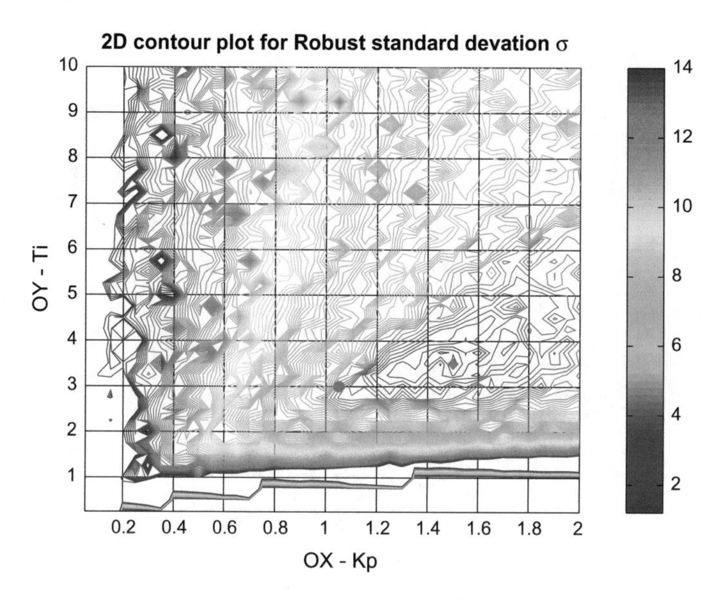

Fig. 11.34 Control quality surface for disturbed G_1(s): σ^{rob}

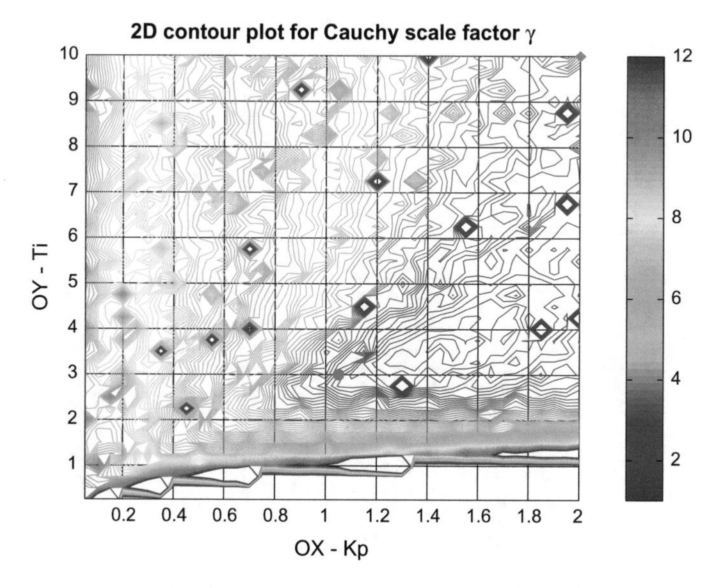

Fig. 11.35 Control quality surface for disturbed G_1(s): Cauchy γ^C

picture. They clearly deteriorate detection. The selection of the best result can be easily biased by the disturbance. This effect has been already noticed and discussed [6, 7]. It is called the *disturbance shadowing* effect.

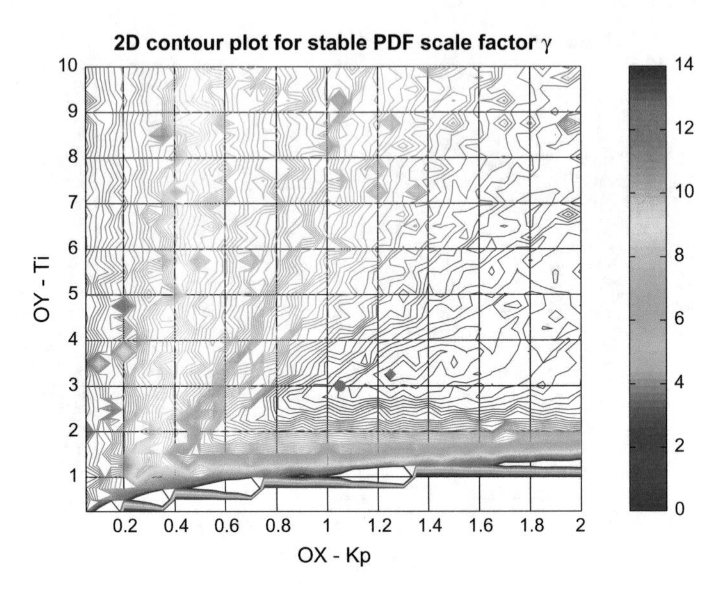

Fig. 11.36 Control quality surface for disturbed $G_1(s)$: stable γ^L

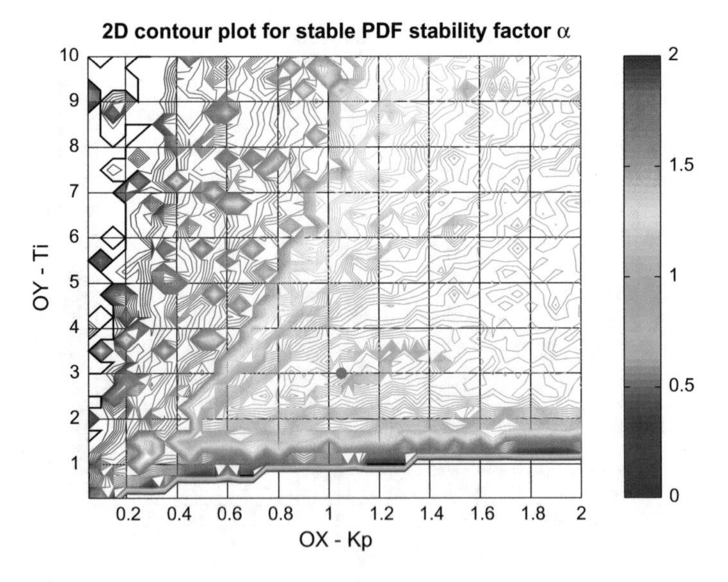

Fig. 11.37 Control quality surface for disturbed $G_1(s)$: stable α (the best value $\alpha = 2.0$)

- Visual inspection expands quantitative analysis. It shows sensitivity of detection. Though some parameters my seem to be good, the shapes of detection surface show that they are ragged so the detection is just random. Simple observation of numbers may be misleading. Results must be verified against disturbances.

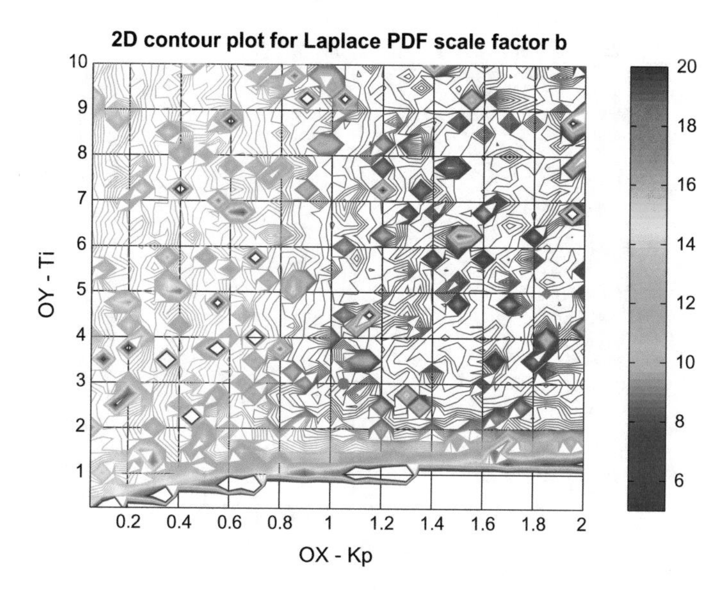

Fig. 11.38 Control quality surface for disturbed $G_1(s)$: Laplace scaling b

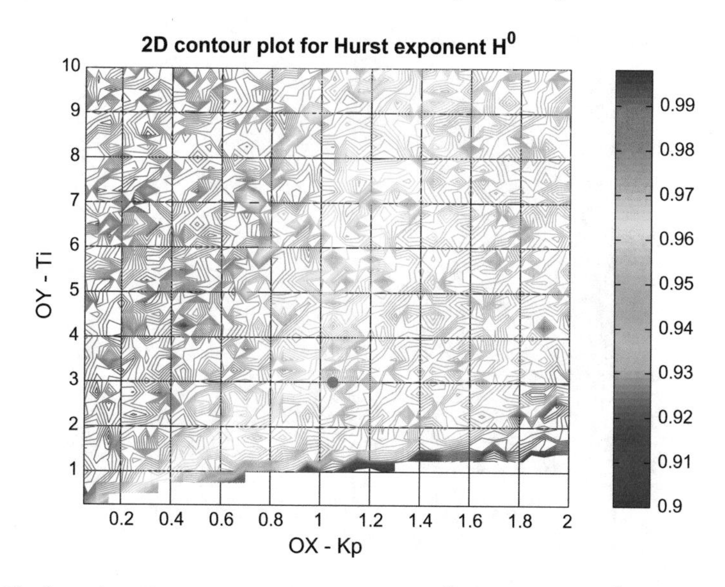

Fig. 11.39 Control quality surface for disturbed $G_1(s)$: H^0 (the best value $H^0 = 0.5$)

Quantitative comparison is not enough. It may give random results. Disturbance shadowing has strong impact on detection. In depth process insight is required.

- We see that MSE (Fig. 11.30) surface is more ragged than IAE (Fig. 11.31). Obviously Quadratic Error QR is close to the MSE. We may also notice that similarly

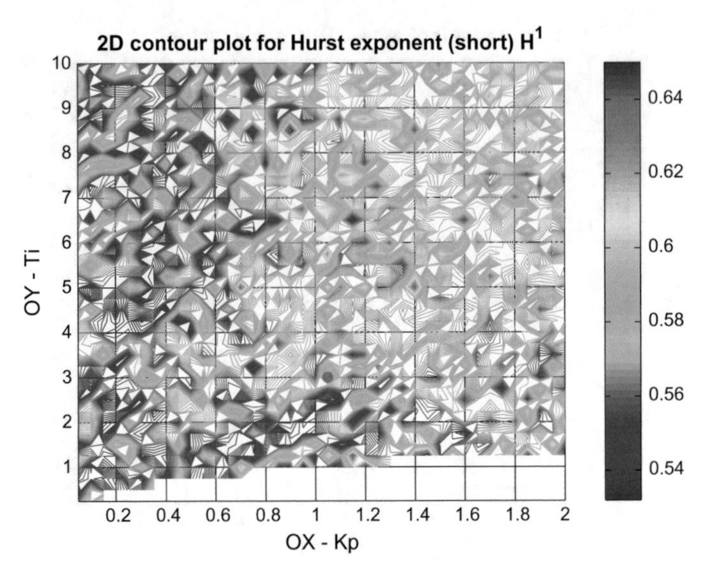

Fig. 11.40 Control quality surface for disturbed $G_1(s)$: H^1 (the best value $H^1 = 0.5$)

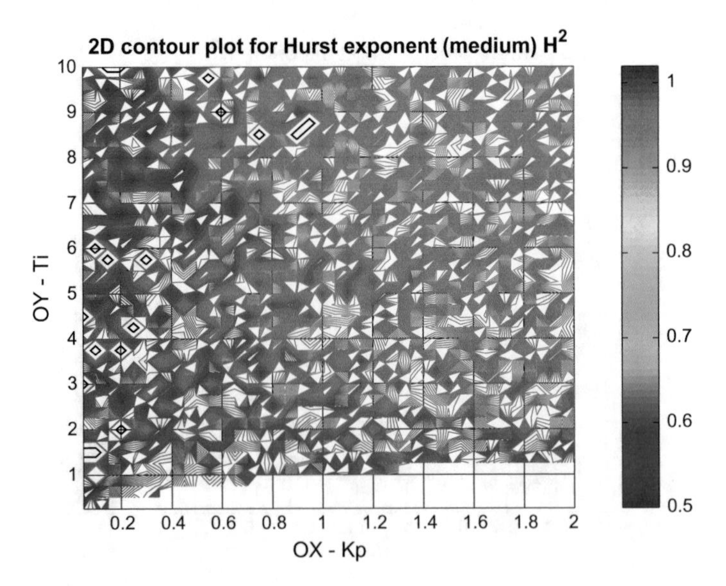

Fig. 11.41 Control quality surface for disturbed $G_1(s)$: H^2 (the best value $H^2 = 0.5$)

to the undisturbed case, Gaussian standard deviation σ (Fig. 11.33) is very close to MSE, while Laplace factor b (Fig. 11.38) is very close to IAE (Fig. 11.32).

- Usage of the robust standard deviation (Fig. 11.34) is more effective. The shape is visibly smoother and the pointed out solution seems to be more reliable.
- Cauchy scale factor γ^C (Fig. 11.35) performs similarly to the robust measures. Interpretation of the stable distribution stability factor α (Fig. 11.37) is not clear.

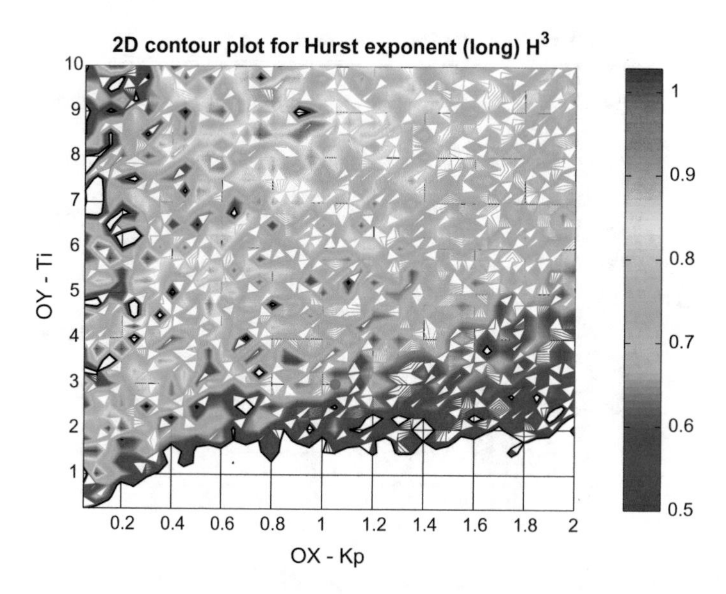

Fig. 11.42 Control quality surface for disturbed $G_1(s)$: H^3 (the best value $H^3 = 0.5$)

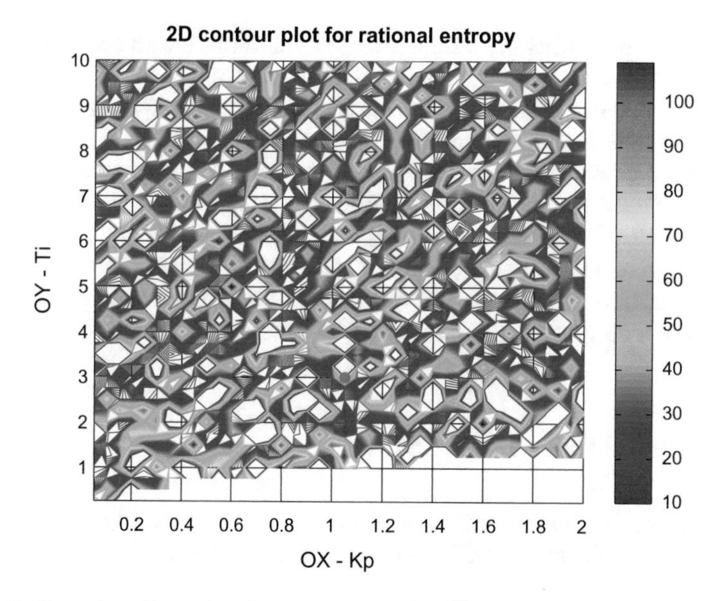

Fig. 11.43 Control quality surface for disturbed $G_1(s)$: H^{RE}

The shape of the surface is close to the scale γ^L. Evidently, the scale γ^L of the stable distribution delivers the most reliable detection pointing out the solution closest (and acceptable) to the optimal one (Figs. 11.36, 11.38 and 11.39).

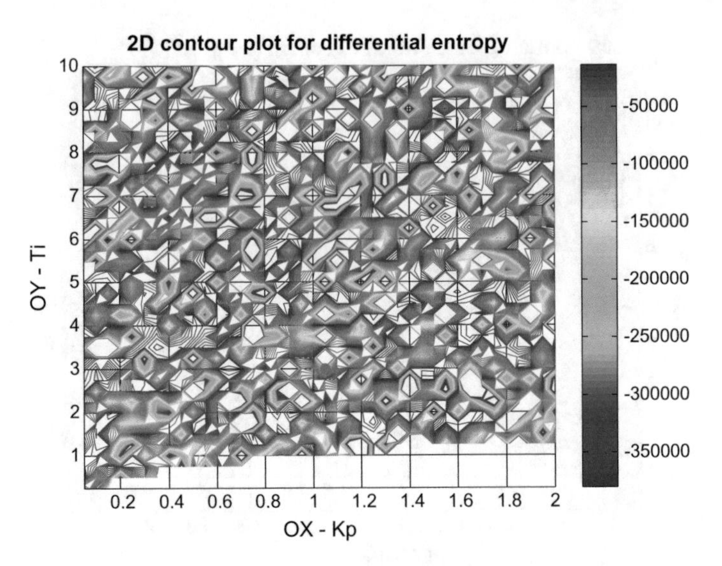

Fig. 11.44 Control quality surface for disturbed $G_1(s)$: H^{DE}

- All the Hurst exponents share similar shape of the surface with not ideally smooth shapes. We also see that the shortest memory Hurst exponent H^1 (Fig. 11.40) is relatively smooth with the shape very close to the scale γ^L of the stable distribution. Its value varies in a very narrow range of $0.55 \div 0.65$ showing similar to the undisturbed case shape. The character of the behavior of the longer memories seems to be veru similar to the undisturbed case (Figs. 11.41 and 11.42).
- Both entropies (Figs. 11.43 and 11.44) are evidently affected by the disturbance and the selection with these measures is doubtful.
- Concluding, the introduction of the **outliers** (fat-tailed disturbance properties) into the control loop generates shadowing effect that makes CPA task questionable.

11.2 Fourth Order Linear System

The analysis starts with the evaluation of the optimal PID controller and is followed by the analysis of the tuning surfaces for all considered CPA indexes.

11.2.1 Optimal Controller Performance

The optimal controller for the $G_2(s)$ transfer function has been designed according to the predefined rules. The evaluated tuning parameters are presented below:

- gain: $k_p^{opt} = 1.19$
- integration: $T_i^{opt} = 1.11$
- derivation: $T_d^{opt} = 0.26$

The graphical representation of the loop setpoint step response is sketched in Fig. 11.45, while Fig. 11.46 shows the disturbance step response. We see that the obtained tuning delivers the response with almost invisible overshoot. The controller output is smooth without any oscillations.

Next, the loop simulation has been run. The time trends (sample of first 120 s) for the selected loop profiles for setpoint, CV and MV are sketched in Fig. 11.47. The figure shows comparison of both undisturbed and disturbed behavior.

Respective control error histogram is shown in Fig. 11.48. It is clearly visible that its shape is not Gaussian with the significant tails. The Histogram can be modeled the most precisely by the α-stable distribution. The analog histogram for the disturbed simulation is sketched in Fig. 11.50.

Finally, the extended range R/S plot has been evaluated (Fig. 11.49). We clearly see that its character is not monofractal. The multi-persistent properties are well reflected with the two crossovers ($cross^1$ and $cross^2$) and three memory ranges described by the respective Hurst exponents (H^1, H^2 and H^3). The analog plot for the disturbed simulation is sketched in Fig. 11.51.

All the evaluated indexes for both undisturbed (nonDist) and disturbed (Dist) loop simulations are summarized in Table 11.3. It is also interested to see what tuning, i.e. controller parameters have been pointed out by each method. Table 11.4 summarizes these results. The table consists only of these measures that can detect the solution, i.e. point out the single result as the extreme index value (minimum or maximum). The settings highlighted show the results which are close enough to the optimal solution $k_p = k_p^{opt} \pm 0.2$ and $T_i = T_i^{opt} \pm 0.4$, i.e. $k_p \in [0.99, 1.39]$ and $T_i \in [0.7, 1.5]$. We notice that none of the indexes is even close to the optimal controller settings. The majority of the indexes are pointing out solutions at the edge of the selected domain associated with rather sluggish than aggressive control. These results are well visualized with the controller tuning surfaces presented in the below sections.

11.2.2 Control Performance Detection Using Step Indexes

The analysis of the performance assessment potential for the considered quality indexes (KPIs) is presented in form of the control quality surfaces. The drawings are prepared according to the same common scheme. The plots are in form of the two dimensional surfaces presenting each KPI as the function of two PID parameters: k_p and T_i. The third parameter, $T_d = 0.26$ is kept constant at its optimal value, so the surface is a 2D cross-section. Due to the lack of place other cross-sections are not presented as they do not bring additional value to the analysis.

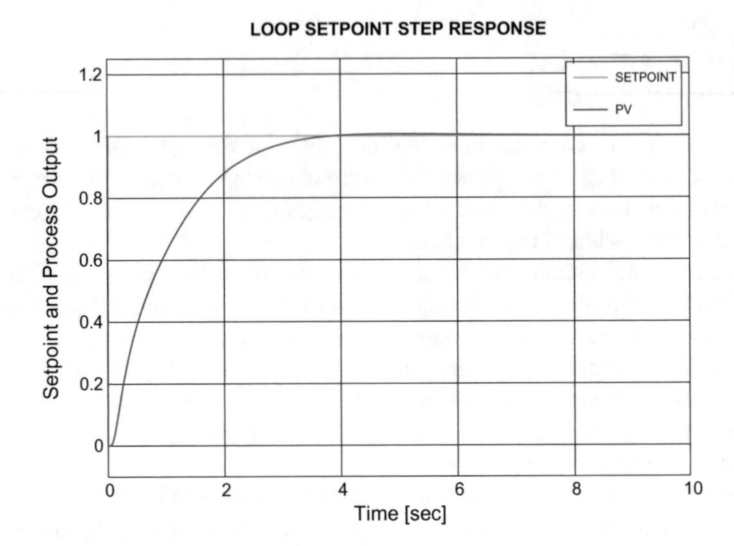

Fig. 11.45 Loop setpoint step response for G_2(s)

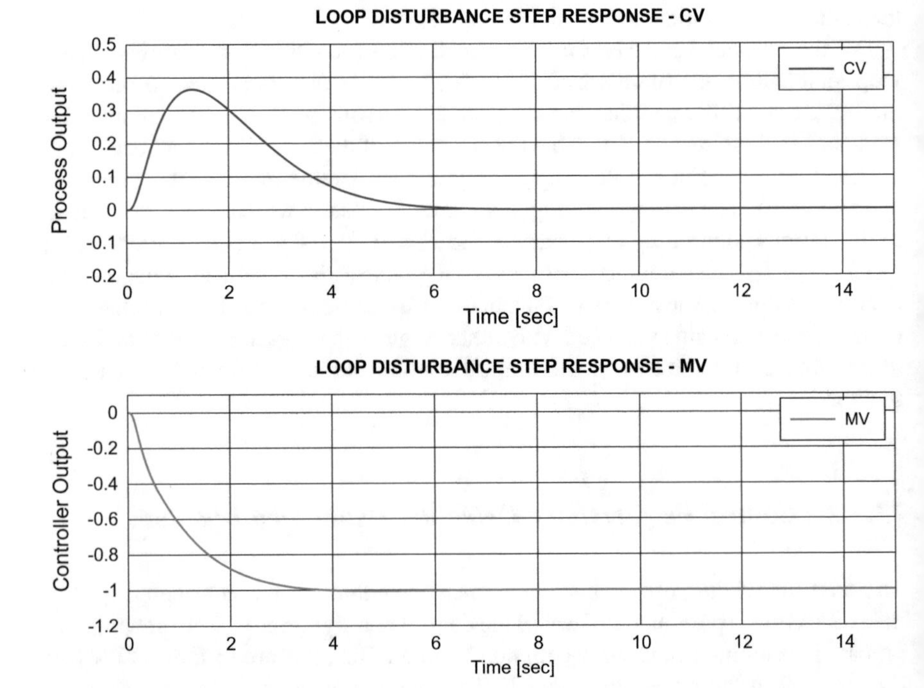

Fig. 11.46 Loop disturbance step response for G_2(s)

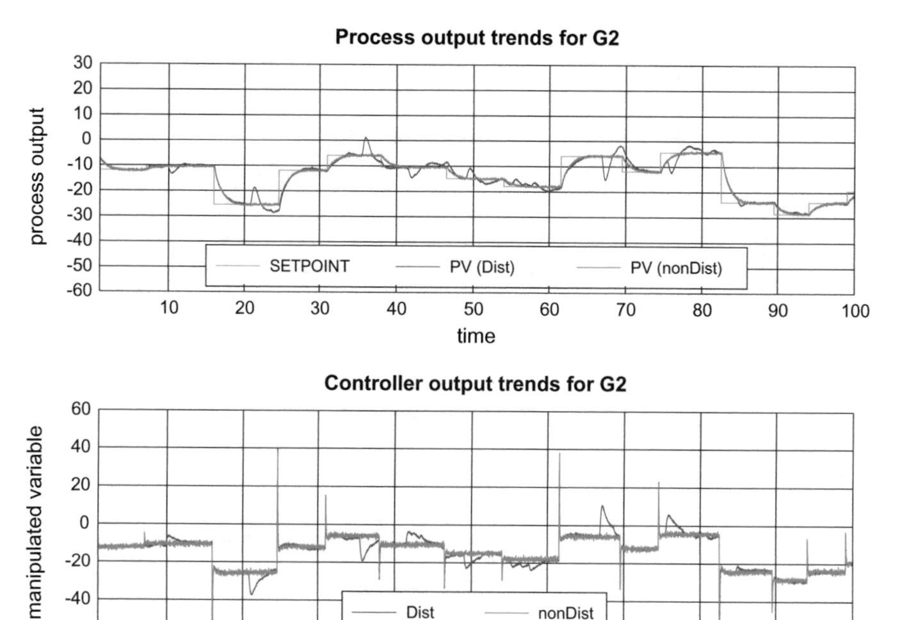

Fig. 11.47 Loop time trends for first 100 s of $G_2(s)$

The position of the found optimal controller is depicted with the magenta circle marker •. There are two variants of the diagram interpretation. The best index is at its minimum value, or there is some optimal value of the index. In the first case the red diamond marker ♦ depicts minimum index position. The black bold contour line for the selected the best value of the KPI is used in the second case.

Step based indexes, i.e. settling time, overshoot, II, AI, OI and RI are evaluated without the disturbance.

Review of the attached drawings for step response indexes enables to bring up the following observations:

- Classical indexes of the settling time (Fig. 11.52) and the overshoot (Fig. 11.53) work fine. It is well seen that the overshoot surface is mostly equal to zero on a large flat domain area. None of the indexes considered independently is able to point out the solution even close to the optimal one.
- Idle Index (Fig. 11.54) Area Index (Fig. 11.55) and the Output Index (not shown) have very similar properties as the overshoot surface. It means that these indexes are not very informative as well.
- Ideal literature value of the Area Index $0.3 < AI < 0.7$ is relatively far away from the optimal controller (Fig. 11.55) and reflects rather sluggish control. Similar

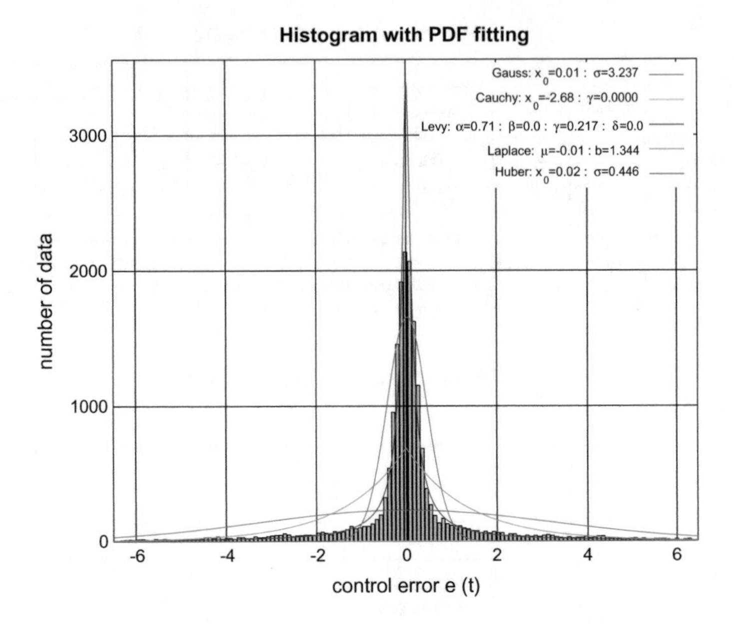

Fig. 11.48 Control error histogram with the fitted different PDFs for $G_2(s)$

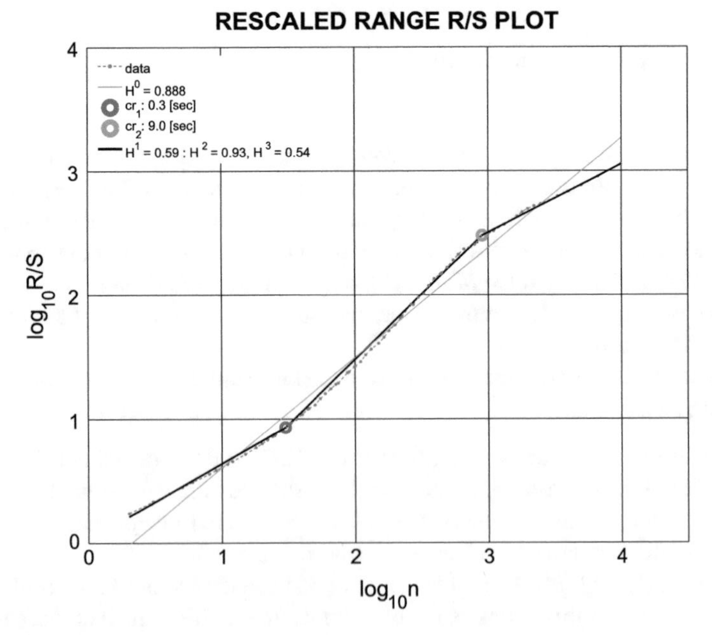

Fig. 11.49 Extended range R/S plot of the control error for $G_2(s)$

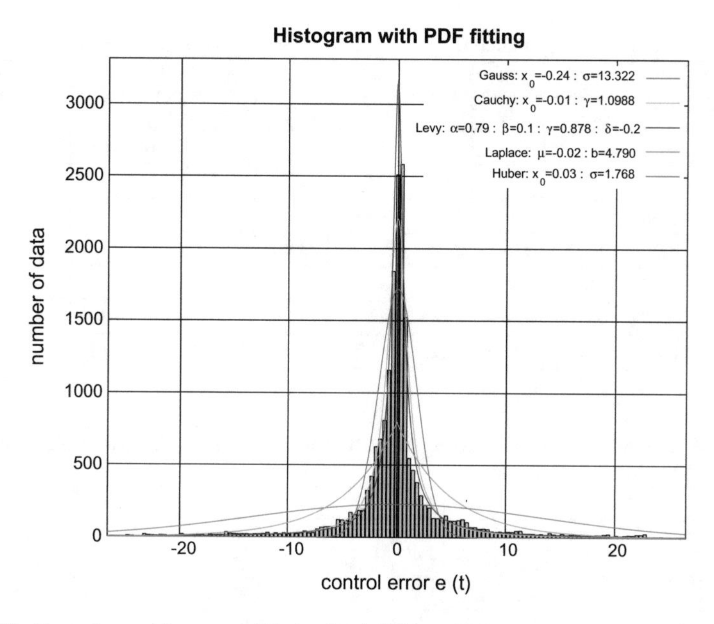

Fig. 11.50 Control error histogram with the fitted different PDFs in disturbed case for $G_2(s)$

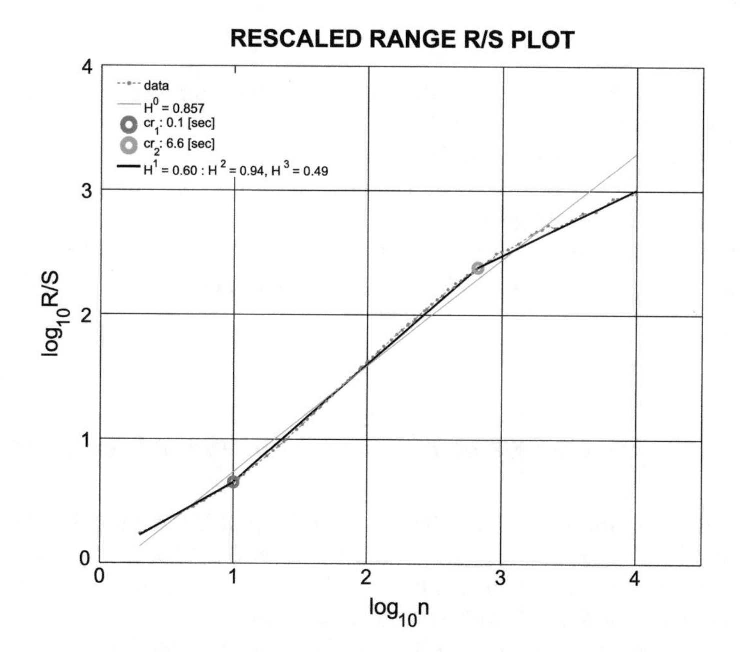

Fig. 11.51 Extended range R/S plot for the control error in disturbed case for $G_2(s)$

Table 11.3 Indexes for the optimal $G_2(s)$ loop controller (undisturbed and disturbed)

	T_s	$\kappa[\%]$	DR	II	AI	OI	RI	MSE	IAE	QE
nonDist	3.39	0.056	0.00	0.41	1.0	0.003	0.44	42.0	2.58	111.4
Dist	–	–	–	–	–	–	–	74.8	3.87	170.7
	σ	γ^C	α	γ^L	b	σ^{rob}				
nonDist	6.48	0.14	0.80	0.12	2.58	0.22				
Dist	8.65	1.30	0.98	1.34	3.87	2.06				
	H^0	H^1	H^2	H^3	cr^1	cr^2	H^{RE}	H^{DE}		
nonDist	0.872	0.614	0.915	0.406	0.30	13.20	41.32	−173230		
Dist	0.854	0.638	0.942	0.569	0.12	3.96	41.19	−156619		

Table 11.4 Controller tuning pointed out by the selected measures for $G_2(s)$—highlighted are the solutions close to the optimal one

		T_s	$\kappa[\%]$	DR	MSE	IAE	QE	σ	γ^C	γ^L	b	σ^{rob}	H^{RE}	H^{DE}
nonDist	k_p	2.00	n/a	n/a	2.00	2.00	1.24	2.00	2.00	2.00	2.00	2.00	0.24	0.24
	T_i	1.20	n/a	n/a	1.20	1.20	10.00	1.20	1.80	1.20	1.20	1.20	10.00	10.00
Dist	k_p	n/a	n/a	n/a	1.92	1.92	1.20	1.92	1.96	2.00	1.92	2.00	0.24	0.24
	T_i	n/a	n/a	n/a	1.20	1.20	4.20	1.20	9.60	1.20	1.20	1.20	7.20	7.20

situation is visible for Output Index, when optimal controller is quite far away from the literature ideal $OI < 0.35$.

- R-Index surface is very scattered (not shown) similarly to the previous transfer function case. It shows that detection ability with RI is very limited and does not work for the considered loop.
- Decay Ratio surface is not evident and the detection ability with DR is very limited and does not work for the considered loop. Due to this fact, the considered surface is not presented.

11.2.3 Control Performance Detection for Undisturbed Loop

The simulation of the loop using predefined setpoint profile is realized in two forms: undisturbed and disturbed one. Thus for such experiment the integral, statistical, fractal and entropy indexes are evaluated and analyzed. Diagrams sketched below present surfaces for the undisturbed simulation mode.

The review of the results obtained for the considered transfer function are consistent with the results for the previous transfer function. Thus, only selected, the most relevant surfaces are sketched for the properties visualization.

Review of the above undisturbed surfaces allows to bring about the following observations:

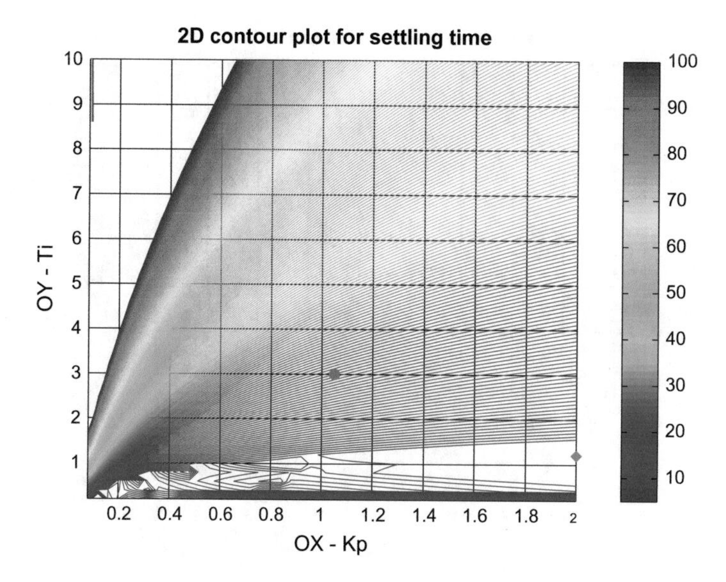

Fig. 11.52 Control quality surface for $G_2(s)$: T_s

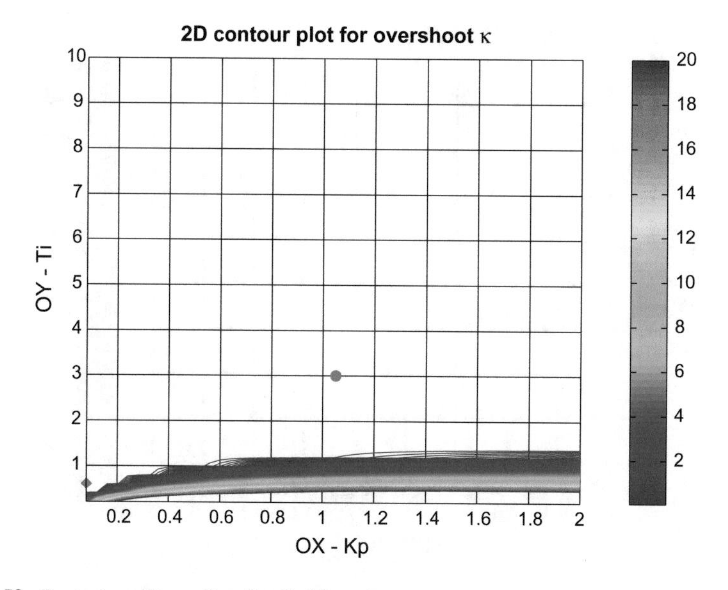

Fig. 11.53 Control quality surface for $G_2(s)$: κ

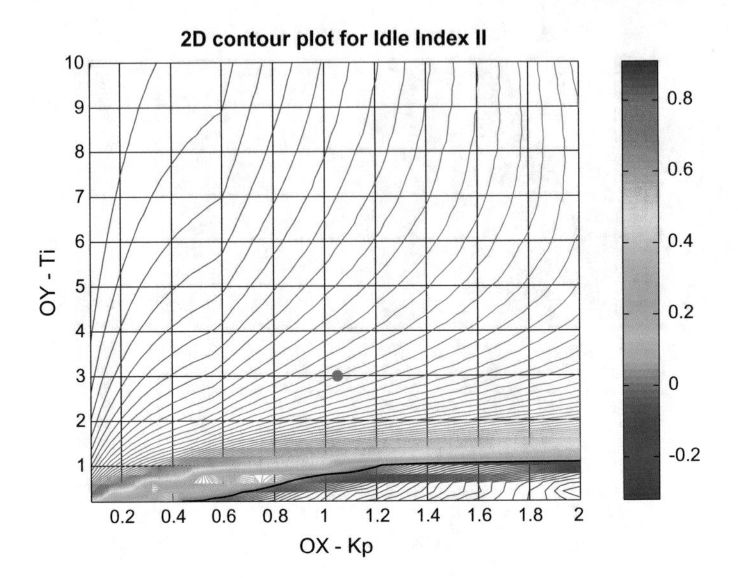

Fig. 11.54 Control quality surface for G_2(s): II (the best value $II = 0$)

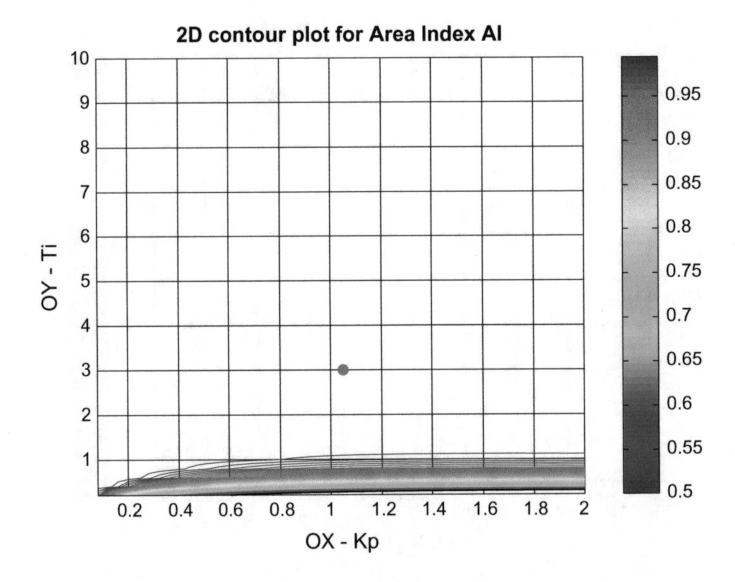

Fig. 11.55 Control quality surface for G_2(s): AI (the best value $AI = 0.5$)

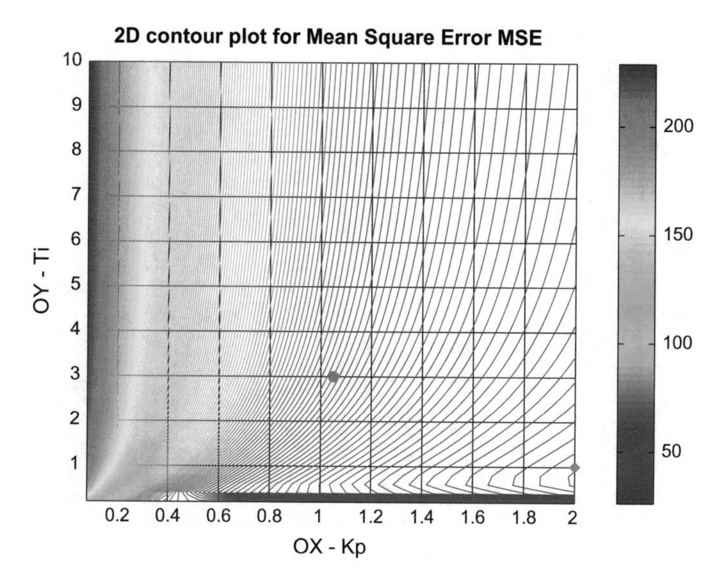

Fig. 11.56 Control quality surface for undisturbed $G_2(s)$: MSE

- Gaussian standard deviation σ, Laplace scale factor b and the IAE (Fig. 11.61) are very close to the MSE (Fig. 11.56), so only two representative surfaces are presented. The solution, which is pointed out is an aggressive one with large gain. It is possible for the selected transfer function as the overshoot is not characterizing the responses.
- Similar results are also visible for the robust standard deviation (Fig. 11.57) and the solution is not even close to the optimal settings.
- Cauchy scale factor γ^C (Fig. 11.58) performs in a different way. In fact it is the only index that points out the solution, which is relatively close to the optimal one (Figs. 11.59, 11.60 and 11.61).
- Interpretation of the stable distribution stability factor α (Fig. 11.62) is not clear. The shape of the surface is close to the scale γ^L.
- All the Hurst exponents share similar shape of the surfaces, however on different levels. The surfaces are also very similar to the MSE and IAE plots and do not bring new information.
- Both entropies share very similar shape (see rational entropy H^{RE} in Fig. 11.63). However, the detection (minimum value) clearly tends towards extremely sluggish control.

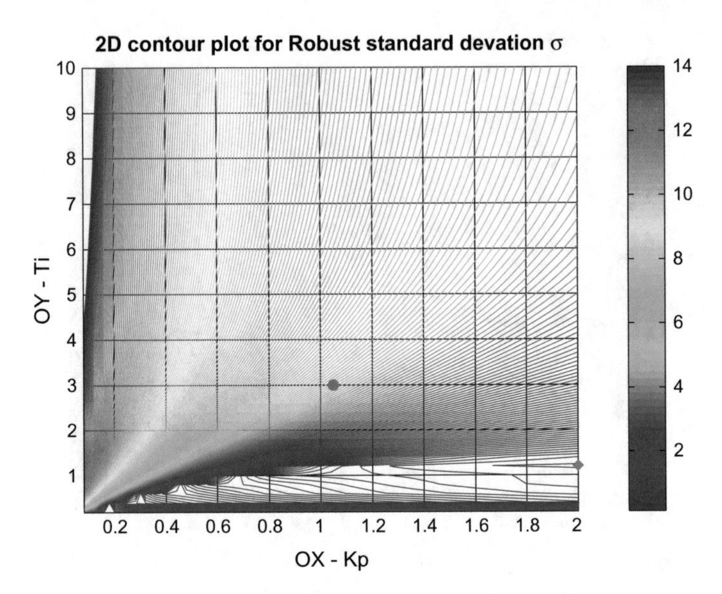

Fig. 11.57 Control quality surface for undisturbed $G_2(s)$: σ^{rob}

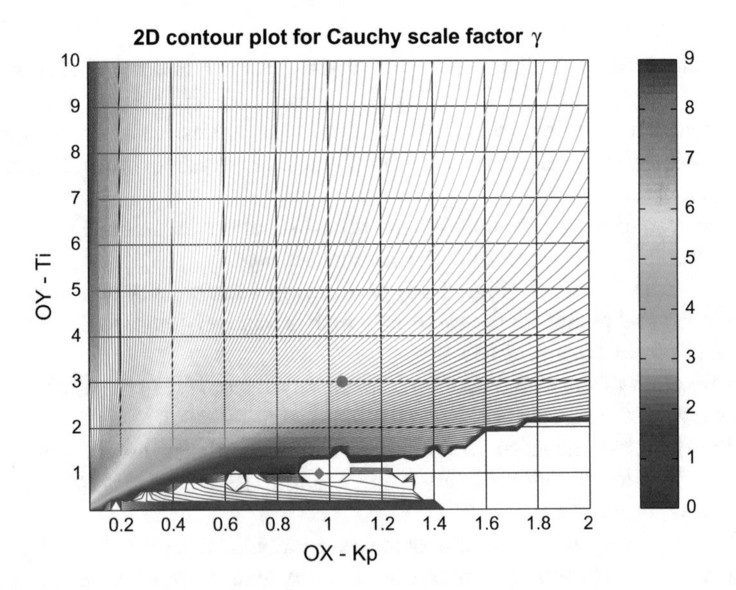

Fig. 11.58 Control quality surface for undisturbed $G_2(s)$: Cauchy γ^C

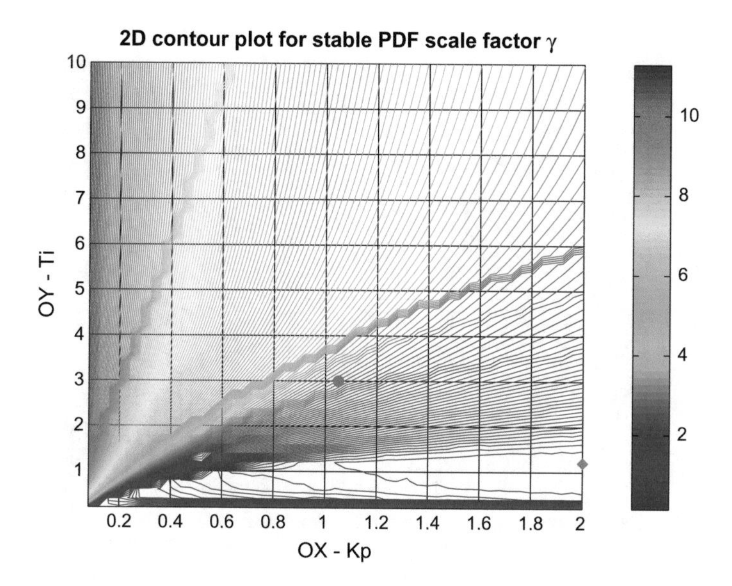

Fig. 11.59 Control quality surface for undisturbed $G_2(s)$: stable γ^L

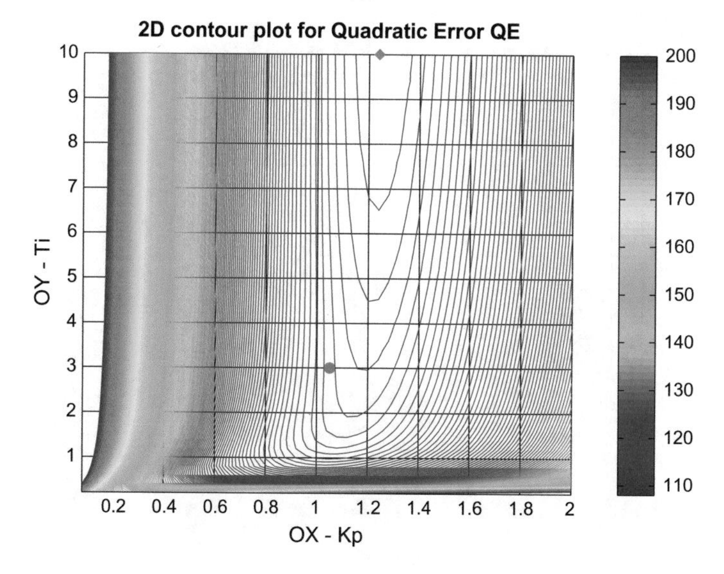

Fig. 11.60 Control quality surface for undisturbed $G_2(s)$: QE

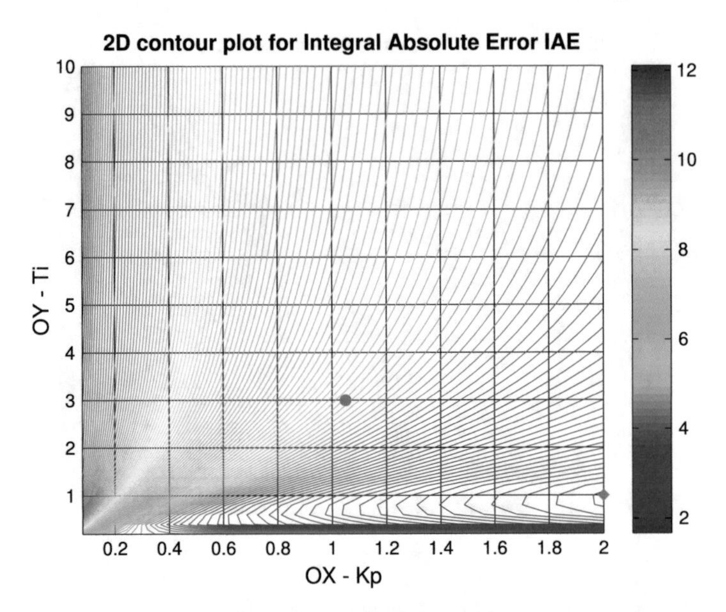

Fig. 11.61 Control quality surface for undisturbed $G_2(s)$: IAE

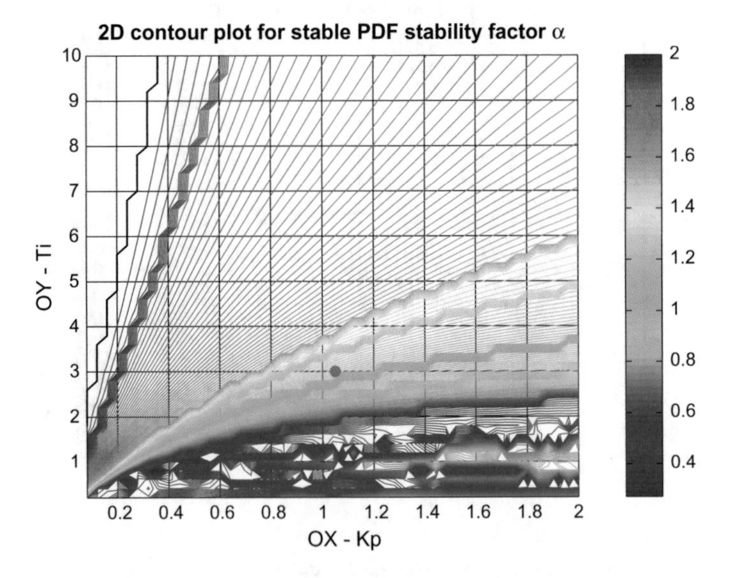

Fig. 11.62 Control quality surface for undisturbed $G_2(s)$: stable α

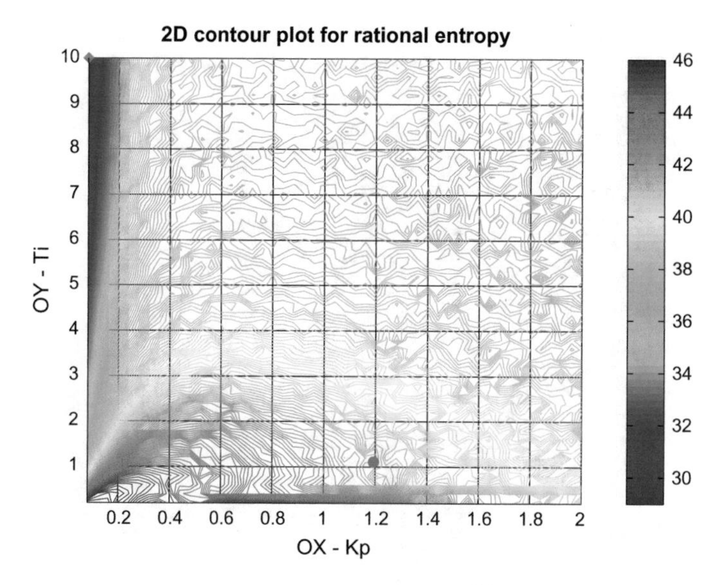

Fig. 11.63 Control quality surface for undisturbed $G_2(s)$: H^{RE}

11.2.4 Control Performance Detection for Disturbed Loop

Next, the control loop is disturbed and the same set of the analyzes is performed. The way of the drawings preparation and surfaces analysis is exactly the same.

Review of the attached drawings enables to bring up the following observations:

- The effect of the disturbance shadowing is also well visible for the considered fourth order linear system. All integral indexes presented highly ragged landscape effectively disabling any rational detection (see MSE surface in Fig. 11.64 and IAE in Fig. 11.65).
- Similarly to the undisturbed case, Gaussian standard deviation σ (Fig. 11.66) is very close to the MSE being highly affected by the disturbances. In contrary, the usage of the robust standard deviation (Fig. 11.67) gives more effective assessment.
- The scale factor for stable distribution seems to be the only index robust to the disturbances and treating reasonably information kept by outliers (Fig. 11.68).
- All the Hurst exponents are also shadowed, however not so much that the other indexes. It is worth to notice that entropies are the most shadowed indexes as for H^{DE} sketched in Fig. 11.69.

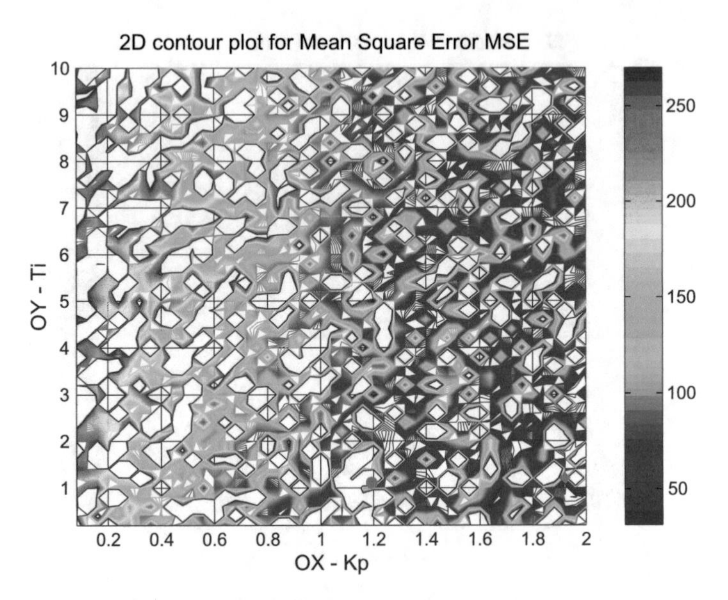

Fig. 11.64 Control quality surface for disturbed $G_2(s)$: MSE

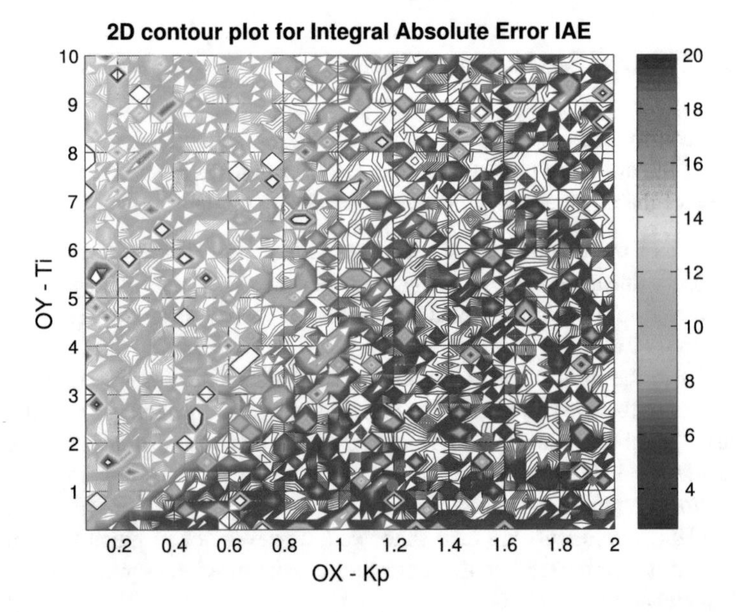

Fig. 11.65 Control quality surface for disturbed $G_2(s)$: IAE

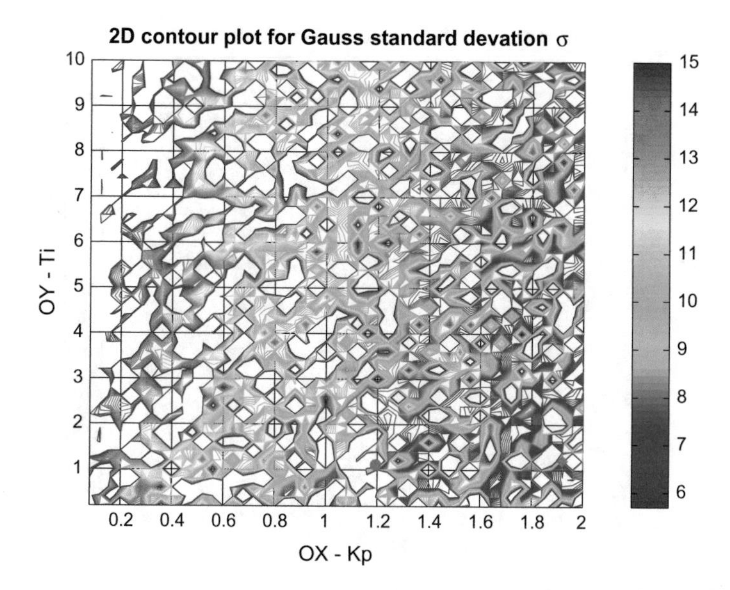

Fig. 11.66 Control quality surface for disturbed $G_2(s)$: Gauss σ

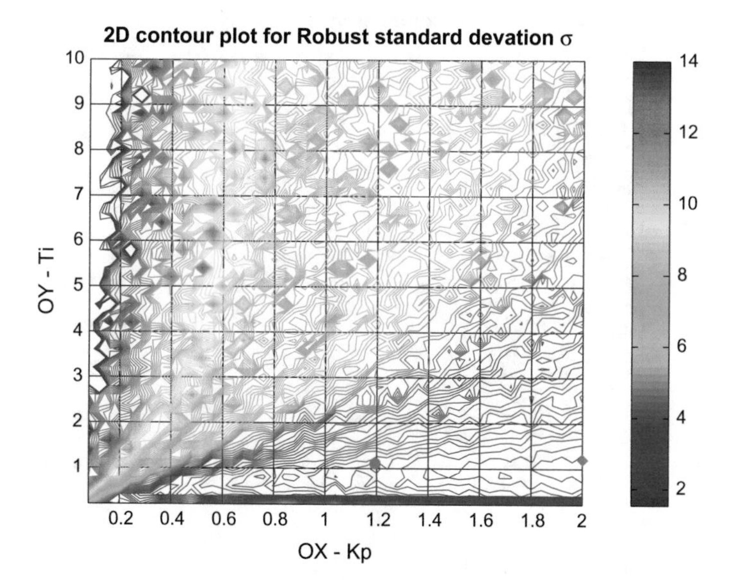

Fig. 11.67 Control quality surface for disturbed $G_2(s)$: σ^{rob}

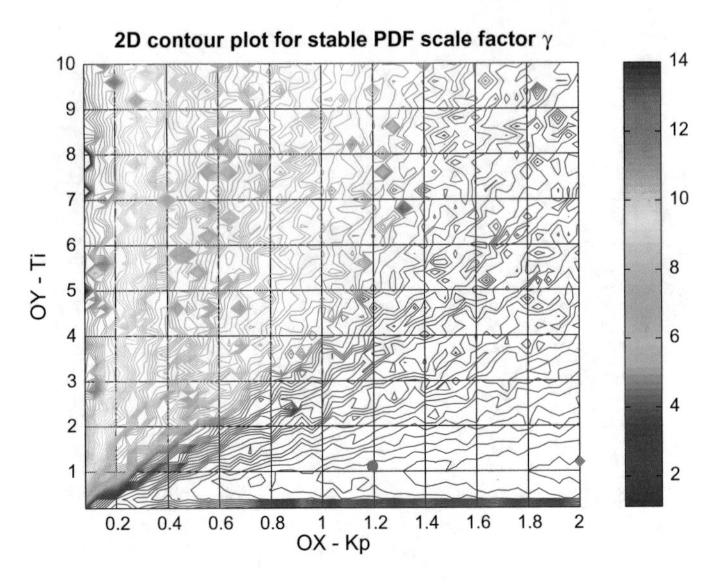

Fig. 11.68 Control quality surface for disturbed $G_2(s)$: stable γ^L

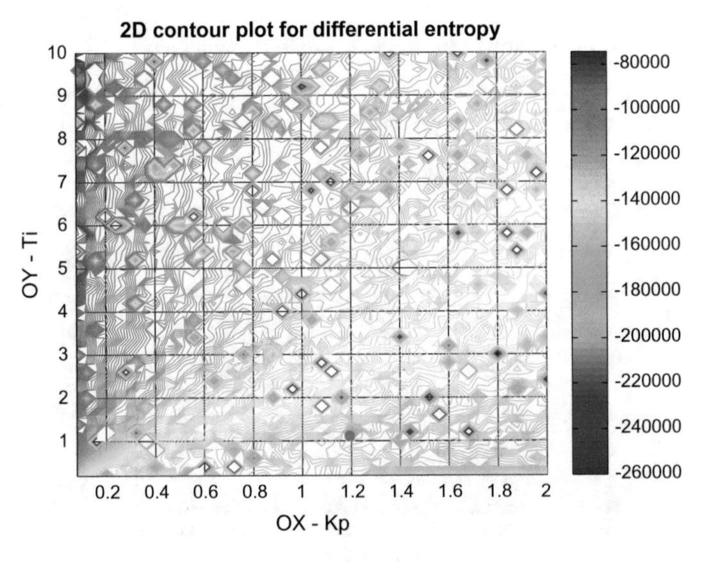

Fig. 11.69 Control quality surface for disturbed $G_2(s)$: H^{DE}

11.3 Nonminimumphase Plant

The analysis starts with the evaluation of the optimal PID controller and is followed by the analysis of the tuning surfaces for all considered CPA indexes.

11.3.1 *Optimal Controller Performance*

The optimal controller for the $G_3(s)$ transfer function has been designed according to the predefined rules. The evaluated tuning parameters are presented below:

- gain: $k_p = 0.885$
- integration: $T_i = 2.481$
- derivation: $T_d = 0.664$

The graphical representation of the loop setpoint step response is sketched in Fig. 11.70, while Fig. 11.71 shows the disturbance step response. We see that the obtained tuning delivers the response with almost invisible overshoot. The controller output is smooth without any oscillations.

Next, the loop simulation has been run. The time trends (sample of first 100 s) for the selected loop profiles for setpoint, CV and MV are sketched in Fig. 11.72. The figure shows comparison of both undisturbed and disturbed behavior.

Respective control error histogram is shown in Fig. 11.73. It is clearly visible that its shape is not Gaussian with the significant tails. The Histogram can be modeled the

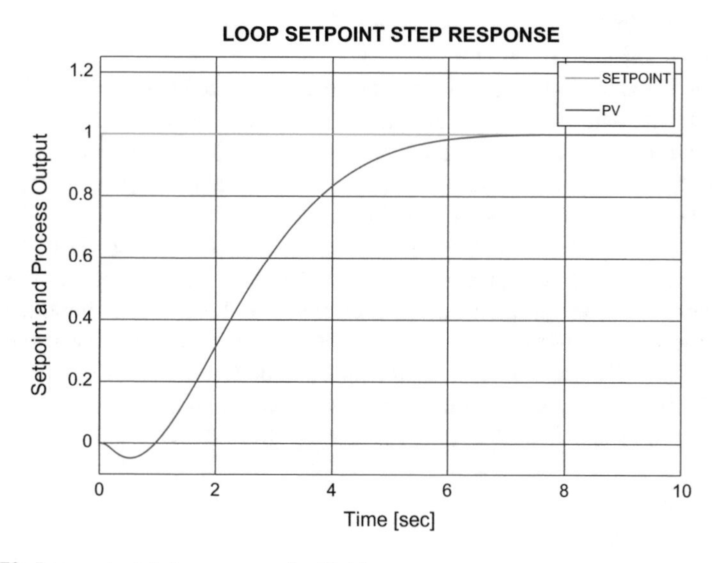

Fig. 11.70 Loop setpoint step response for $G_3(s)$

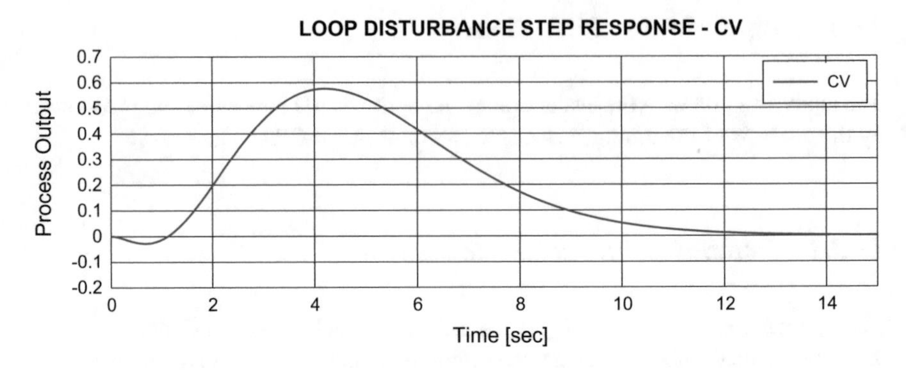

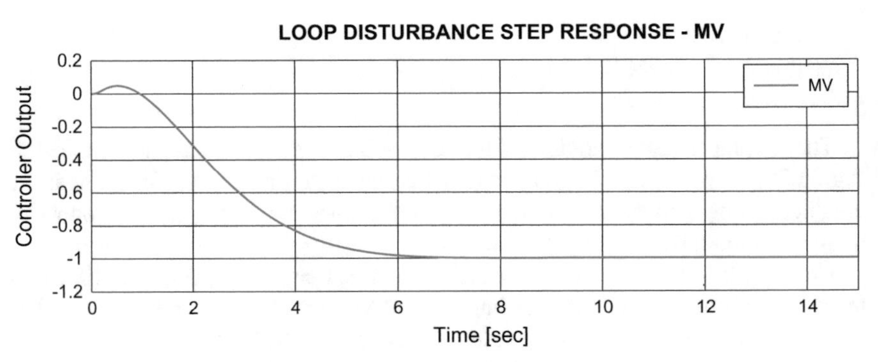

Fig. 11.71 Loop disturbance step response for $G_3(s)$

most precisely by the α-stable distribution. The analog histogram for the disturbed simulation is sketched in Fig. 11.75.

Finally, the extended range R/S plot has been evaluated (Fig. 11.74). We clearly see that its character is not monofractal. The multi-persistent properties are well reflected with the two crossovers ($cross^1$ and $cross^2$) and three memory ranges described by the respective Hurst exponents (H^1, H^2 and H^3). The analog plot for the disturbed simulation is sketched in Fig. 11.76.

All evaluated indexes for undisturbed (nonDist) and disturbed (Dist) loops are summarized in Table 11.5. It is also interested to see what tuning, i.e. controller parameters have been pointed out by each method. Table 11.6 summarizes these results. The table consists only of these measures that can detect the solution, i.e. point out the single result as the extreme index value (minimum or maximum). The settings highlighted show the results which are close enough to the optimal solution $k_p = k_p^{opt} \pm 0.25$ and $T_i = T_i^{opt} \pm 0.32$, i.e. $k_p \in [0.64, 1.14]$ and $T_i \in [2.16, 2.80]$.

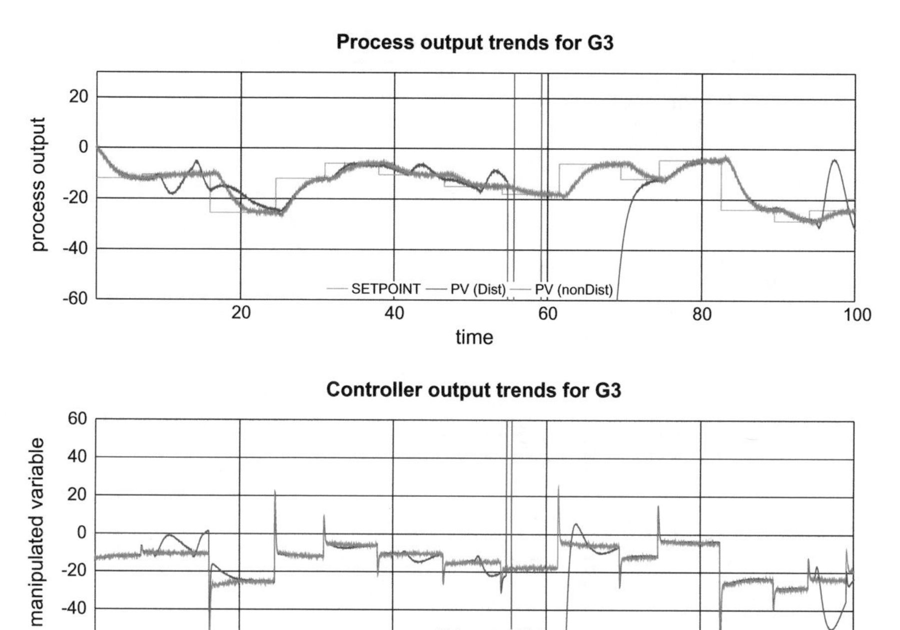

Fig. 11.72 Loop time trends for first 100 s for $G_3(s)$

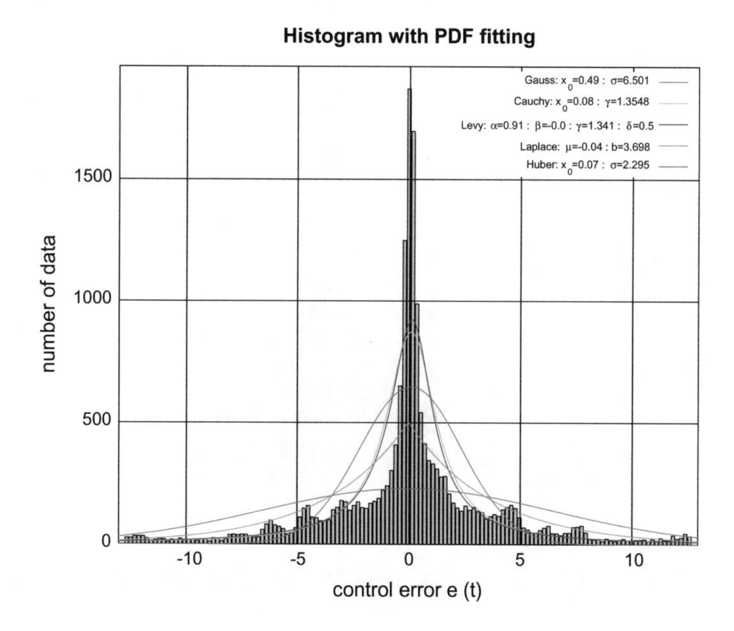

Fig. 11.73 Control error histogram with the fitted different PDFs for $G_3(s)$

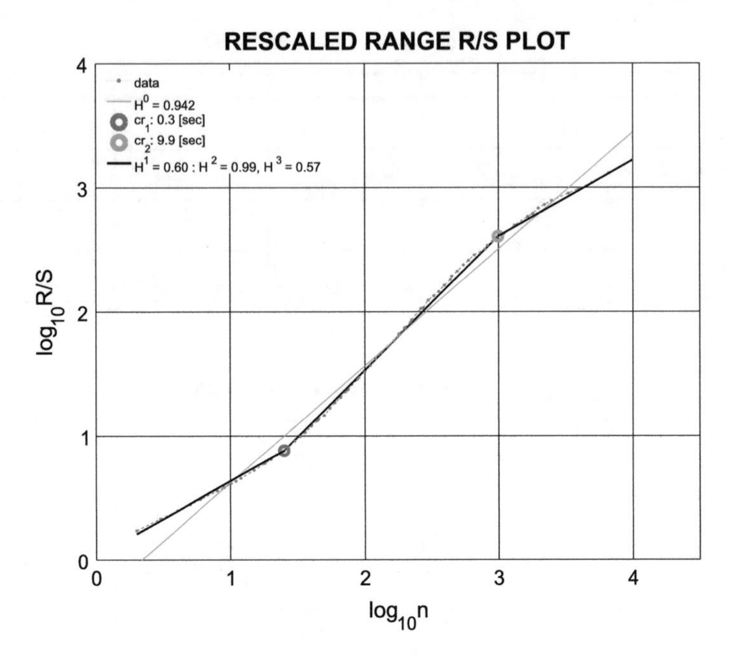

Fig. 11.74 Extended range R/S plot of the control error for $G_3(s)$

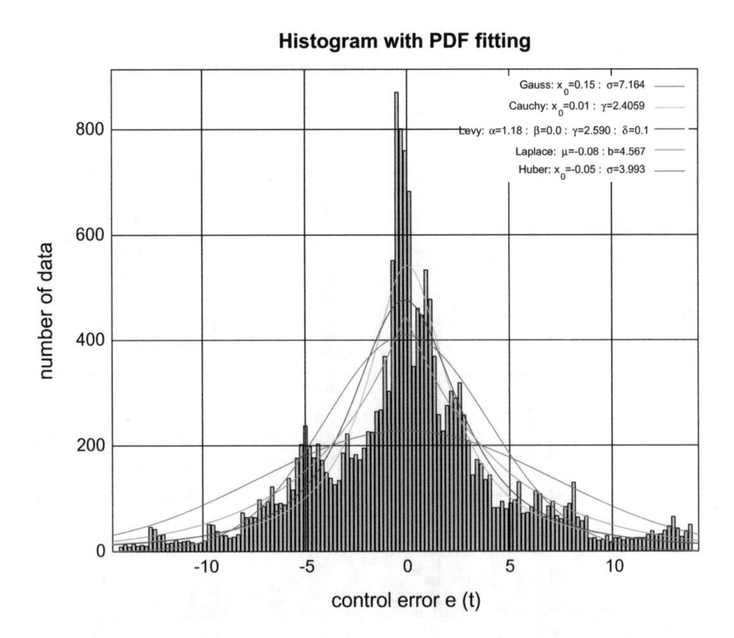

Fig. 11.75 Control error histogram with the fitted different PDFs in disturbed case for $G_3(s)$

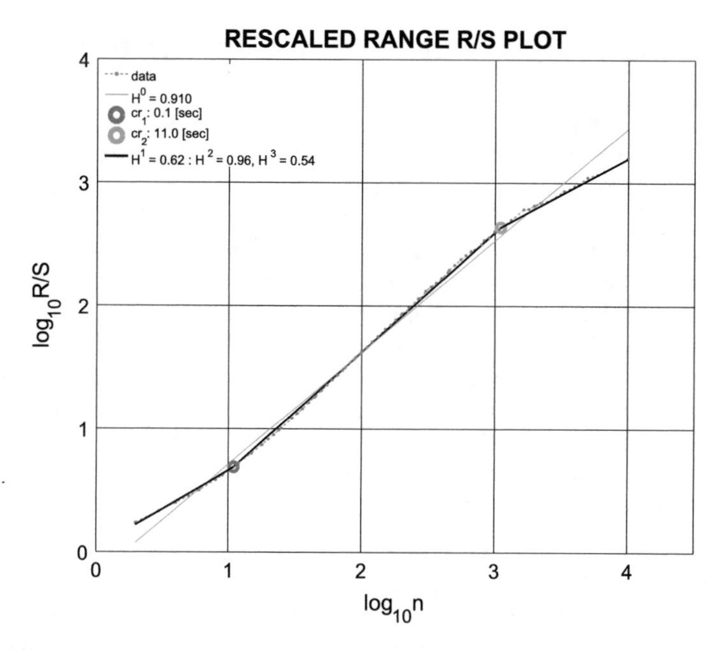

Fig. 11.76 Extended range R/S plot for the control error of disturbed case for $G_3(s)$

Table 11.5 Indexes for the optimal $G_3\,(s)$ loop controller

	T_s	$\kappa[\%]$	DR	II	AI	OI	RI	MSE	IAE	QE
nonDist	6.280	0.024	0.000	−0.087	1.000	0.006	0.611	160.265	6.968	237.741
Dist	–	–	–	–	–	–	–	161.279	7.349	239.653
	σ	γ^C	α	γ^L	b	σ^{rob}				
nonDist	12.659	1.343	0.379	0.347	6.988	2.310				
Dist	12.700	2.434	0.840	2.123	7.355	3.934				
	H^0	H^1	H^2	H^3	cr^1	cr^2	H^{RE}	H^{DE}		
nonDist	0.923	0.614	0.967	0.481	0.22	13.20	48.36	−152107		
Dist	0.900	0.622	0.953	0.410	0.10	13.20	49.49	−147822		

Table 11.6 Controller tuning pointed out by the selected measures for $G_3\,(s)$—the highlighted solutions are close to the optimal one

		T_s	$\kappa[\%]$	DR	MSE	IAE	QE	σ	γ^C	γ^L	b	σ^{rob}	H^{RE}	H^{DE}
nonDist	k_p	**0.96**	n/a	n/a	1.72	1.48	0.68	1.72	1.12	**0.92**	1.48	**1.12**	0.20	0.20
	T_i	**2.5**	n/a	n/a	10.0	3.2	10.0	10.0	2.8	**2.5**	3.2	**2.8**	10.0	10.0
Dist	k_p	n/a	n/a	n/a	1.76	1.72	0.60	1.76	1.80	**1.16**	1.72	**1.16**	0.20	0.20
	T_i	n/a	n/a	n/a	9.20	4.50	9.80	9.20	10.0	**2.80**	4.50	**2.80**	10.00	10.00

11.3.2 Control Performance Detection Using Step Indexes

The analysis of the performance assessment potential for the considered KPIs is presented in form of the control quality surfaces. The drawings are prepared according to the same common scheme. The plots are in form of the two dimensional surfaces presenting each KPI as the function of two PID parameters: k_p and T_i. The third parameter, $T_d = 0.66$ is kept constant at its optimal value, so the surface is a 2D cross-section. Due to the lack of place other cross-sections are not presented as they do not bring additional value to the analysis.

The position of the optimal controller is depicted with the magenta circle •. There are two possible variants of the interpretation. The best index is at its minimum value, or there is many optimal values. In the first case the red diamond ♦ depicts index position. The black bold contour line for the best values of the KPI is used in the second case. Step based indexes are evaluated without the disturbance.

Review of the drawings enables to bring up the following observations:

- Classical indexes of the settling time (Fig. 11.77) and the overshoot (Fig. 11.78) work fine. It is clearly seen that the calculated optimal controller $R3_{PID}^{opt}(s)$ is very close to the minimum index one. On the other hand, the overshoot treated independently does not enable to find single solution, but the optimal solution lies on the zero overshoot contour line.
- Optimal controller lies on the contour line depicting ideal II value (Fig. 11.79).
- Ideal literature value of the Area Index $0.3 < AI < 0.7$ is relatively far away from the optimal controller (Fig. 11.80) and reflects rather sluggish control. Similar

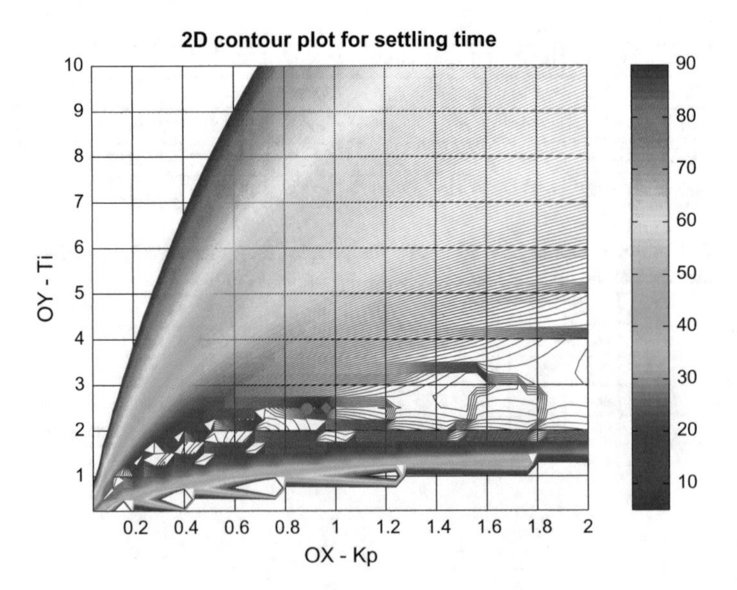

Fig. 11.77 Control quality surface for $G_3(s)$: T_s

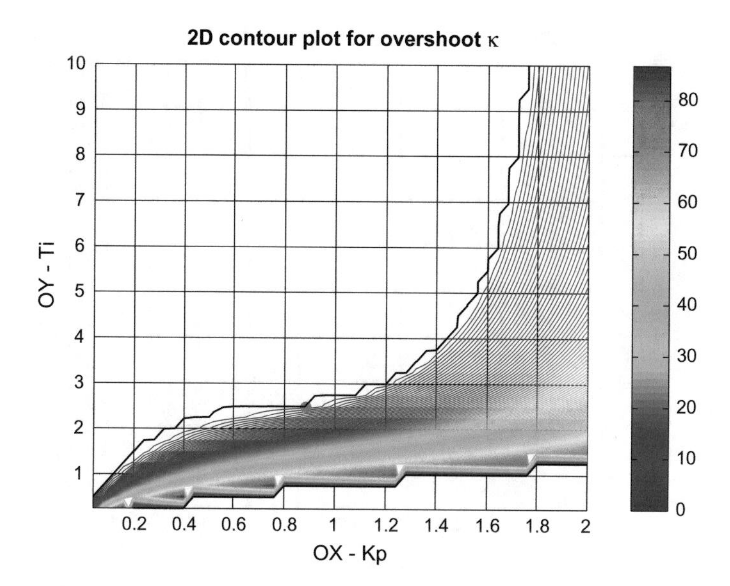

Fig. 11.78 Control quality surface for $G_3(s)$: κ

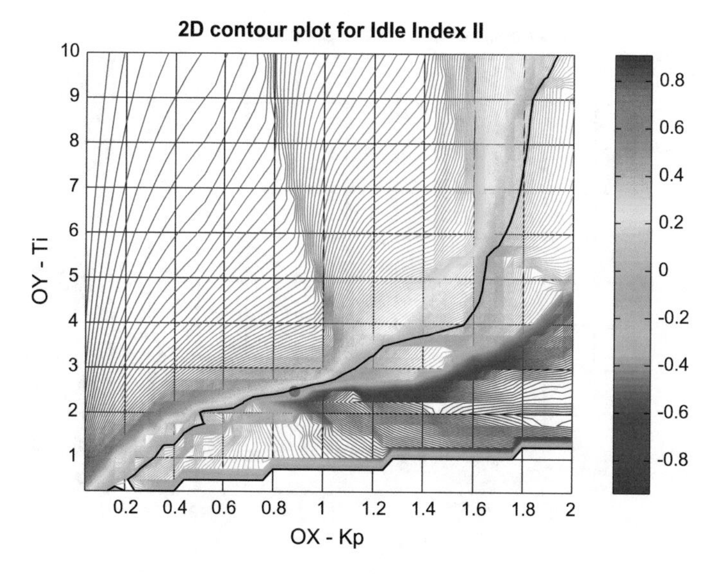

Fig. 11.79 Control quality surface for $G_3(s)$: II (the best value $II = 0$)

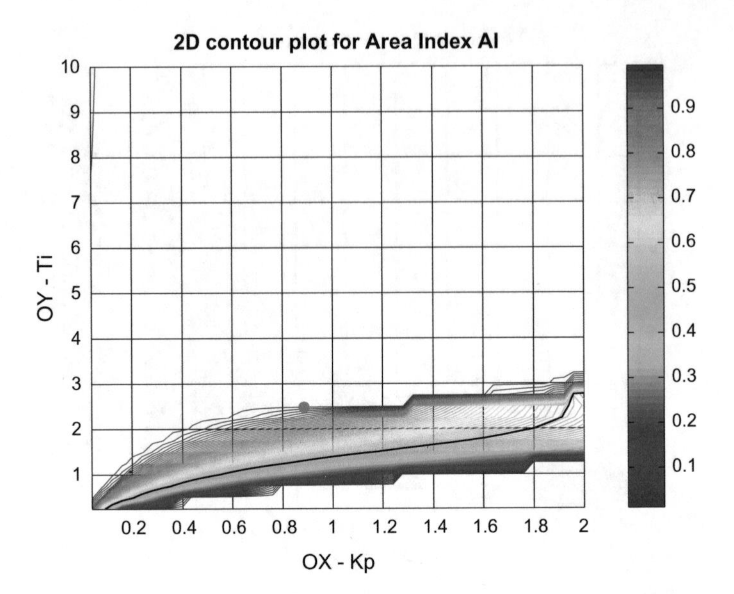

Fig. 11.80 Control quality surface for G_3(s): AI (the best value $AI = 0.5$)

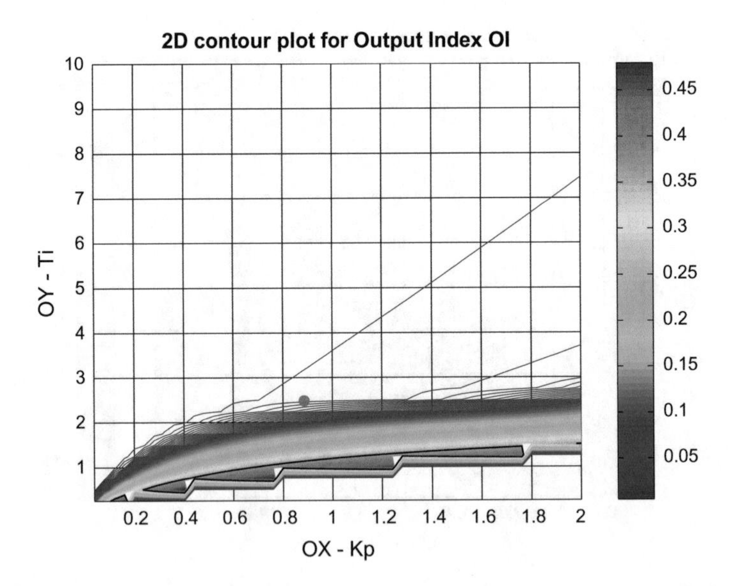

Fig. 11.81 Control quality surface for G_3(s): OI (the best value $OI = 0.35$)

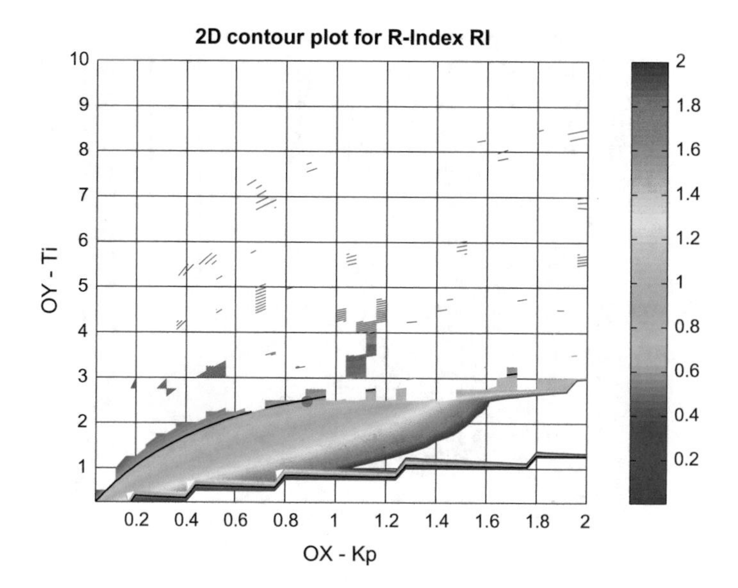

Fig. 11.82 Control quality surface for $G_3(s)$: RI (the best value $RI = 0.6$)

situation is visible for the Output Index (Fig. 11.81), when optimal controller $OI \approx 0.0$ is quite far away from the literature ideal $OI < 0.35$.

- R-Index surface is not smooth and very scattered (Fig. 11.82). Nonetheless the optimal controller hits optimal $RI = 0.6$ contour line.
- Decay Ratio surface is also not evident, so its plot is not included.

11.3.3 Control Performance Detection for Undisturbed Loop

The simulation of the loop using predefined setpoint profile is realized in two forms: undisturbed and disturbed one. Thus for such experiment the integral, statistical, fractal and entropy indexes are evaluated and analyzed. Diagrams sketched below present surfaces for the undisturbed simulation mode.

Review of the above undisturbed surfaces allows to bring about the following observations:

- Gaussian standard deviation σ (Fig. 11.85) is very close to the MSE (Fig. 11.83), while Laplace scale factor b is very close to the IAE (Fig. 11.84). These shapes are almost identical. However, the detected results differ significantly. Square error and σ tend to the sluggish and non-overshoot control highly disliking aggressiveness in tuning. On the other hand absolute error and Laplace scale factor b are more realistic and closer to the expectations. Quadratic Error QR is close to the MSE lies somewhere between depicting extremely sluggish control.

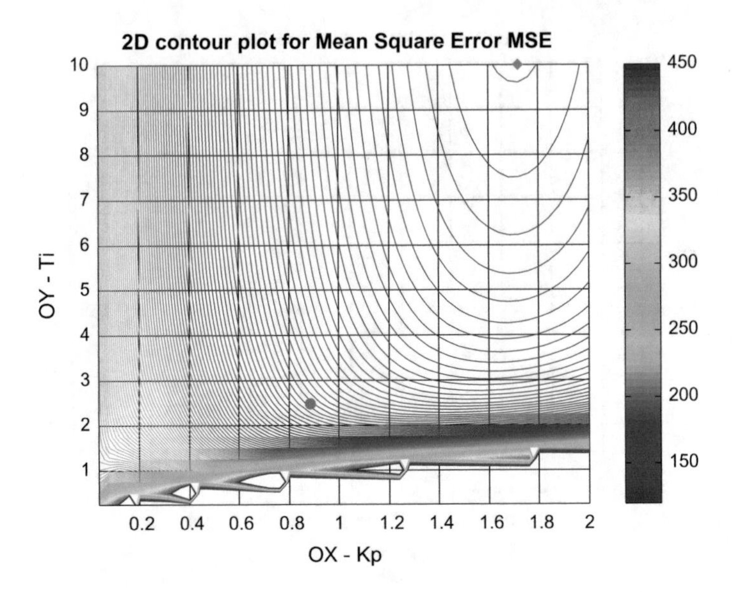

Fig. 11.83 Control quality surface for undisturbed $G_3(s)$: MSE

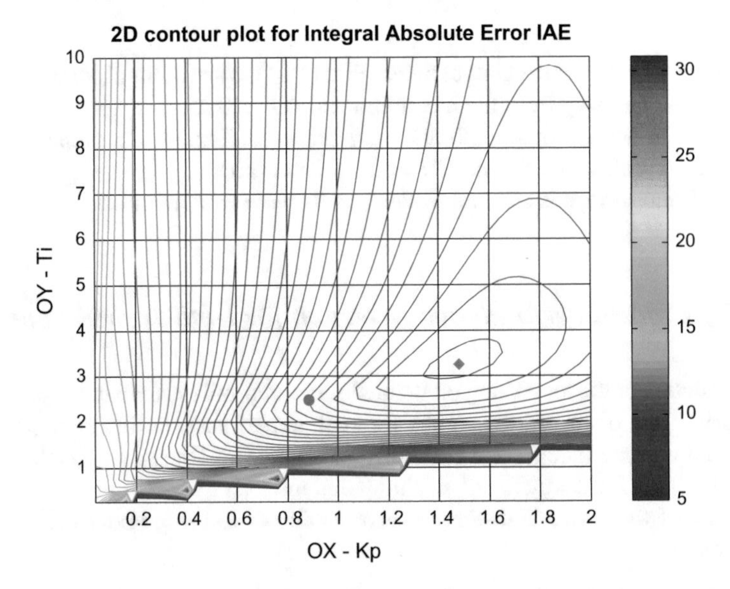

Fig. 11.84 Control quality surface for undisturbed $G_3(s)$: IAE

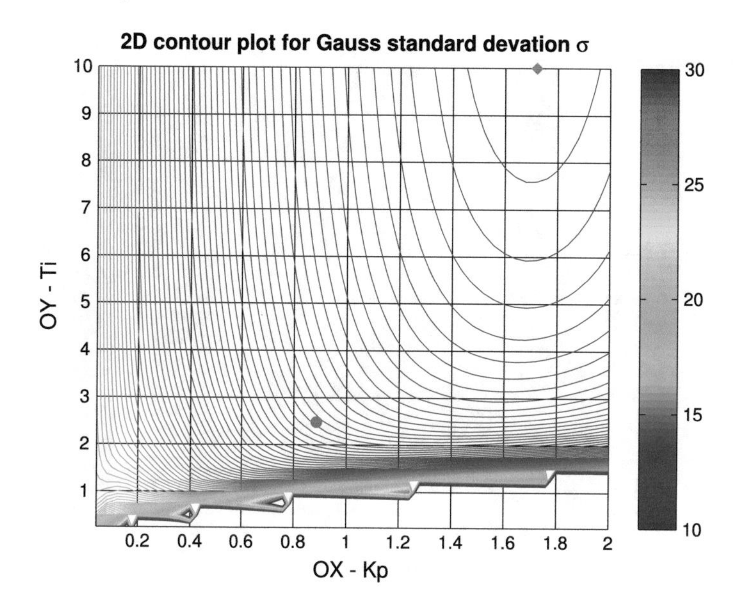

Fig. 11.85 Control quality surface for undisturbed $G_3(s)$: Gauss σ

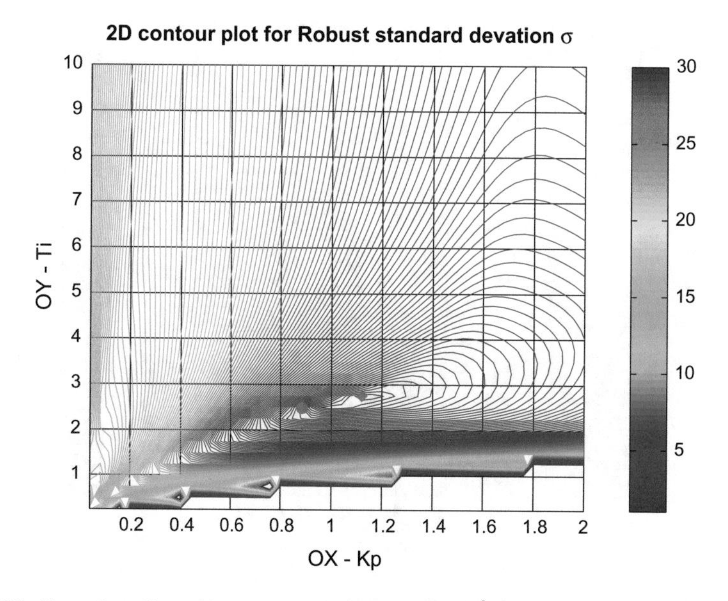

Fig. 11.86 Control quality surface for undisturbed $G_3(s)$: σ^{rob}

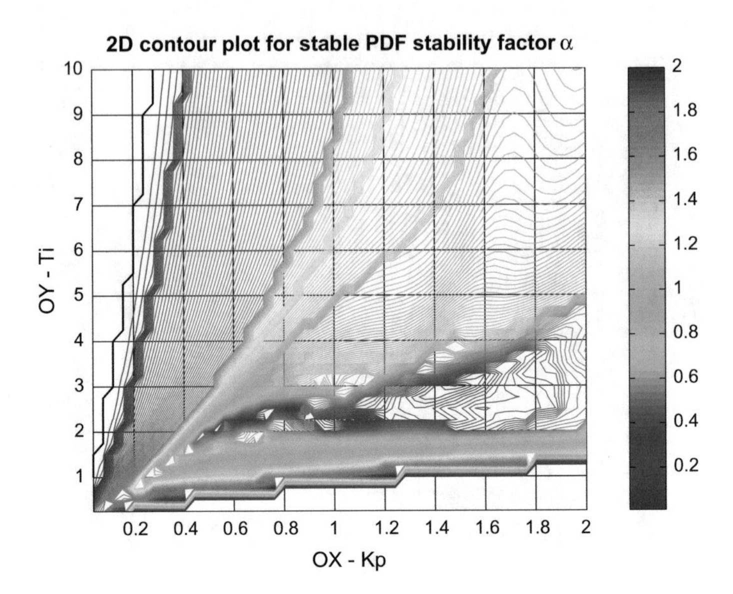

Fig. 11.87 Control quality surface for undisturbed $G_3(s)$: stable α (the best value $\alpha = 2.0$)

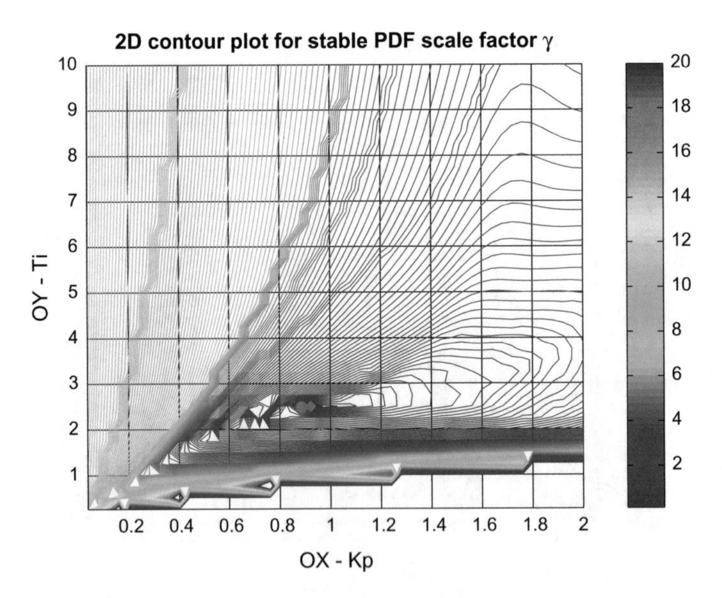

Fig. 11.88 Control quality surface for undisturbed $G_3(s)$: stable γ^L

- Usage of the robust standard deviation (Fig. 11.86) seems to be effective with the best found controller relatively close to the optimal one. Cauchy scale factor γ^C performs similarly.
- Detection with the scale factor of the stable distribution (Fig. 11.88) is the accurate (it points out almost the same point) and seems to be the best KPI. Interpretation

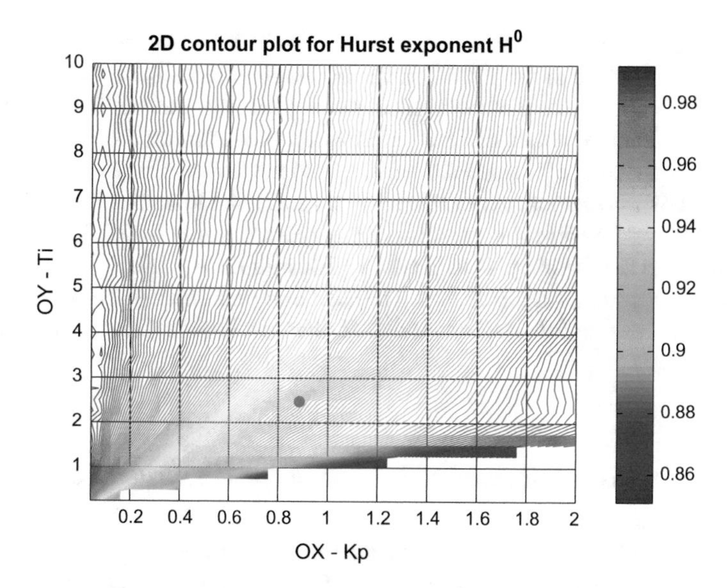

Fig. 11.89 Control quality surface for undisturbed $G_3(s)$: H^0

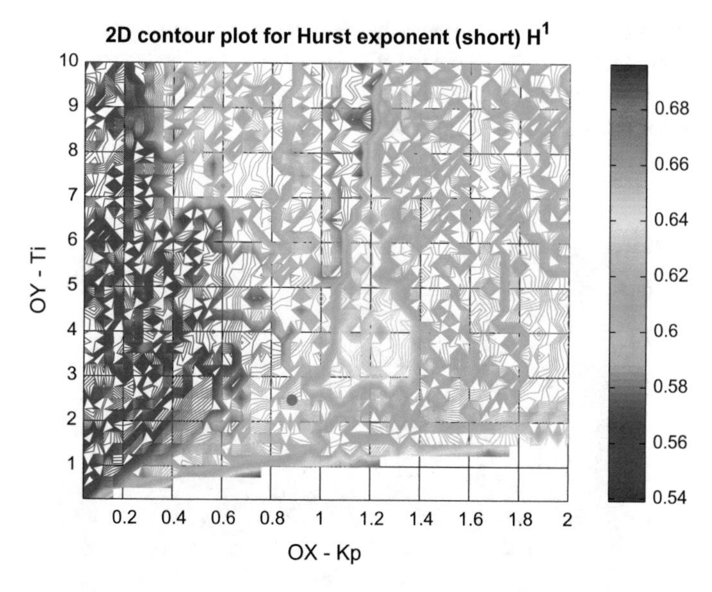

Fig. 11.90 Control quality surface for undisturbed $G_3(s)$: H^1

of the stable distribution stability factor α (Fig. 11.87) is not clear. The shape of the surface is close to the γ^L. However the solution tends towards lower values, which according to the literature reflect rather aggressive tuning. It rather informs that the outliers exist than delivers information about control quality.

- All the Hurst exponents share similar shape of the surfaces, however on different levels. The short memory exponent H^1 (Fig. 11.89) should be the most informative. The value for the optimal control is not as expected, i.e. $H \approx 0.5$. The controller, according to the literature, is clearly sluggish with $H^1 \to 1.0$.
- The entropies share similar shapes. The detection (minimum value) tends towards unreasonably sluggish control and is rather useless in the assessment task (Fig. 11.90).

11.3.4 Control Performance Detection for Disturbed Loop

Next, the control loop is disturbed and the same set of the analyzes is performed. The way of the drawings preparation and surfaces analysis is exactly the same.

Review of the disturbed control drawings brings up the following observations:

- The majority of the surfaces witness the disturbance shadowing effect. It is disturbing proper detection for MSE, IAE, QE, Gaussian σ, Laplace b and both entropies. Only two examples of the Gaussian σ (Fig. 11.91) and Laplace b (Fig. 11.92) are sketched for the visualization purposes.

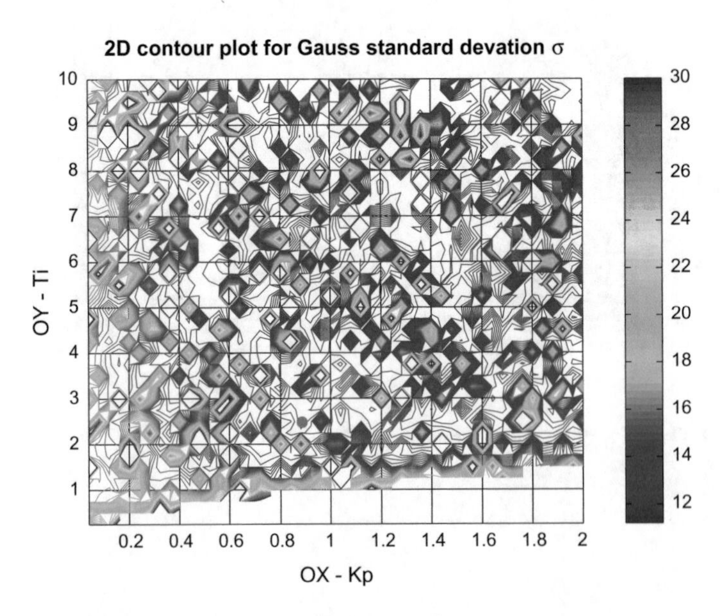

Fig. 11.91 Control quality surface for disturbed $G_3(s)$: Gauss σ

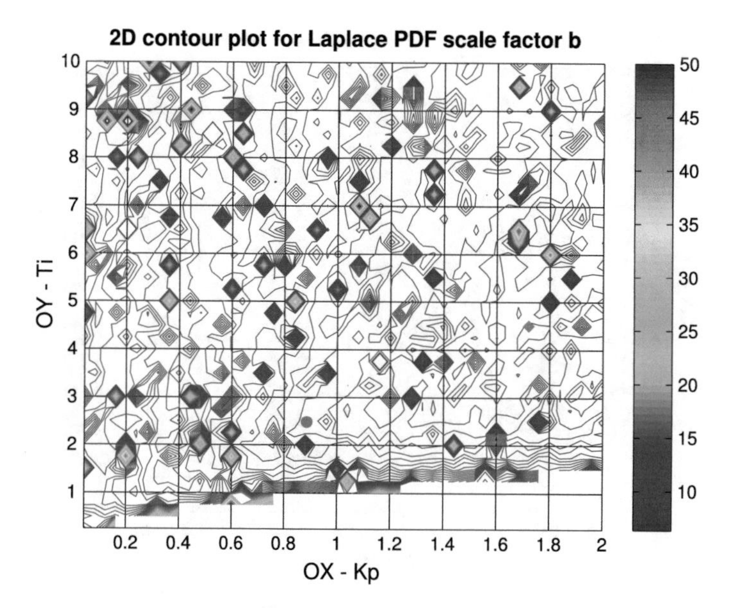

Fig. 11.92 Control quality surface for disturbed $G_3(s)$: Laplace b

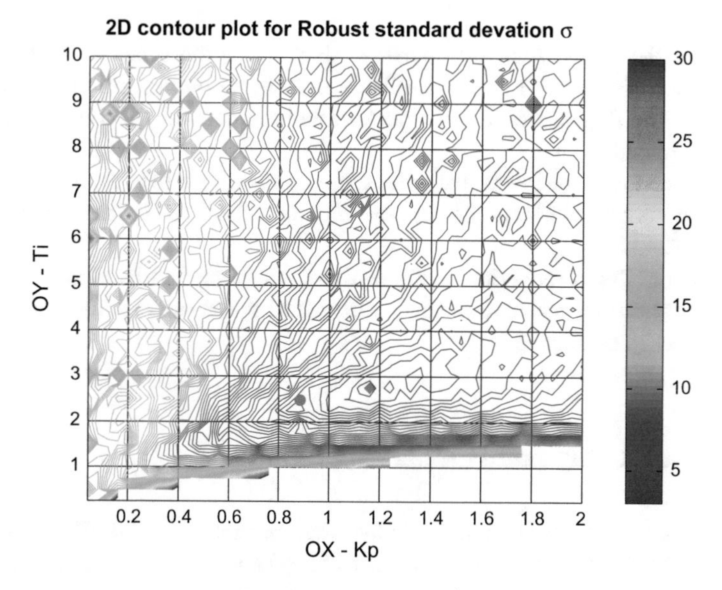

Fig. 11.93 Control quality surface for disturbed $G_3(s)$: σ^{rob}

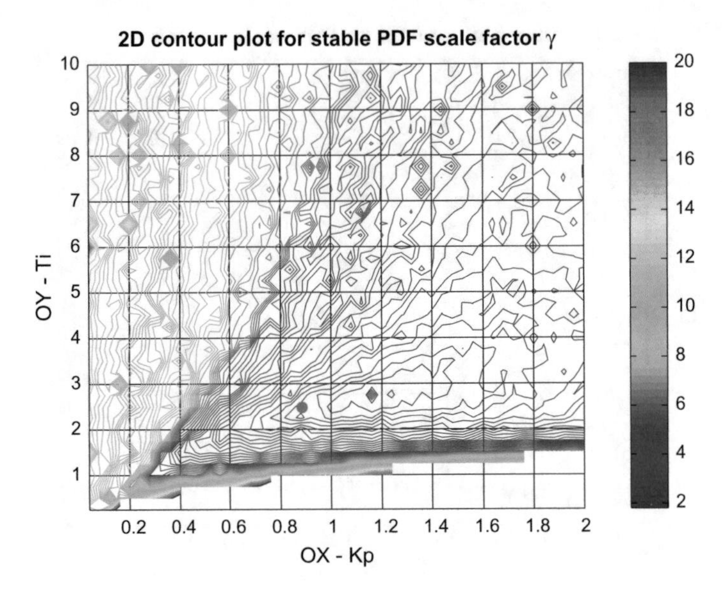

Fig. 11.94 Control quality surface for disturbed $G_3(s)$: stable γ

- Usage of the robust standard deviation (Fig. 11.93) the scale γ^L of the stable distribution (Fig. 11.94) delivers reliable detection pointing out the acceptable solution. The shapes are visibly smoother and the pointed out solution seems to be more reliable.
- Fractal measures are indecisive with some shadowing effect. Although the surfaces are smoother than for the classical measures the detection is not evident and indecisive.
- Concluding, the introduction of the **outliers** (as fat-tailed disturbance property) into the control loop generates shadowing effect that makes the most of the CPA approaches questionable.

11.4 Time-Delay and Double Lag Plant

The analysis of time-delay double lag plant starts with the optimal PID controller. It is followed by the review of tuning surfaces for all considered CPA indexes.

11.4.1 Optimal Controller Performance

The optimal controller for the $G_4(s)$ transfer function has been designed according to the predefined rules. The evaluated tuning parameters are presented below:

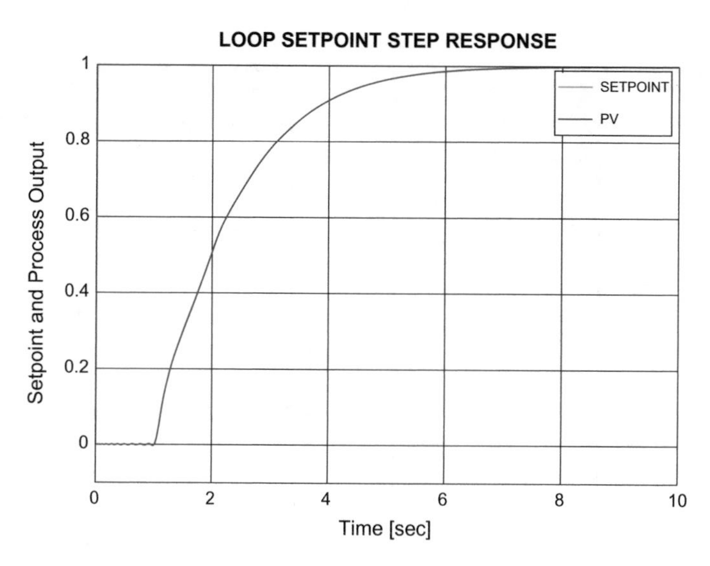

Fig. 11.95 Loop setpoint step response for $G_4(s)$

- gain: $k_p = 0.27$
- integration: $T_i = 0.61$
- derivation: $T_d = 0.21$

The graphical representation of the loop setpoint step response is sketched in Fig. 11.95, while Fig. 11.96 shows the disturbance step response. We see that the obtained tuning delivers the response with almost invisible overshoot. The controller output is smooth without any oscillations. Further, the loop simulation has been run. The time trends (sample of first 100 s) for the selected loop profiles for setpoint, CV and MV (both undisturbed and disturbed) are sketched in Fig. 11.97.

Respective control error histogram is shown in Fig. 11.98. It is clearly visible that its shape is not Gaussian with the significant tails. The Histogram can be modeled the most precisely by the α-stable distribution. The analog histogram for the disturbed simulation is sketched in Fig. 11.100.

Finally, the extended range R/S plot has been evaluated (Fig. 11.99). We clearly see that its character is not monofractal. The multi-persistent properties are well reflected with the two crossovers ($cross^1$ and $cross^2$) and three memory ranges described by the respective Hurst exponents (H^1, H^2 and H^3). The analog plot for the disturbed simulation is sketched in Fig. 11.101.

Control error histogram is shown in Fig. 11.100. The non-Gaussian shape with the significant tails is observed. The Histogram can be modeled the most precisely by the α-stable distribution. Finally, the extended range R/S plot has been evaluated (Fig. 11.99). We clearly see multifractal character, well reflected with the two crossovers ($cross^1$ and $cross^2$) and three memory ranges described by the respective Hurst exponents (H^1, H^2 and H^3).

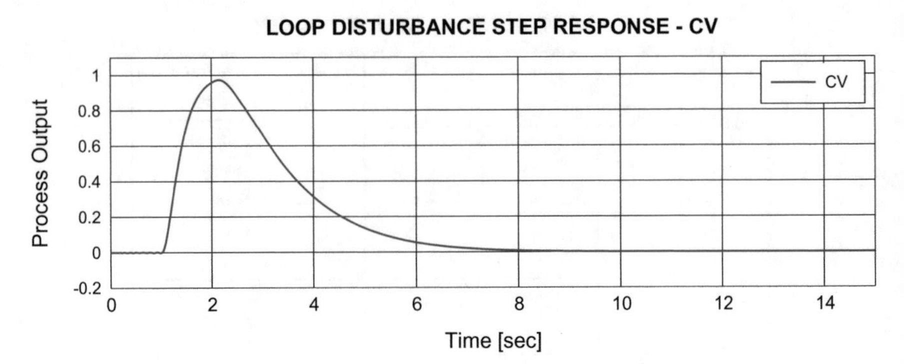

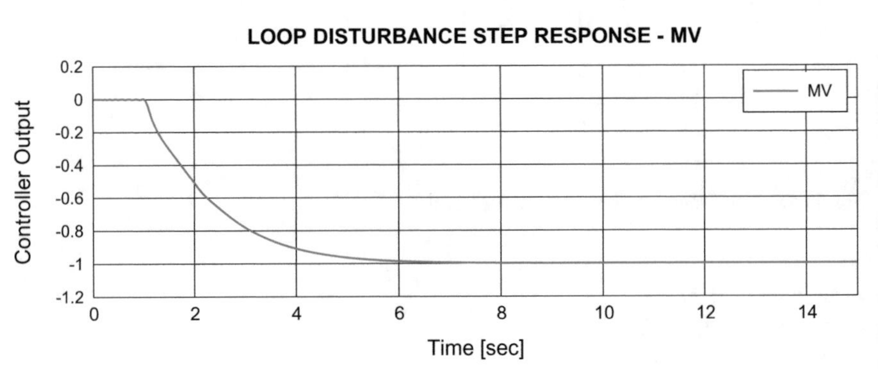

Fig. 11.96 Loop disturbance step response for $G_4(s)$

All the evaluated indexes for undisturbed (nonDist) and disturbed (Dist) simulations are summarized in Table 11.7. It is also interested to see what controller parameters have been selected by each method. Table 11.8 summarizes these results. The table consists only of these measures that can detect the solution, i.e. point out the single result as the extreme index value (minimum or maximum). The settings highlighted show the results, which are close enough to the optimal solution $k_p \in [0.15, 0.45]$ and $T_i \in [0.5, 0.7]$.

11.4.2 Control Performance Detection Using Step Indexes

The performance assessment potential analysis for the KPIs is presented using control quality surfaces. The drawings are prepared according to the same common scheme. The plots are in form of the two dimensional surfaces presenting each KPI as the function of two PID parameters: k_p and T_i. The third parameter, $T_d = 0.21$ is kept constant at its optimal value, so the surface is a 2D cross-section. Other cross-sections are not presented, as they do not bring any additional value to the analysis.

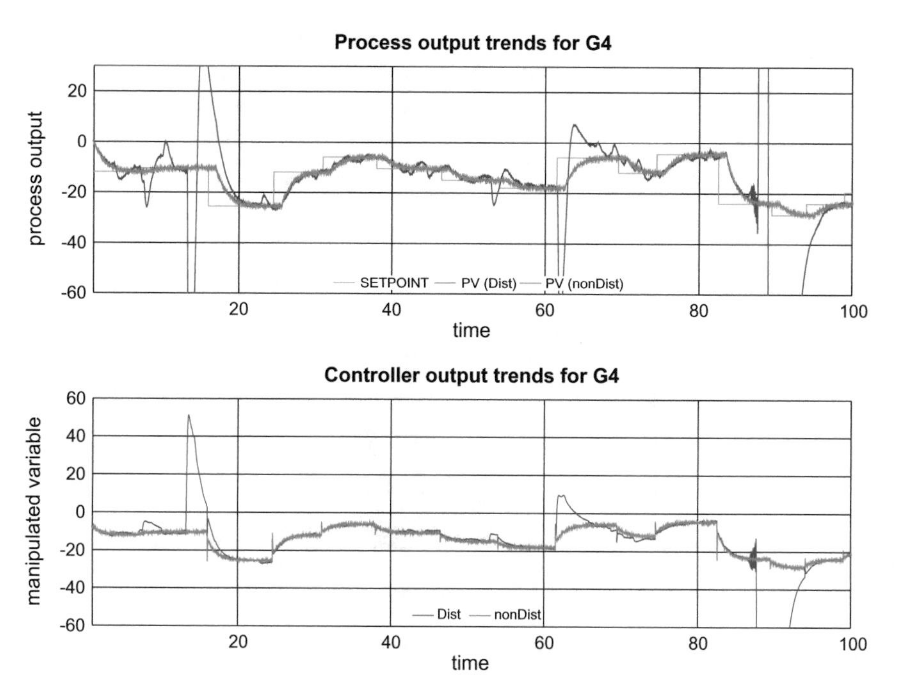

Fig. 11.97 Loop time trends for first 100 s of $G_4(s)$

The position of the found optimal controller is depicted with the magenta circle •. There are two variants for the diagram interpretation: the best index is at its minimum value or there is optimal set of the indexes. In the first case the red diamond ♦ depicts minimum index position. The black bold contour line for the best KPI values is used in the second case. Step based indexes are evaluated without the disturbance.

Review of the drawings enables to bring up the following observations:

- Classical indexes of the settling time (Fig. 11.102) and the overshoot (Fig. 11.103) work fine. It is clearly seen that the calculated optimal controller $R4^{opt}_{PID}(s)$ is relatively close to the minimum index one. On the other hand, the overshoot treated independently does not enable to find single solution, but the optimal solution lies on the zero overshoot contour line.
- Optimal controller lies on the contour line depicting ideal II value (Fig. 11.104).
- Output Index (Fig. 11.105) shows smooth and reasonable shape, though optimal controller $OI \approx 0.0$ is not close to the literature suggestion $OI < 0.35$.
- The Area Index $0.3 < AI < 0.7$ seems to be unreadable in the considered case and the plot is completely uninformative. Similar situation is visible R-Index surface being very scattered. Their plots ir thus not included.
- Decay Ratio surface is also not evident, so its plot is not included.

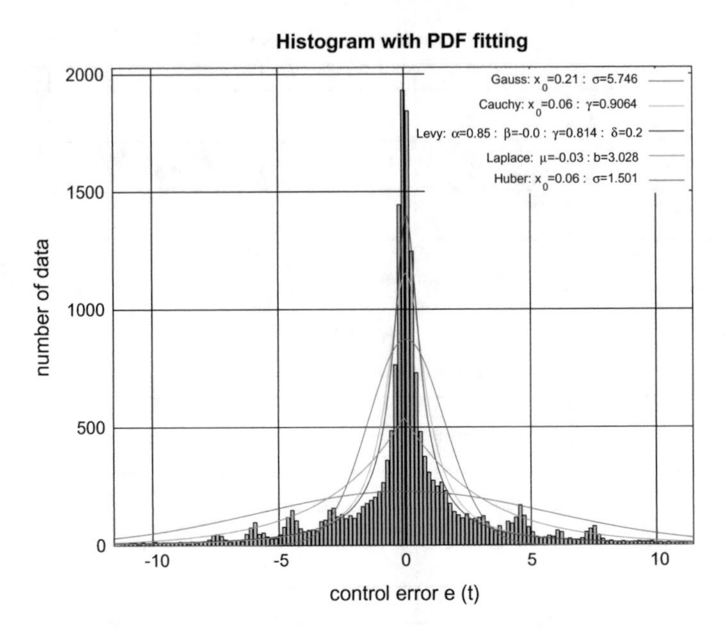

Fig. 11.98 Control error histogram with the fitted different PDFs for $G_4(s)$

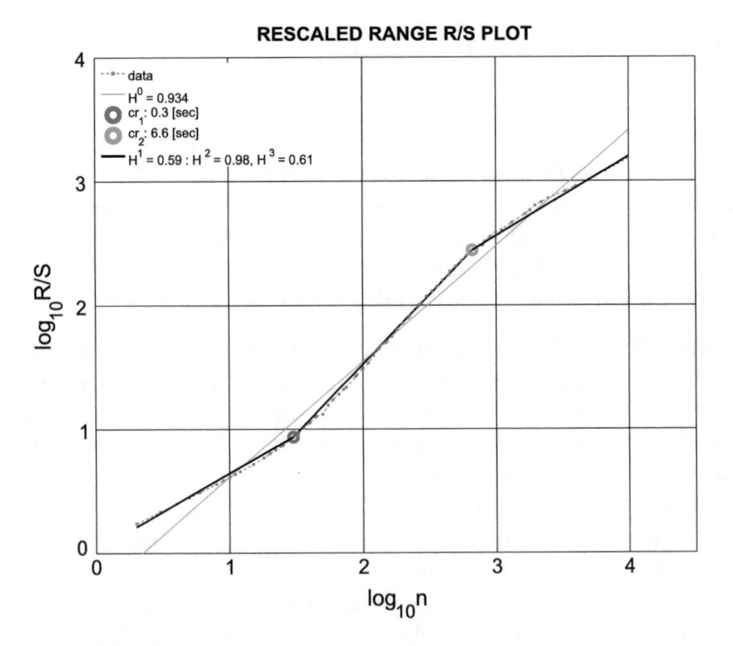

Fig. 11.99 Extended range R/S plot of the control error for $G_4(s)$

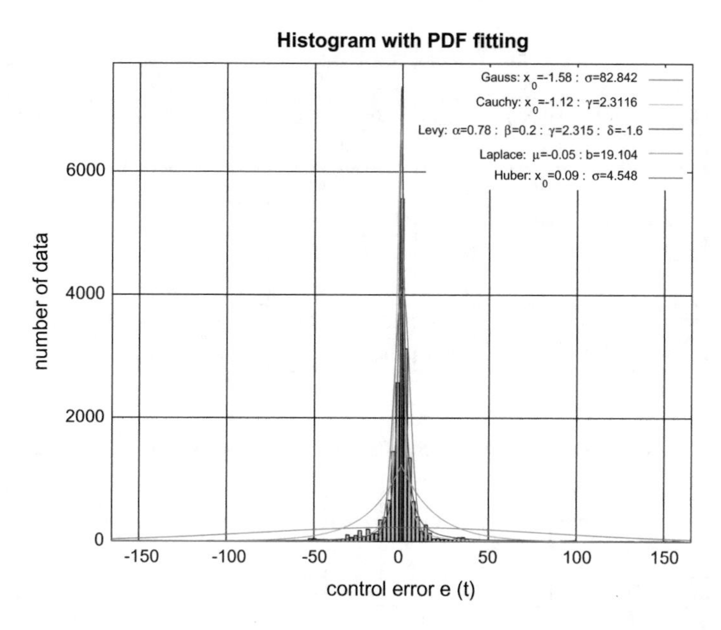

Fig. 11.100 Control error histogram with the fitted different PDFs in disturbed case $G_4(s)$

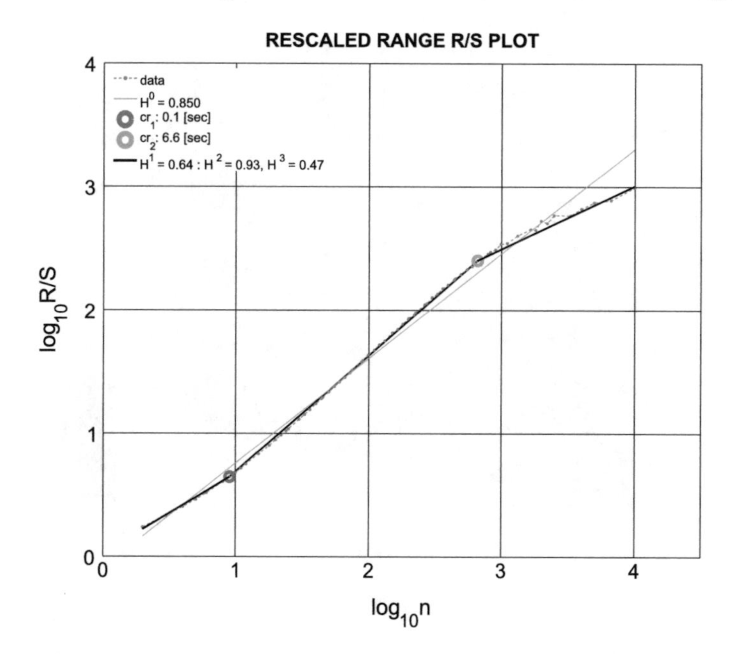

Fig. 11.101 Extended range R/S plot for the control error of disturbed case for $G_4(s)$

Table 11.7 Indexes for the optimal $G_4\,(s)$ loop controller

	T_s	$\kappa[\%]$	DR	II	AI	OI	RI	MSE	IAE	QE
nonDist	6.290	0.000	0.000	0.423	1.000	0.000	0.479	128.093	5.800	172.119
Dist	–	–	–	–	–	–	–	129.113	6.137	173.277
	σ	γ^C	α	γ^L	b	σ^{rob}				
nonDist	11.318	0.873	0.476	0.311	5.815	1.467				
Dist	11.363	1.813	0.709	1.671	6.141	2.873				
	H^0	H^1	H^2	H^3	cr^1	cr^2	H^{RE}	H^{DE}		
nonDist	0.916	0.622	0.961	0.540	0.30	8.25	42.95	−149965		
Dist	0.880	0.645	0.940	0.592	0.10	6.60	46.17	−144486		

Table 11.8 Tuning selected by the measures for $G_4\,(s)$—the highlighted solutions are close to the optimal one

		T_s	$\kappa[\%]$	DR	MSE	IAE	QE	σ	γ^C	γ^L	b	σ^{rob}	H^{RE}	H^{DE}
nonDist	k_p	0.42	n/a	n/a	0.66	0.54	0.66	0.66	0.66	**0.45**	0.54	**0.45**	0.15	0.15
	T_i	0.70	n/a	n/a	1.10	0.80	4.00	1.10	1.00	**0.70**	0.80	**0.70**	4.00	4.00
Dist	k_p	1.20	n/a	n/a	0.66	0.54	0.66	0.66	0.87	0.54	0.54	0.54	0.15	0.15
	T_i	4.00	n/a	n/a	1.10	0.80	3.10	1.10	2.30	0.80	0.80	0.80	3.80	3.80

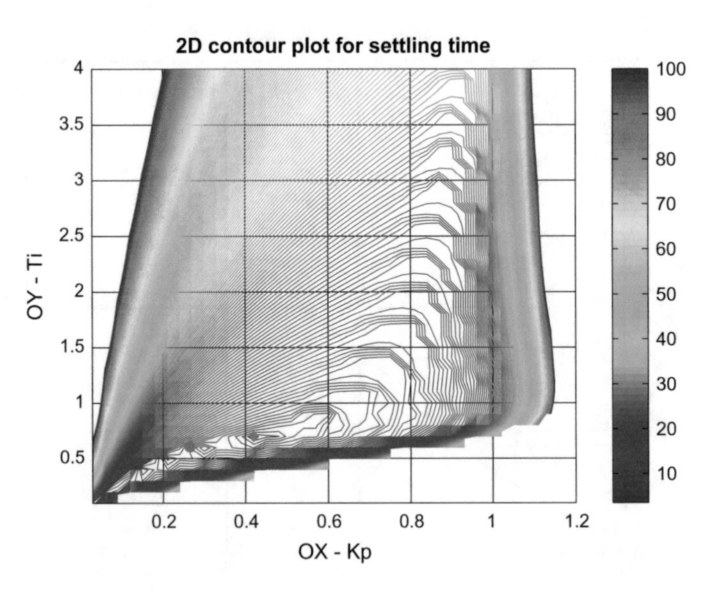

Fig. 11.102 Control quality surface for $G_4(s)$: T_s

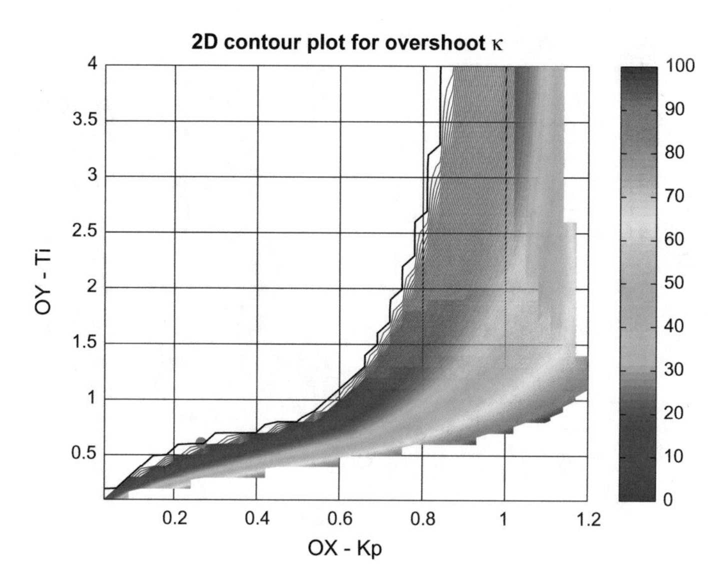

Fig. 11.103 Control quality surface for $G_4(s)$: κ

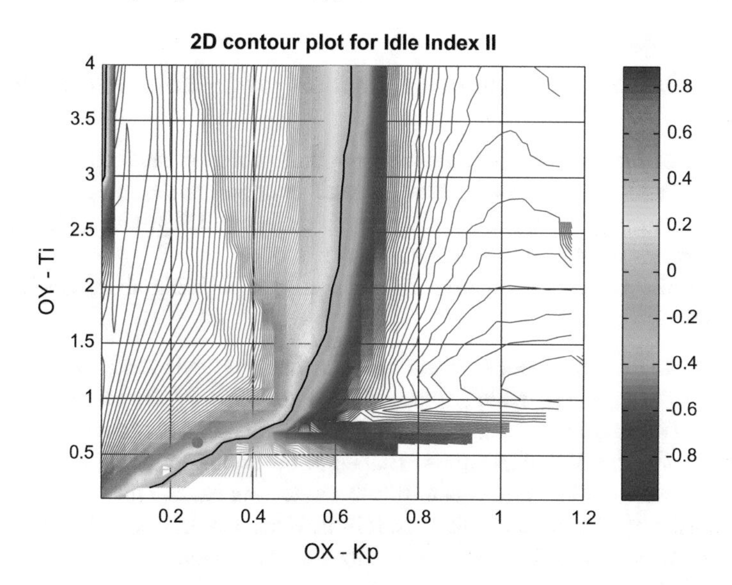

Fig. 11.104 Control quality surface for $G_4(s)$: II (the best value $II = 0$)

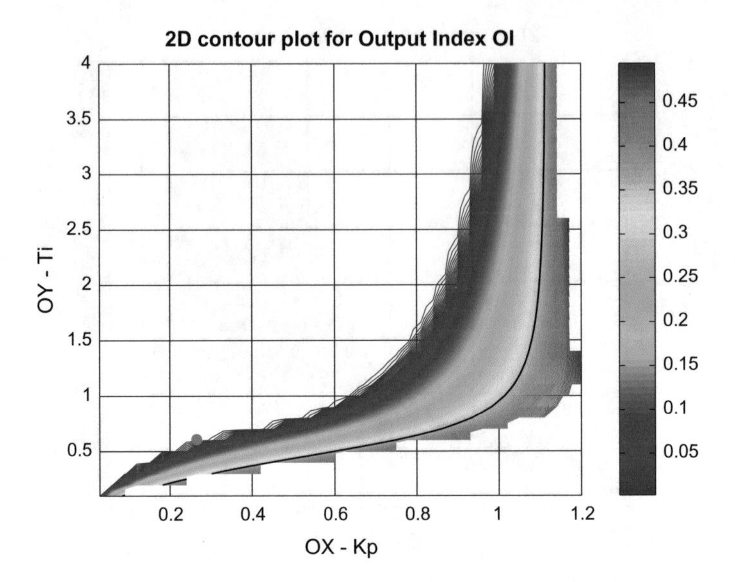

Fig. 11.105 Control quality surface for $G_4(s)$: OI (the best value $OI = 0.35$)

11.4.3 Control Performance Detection for Undisturbed Loop

Simulation of the loop using predefined setpoint profile is realized in two forms: undisturbed and disturbed one. Thus for such experiment the integral, statistical, fractal and entropy indexes are evaluated and analyzed. Diagrams sketched below present surfaces for the undisturbed simulation mode.

Review of the above undisturbed surfaces allows to bring about the following observations:

- Similarly to all the previous cases Gaussian standard deviation σ (Fig. 11.106) is very close to the MSE, while Laplace scale factor b is very close to the IAE and to the robust standard deviation (Fig. 11.107). Both fat tailed scaling factors (Cauchy and α-stable PDF) exhibit similar shapes to the robust standard deviation. The α-stable scaling (Fig. 11.108) is the closest detecting exactly the same solution to the robust σ^{rob}. Additionally, this result is the closest to the optimal PID controller setting. Thus, only both standard deviations are presented.
- Quadratic Error as in all the previous cases points out the distant solution, a sluggish one, with the surface shame also similar as previously.
- Usage of the robust standard deviation (Fig. 11.107) seems to be effective with the best found controller relatively close to the optimal one. Cauchy scale factor γ^C performs similarly.
- Interpretation of the stable distribution stability factor α is unclear. It rather only informs that the outliers exist, than delivers information about control quality.

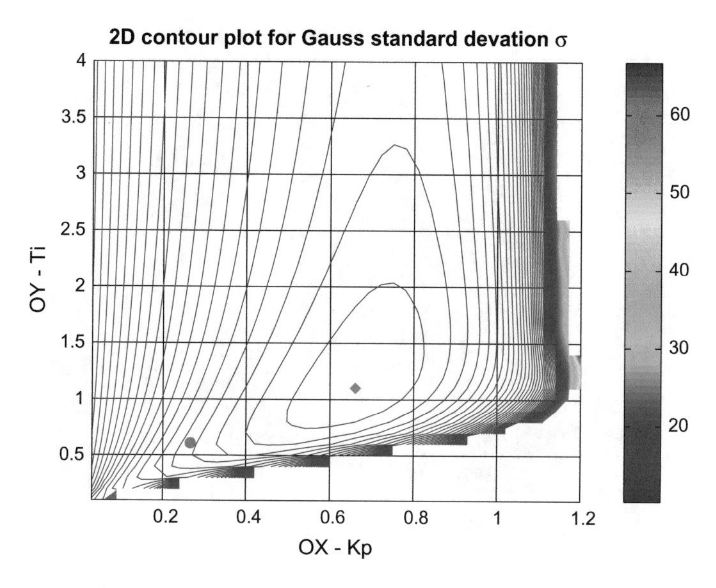

Fig. 11.106 Control quality surface for undisturbed $G_4(s)$: Gauss σ

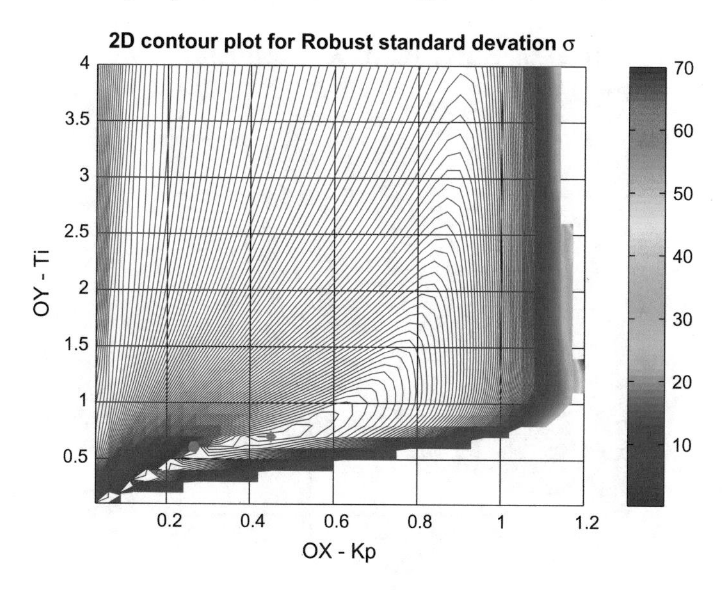

Fig. 11.107 Control quality surface for undisturbed $G_4(s)$: σ^{rob}

- All the Hurst exponents share similar shape of the surfaces, however on different levels. The short memory exponent H^1 (Fig. 11.109) should be the most informative. The value for the optimal control is not as expected, i.e. $H \approx 0.5$. The controller, according to the literature, is sluggish.

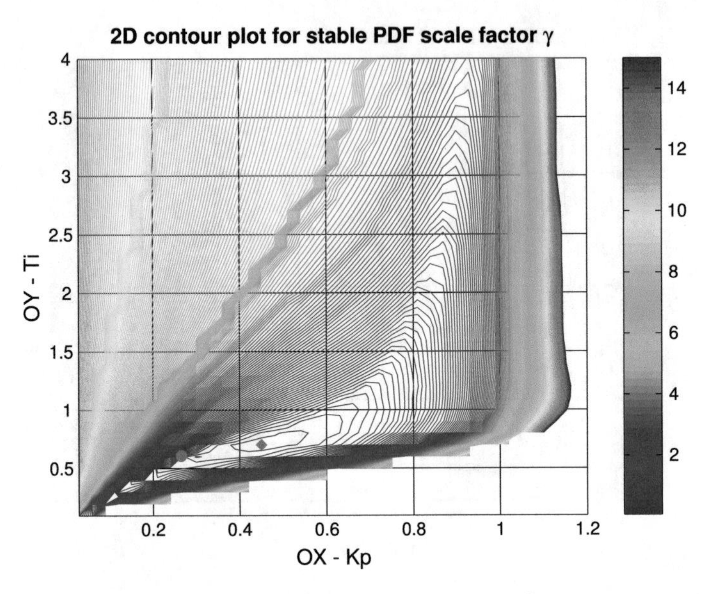

Fig. 11.108 Control quality surface for undisturbed $G_4(s)$: stable γ^L

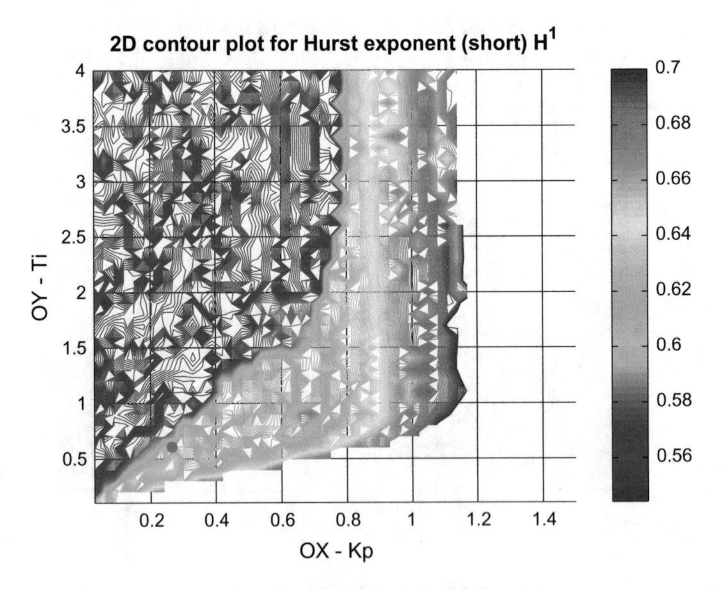

Fig. 11.109 Control quality surface for undisturbed $G_4(s)$: H^1

- The entropies share very similar shape. The detection tends towards extremely sluggish control and thus is rather useless for the assessment task.

11.4.4 Control Performance Detection for Disturbed Loop

Next, the control loop is disturbed and the same set of the analyzes is performed. The way of the drawings preparation and surfaces analysis is exactly the same. Review of the attached disturbed control drawings brings up the following observations:

- Actually, we may distinguish to types of the evaluated control surfaces. Part of them seems to be strongly affected by the disturbances showing "ragged" surface landscape, while the shape is smooth (unaffected by the disturbance *shadowing* effect) in the second group of the diagrams. The group of the disturbance sensitive measures consists of MSE, IAE, QE, normal standard deviation, Laplace scale factor and both entropies. On the other hand, the robust standard deviation, Cauchy and α-stable distribution factors together with all the Hurst exponents are robust against loop disturbances, mainly the **outliers**.
- Two surfaces for each of the groups are sketched as the representatives. Gaussian standard deviation (Fig. 11.110) and Laplace b (Fig. 11.111) surfaces are shown to present the existence of the *shadowing* effect for the above-mentioned measures. On the other hand, robust standard deviation (Fig. 11.112) and the scale factor γ^L of the stable distribution (Fig. 11.113) are sketched for the visualization purposes of the robust surfaces.

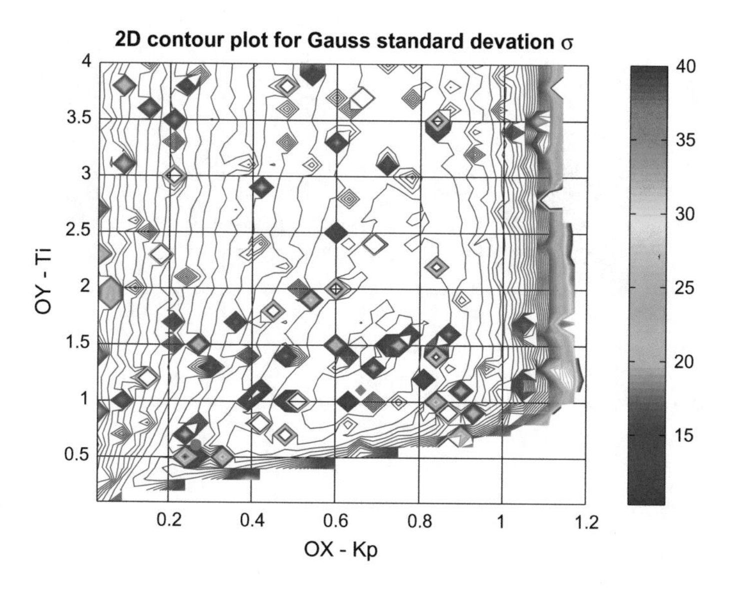

Fig. 11.110 Control quality surface for disturbed $G_4(s)$: Gauss σ

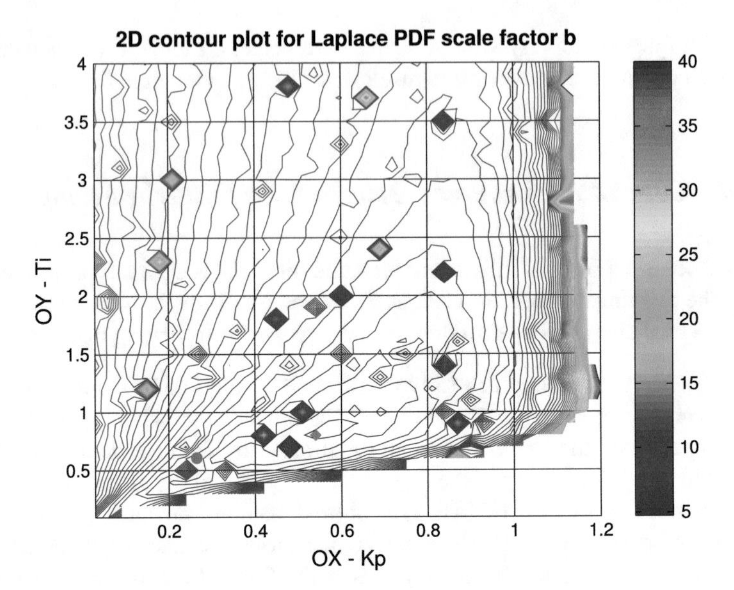

Fig. 11.111 Control quality surface for disturbed $G_4(s)$: Laplace b

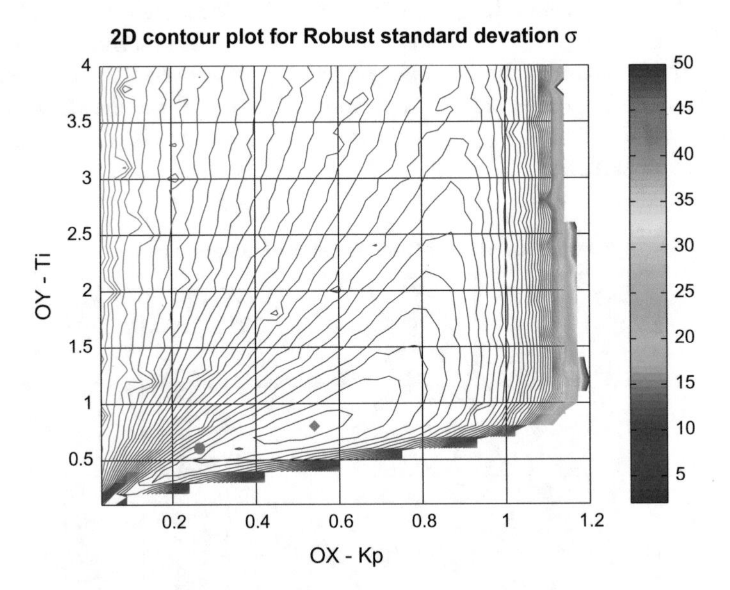

Fig. 11.112 Control quality surface for disturbed $G_4(s)$: σ^{rob}

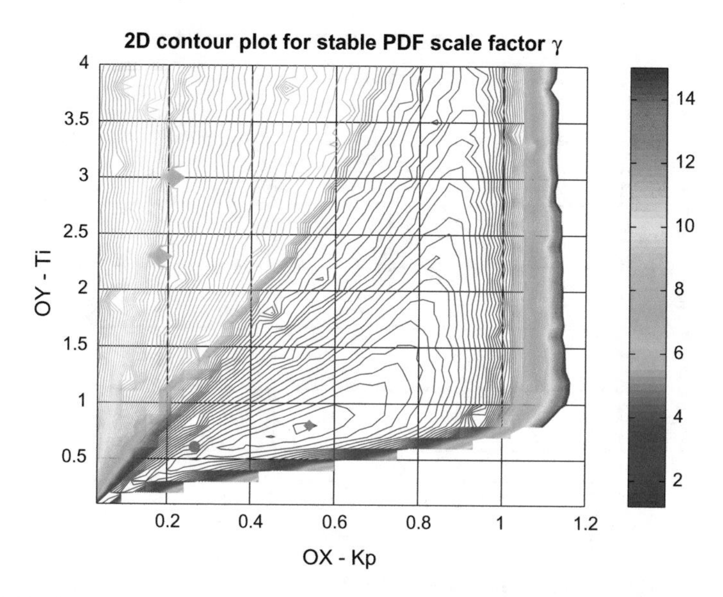

Fig. 11.113 Control quality surface for disturbed $G_4(s)$: stable γ

- We also see that the detection for the disturbance robust indexes is slightly biased in comparison with the undisturbed case, however looks acceptable.
- It is well seen that the introduction of the **outliers** (fat-tailed disturbance properties) into the control loop generates *shadowing* effect that makes the most of the CPA approaches questionable.

11.5 Oscillatory Transfer Function

The analysis starts with the evaluation of the optimal PID controller and is followed by the analysis of the tuning surfaces for all considered CPA indexes.

11.5.1 Optimal Controller Performance

The optimal controller for the $G_5(s)$ transfer function has been designed according to the predefined rules. The evaluated tuning parameters are presented below:

- gain: $k_p = 0.13$
- integration: $T_i = 0.26$
- derivation: $T_d = 0.08$

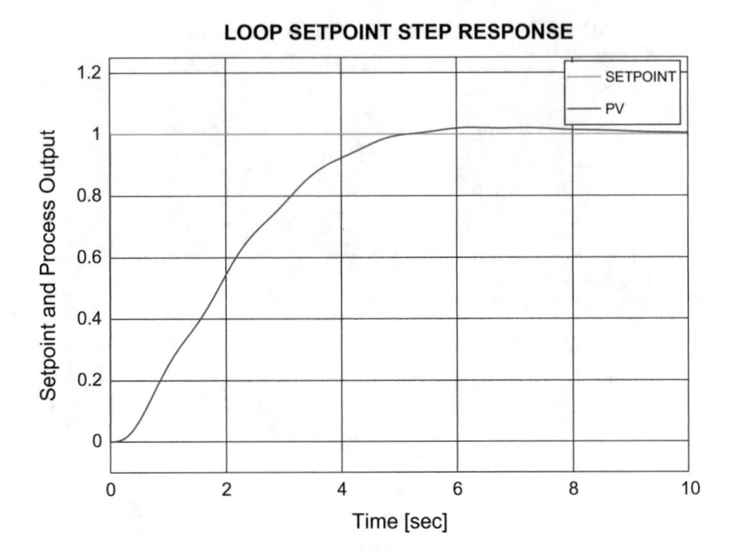

Fig. 11.114 Loop setpoint step response for $G_5(s)$

The graphical representation of the loop setpoint step response is sketched in Fig. 11.114, while Fig. 11.115 shows the disturbance step response. We see that the obtained tuning delivers the response with almost invisible overshoot. The controller output is smooth without any oscillations.

Next, the loop simulation has been run. The time trends (sample of first 120 s) for the selected loop profiles for setpoint, CV and MV are sketched in Fig. 11.116. The figure shows comparison of both undisturbed and disturbed behavior.

Respective control error histogram is shown in Fig. 11.117. It is clearly visible that its shape is not Gaussian with the significant tails. The Histogram can be modeled the most precisely by the α-stable distribution. The analog histogram for the disturbed simulation is sketched in Fig. 11.119.

Finally, the extended range R/S plot has been evaluated (Fig. 11.118). We clearly see that its character is not monofractal. The multi-persistent properties are well reflected with the two crossovers ($cross^1$ and $cross^2$) and three memory ranges described by the respective Hurst exponents (H^1, H^2 and H^3). The analog plot for the disturbed simulation is sketched in Fig. 11.120.

All the evaluated indexes for both undisturbed (nonDist) and disturbed (Dist) loop simulations are summarized in Table 11.9. It is also interested to see what tuning, i.e. controller parameters have been pointed out by each method. Table 11.10 summarizes these results. The table consists only of these measures that can detect the solution, i.e. point out the single result as the extreme index value (minimum or maximum). We see that none of the indexes is able to point out the solution close to the one identified with the applied optimization method.

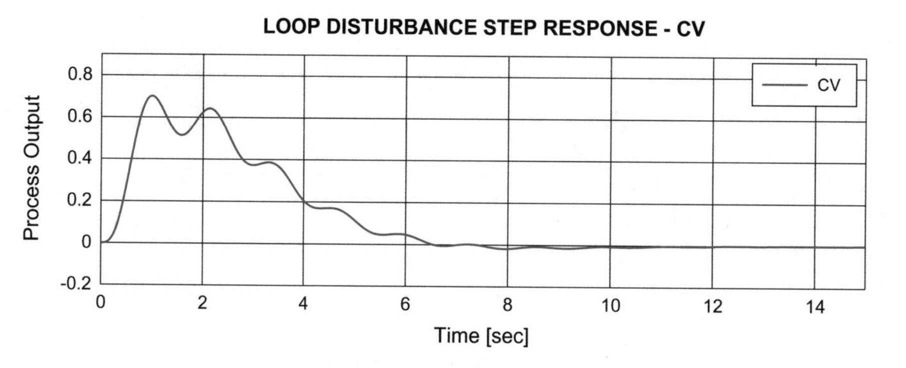

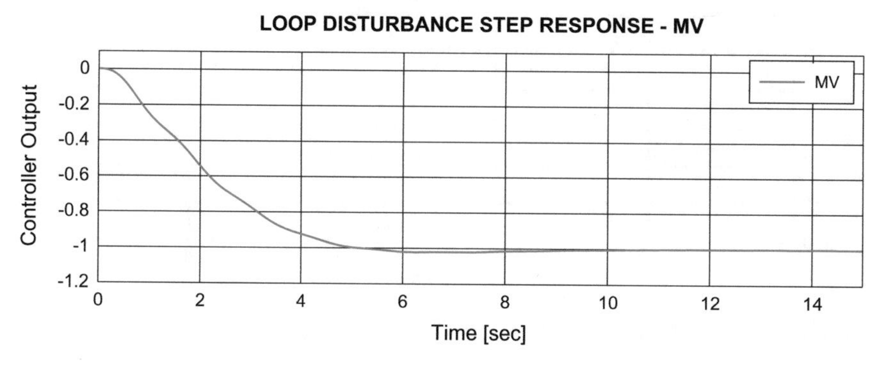

Fig. 11.115 Loop disturbance step response for $G_5(s)$

Table 11.9 Indexes for the optimal $G_5\,(s)$ loop controller

	T_s	$\kappa[\%]$	DR	II	AI	OI	RI	MSE	IAE	QE
nonDist	8.860	2.200	0.000	0.120	0.978	0.021	0.191	104.011	5.284	149.941
Dist	–	–	–	–	–	–	–	105.120	5.637	151.637
	σ	γ^C	α	γ^L	b	σ^{rob}				
nonDist	10.198	0.861	0.433	0.618	5.289	1.381				
Dist	10.253	1.796	0.729	1.658	5.638	2.855				
	H^0	H^1	H^2	H^3	cr^1	cr^2	H^{RE}	H^{DE}		
nonDist	0.931	0.592	0.983	0.564	0.25	9.00	45.44	−148350		
Dist	0.837	0.606	0.913	0.367	0.06	13.2	46.19	−145984		

Table 11.10 Controller tuning pointed out by the selected measures for $G_5\,(s)$—the highlighted solutions are close to the optimal one

		T_s	$\kappa[\%]$	DR	MSE	IAE	QE	σ	γ^C	γ^L	b	σ^{rob}	H^{RE}	H^{DE}
nonDist	k_p	0.38	n/a	n/a	0.74	0.64	0.82	0.74	0.44	0.38	0.64	0.44	0.02	0.02
	T_i	0.70	n/a	n/a	0.40	0.40	0.90	0.40	0.60	0.60	0.30	0.40	2.00	2.00
Dist	k_p	n/a	n/a	n/a	0.72	0.64	0.74	0.72	0.84	0.50	0.64	0.50	0.12	0.12
	T_i	n/a	n/a	n/a	0.40	0.30	0.60	0.40	2.00	0.50	0.30	0.50	2.00	2.00

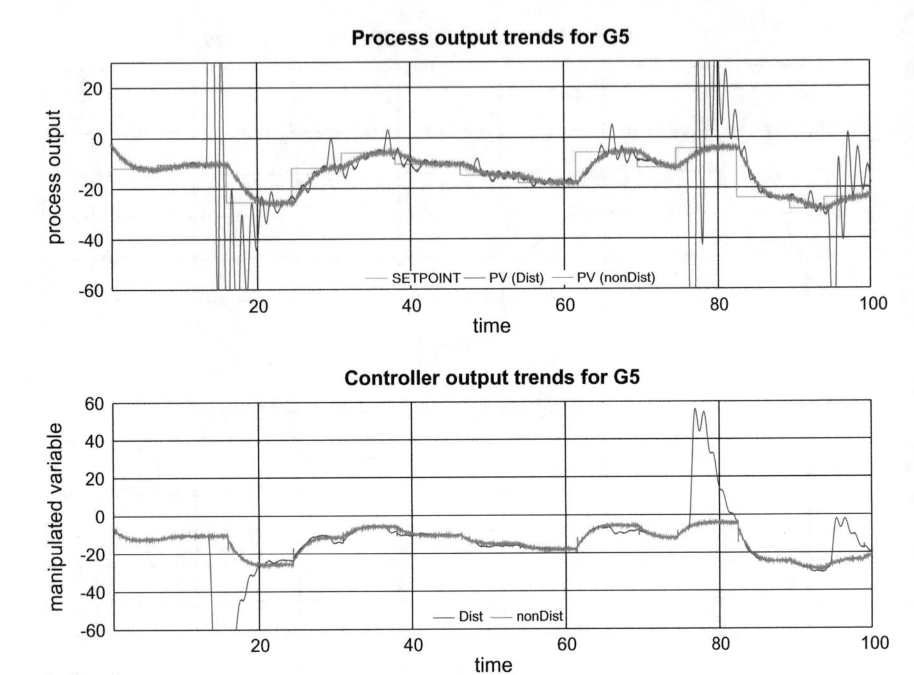

Fig. 11.116 Loop time trends of first 100 s for $G_5(s)$

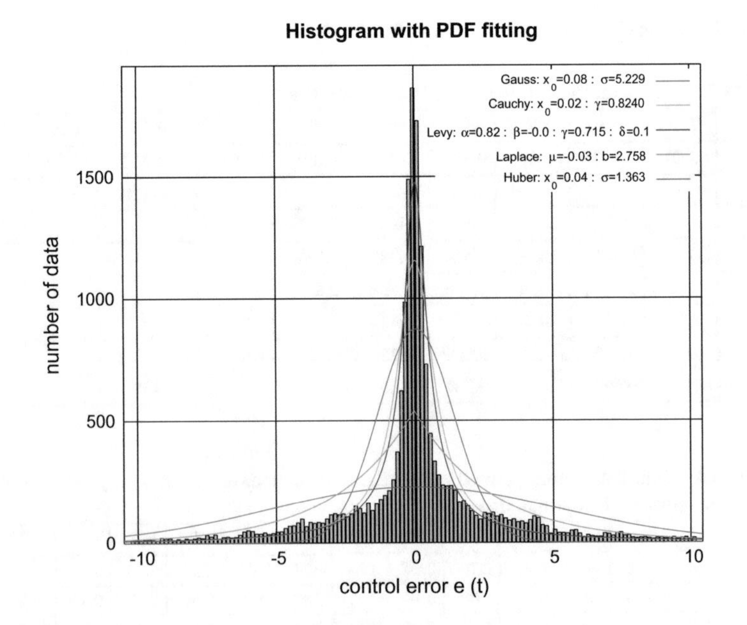

Fig. 11.117 Control error histogram with the fitted different PDFs for $G_5(s)$

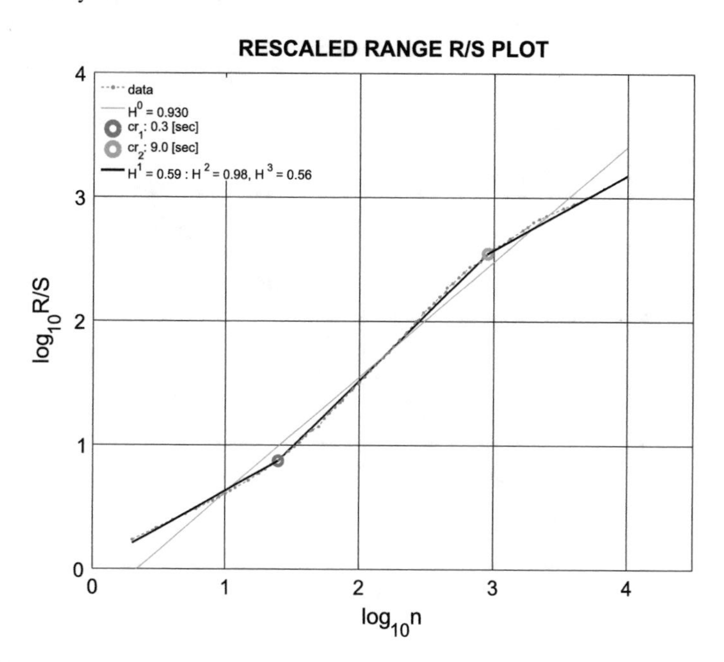

Fig. 11.118 Extended range R/S plot of the control error for $G_5(s)$

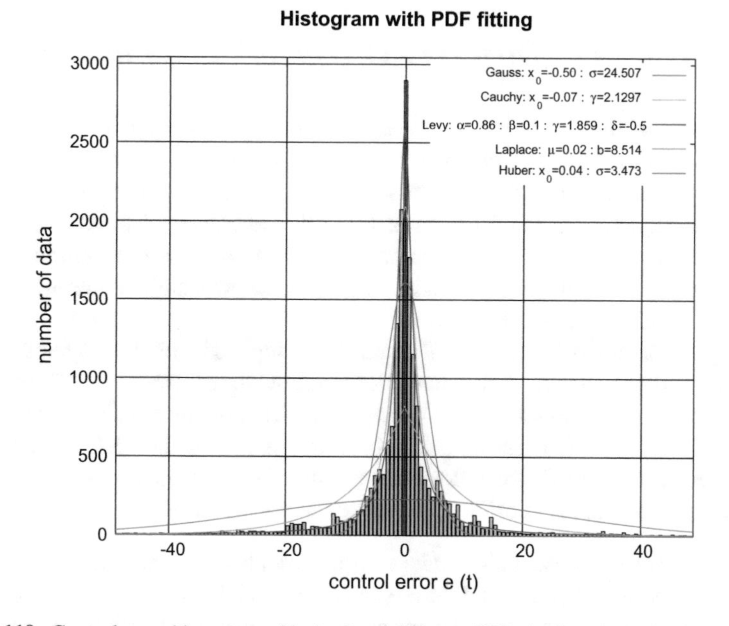

Fig. 11.119 Control error histogram with the fitted different PDFs in disturbed case for $G_5(s)$

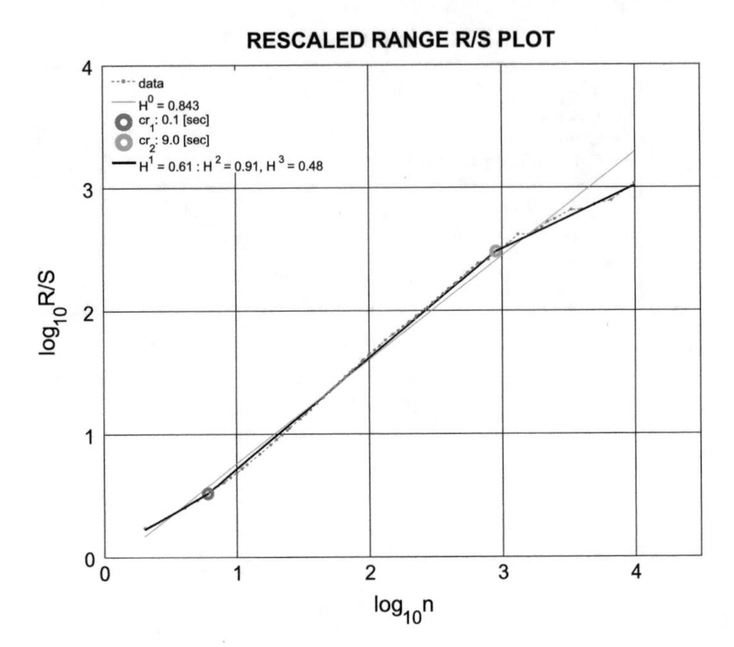

Fig. 11.120 Extended range R/S plot for the control error in disturbed case for $G_5(s)$

11.5.2 Control Performance Detection Using Step Indexes

The analysis of the performance assessment potential for the considered KPIs is presented in form of the control quality surfaces. The drawings are prepared according to the same common scheme. The plots are in form of the two dimensional surfaces presenting each KPI as the function of two PID parameters: k_p and T_i. The third parameter, $T_d = 0.08$ is kept constant at its optimal value, so the surface is a 2D cross-section. Due to the lack of place other cross-sections are not presented as they do not bring additional value to the analysis.

The position of the found optimal controller is depicted with the magenta circle •. There are two variants for the diagram interpretation: the best index is at its minimum value or there is optimal set of the indexes. In the first case the red diamond ♦ depicts minimum index position. The black bold contour line for the best KPI values is used in the second case. Step based indexes are evaluated without the disturbance. Review of the drawings enables to bring up the following observations:

- Four the most representative indexes are presented. The surfaces of the basic settling time (Fig. 11.121) and the overshoot (Fig. 11.122) seems to work fine. The only effect that we see is that the optimal solution identified with the optimization algorithm has quite significant overshoot, especially in comparison with all the previous cases. It might be one of the reasons that none of the detected tunings is close enough to the optimal one.

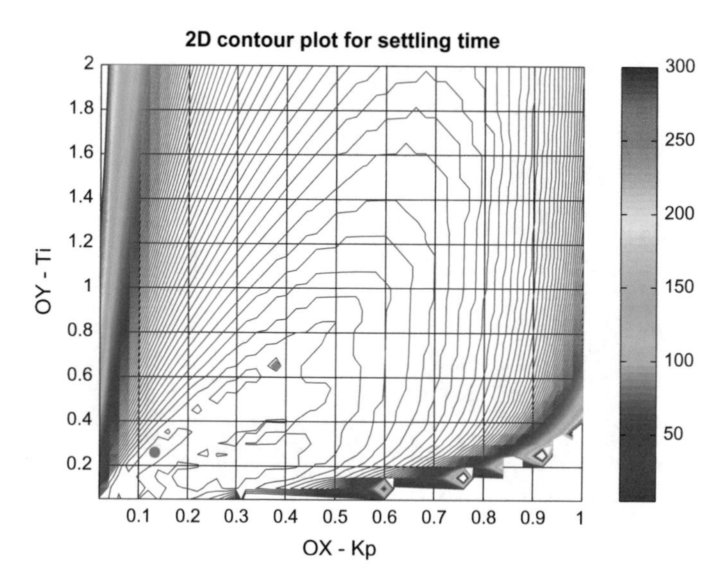

Fig. 11.121 Control quality surface for $G_5(s)$: T_s

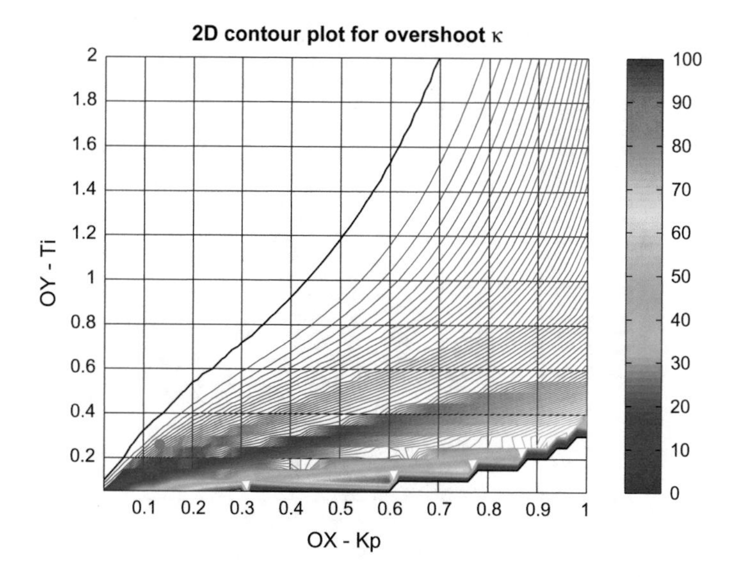

Fig. 11.122 Control quality surface for $G_5(s)$: κ

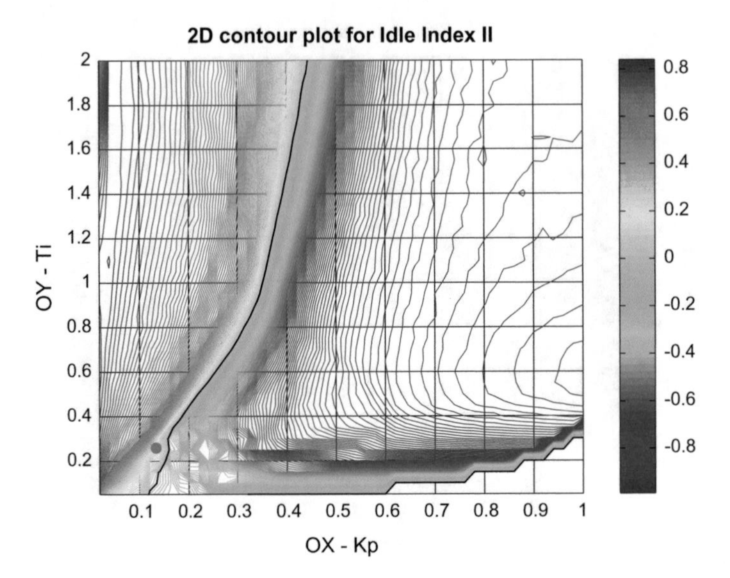

Fig. 11.123 Control quality surface for $G_5(s)$: II (the best value $II = 0$)

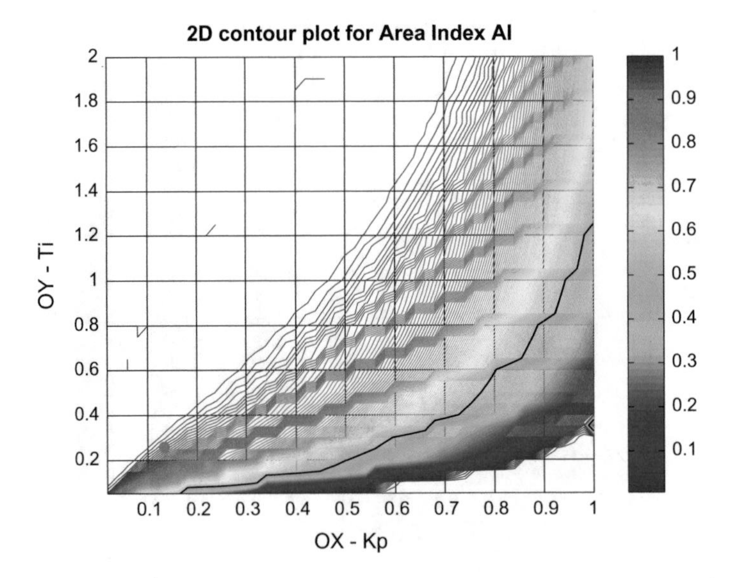

Fig. 11.124 Control quality surface for $G_5(s)$: AI (the best value $AI = 0.5$)

- Optimal controller lies relatively close to the contour line depicting ideal II value (Fig. 11.123). It confirms that II index seems to be a reliable one.
- The character of the surfaces obtained for the AI and OI is very similar. Ideal literature value of the Area Index $0.3 < AI < 0.7$ is relatively distant from the optimal controller (Fig. 11.124) and reflects rather sluggish control. Similar situation is visible for the Output Index (not presented but almost the same in character as AI), when optimal controller $OI \approx 0.0$ is distant from the literature $OI < 0.35$.
- R-Index surface is not smooth and very scattered. Decay Ratio surface is also not evident, so the respective plots are not included.

11.5.3 Control Performance Detection for Undisturbed Loop

The simulation of the loop using predefined setpoint profile is realized in two forms: undisturbed and disturbed one. Thus for such experiment the integral, statistical, fractal and entropy indexes are evaluated and analyzed. Diagrams sketched below present surfaces for the undisturbed simulation mode.

Review of the above undisturbed surfaces brings about the following observations:

- Next, sketched plots present representative selection if the surfaces. First of all, they are all smooth. Secondly, the detected results are not in the close proximity of the optimally identified tuning. Gaussian standard deviation σ is very close to the MSE (Fig. 11.126), while Laplace scale factor b (Fig. 11.129) is very close to

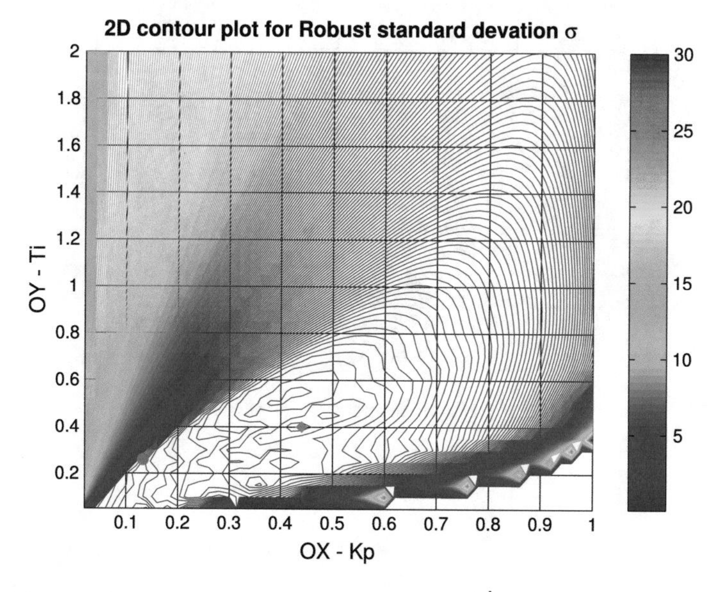

Fig. 11.125 Control quality surface for undisturbed $G_5(s)$: σ^{rob}

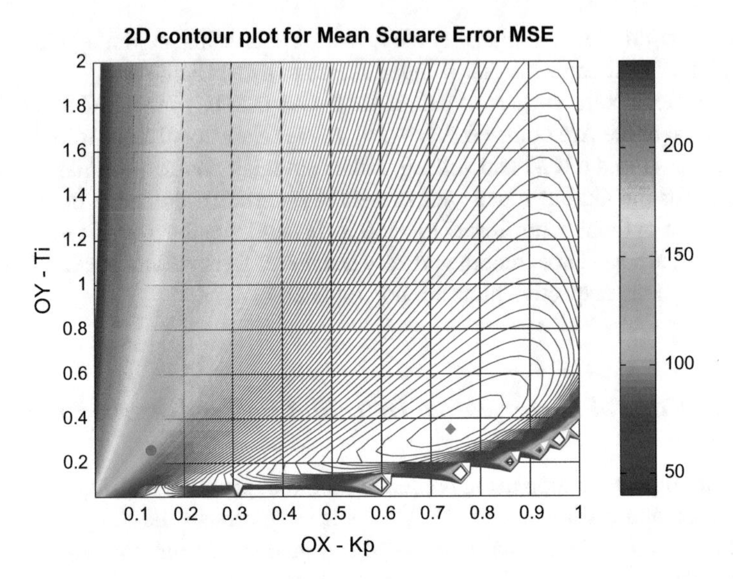

Fig. 11.126 Control quality surface for undisturbed $G_5(s)$: MSE

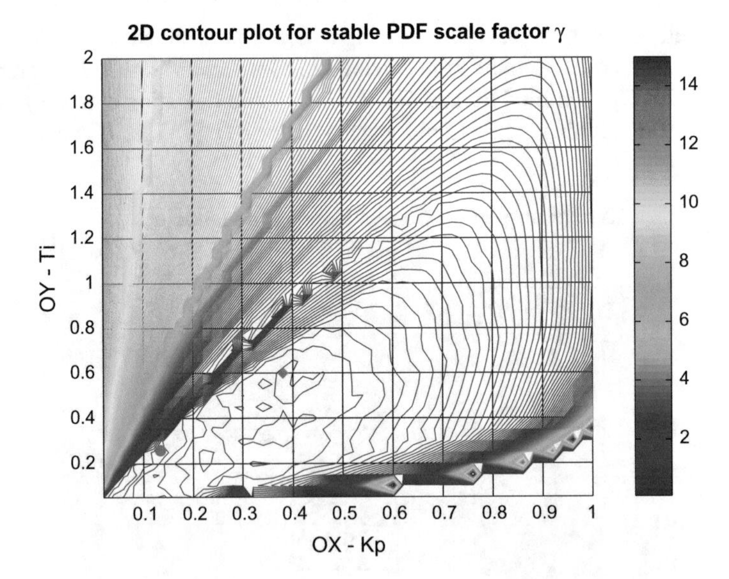

Fig. 11.127 Control quality surface for undisturbed $G_5(s)$: stable γ

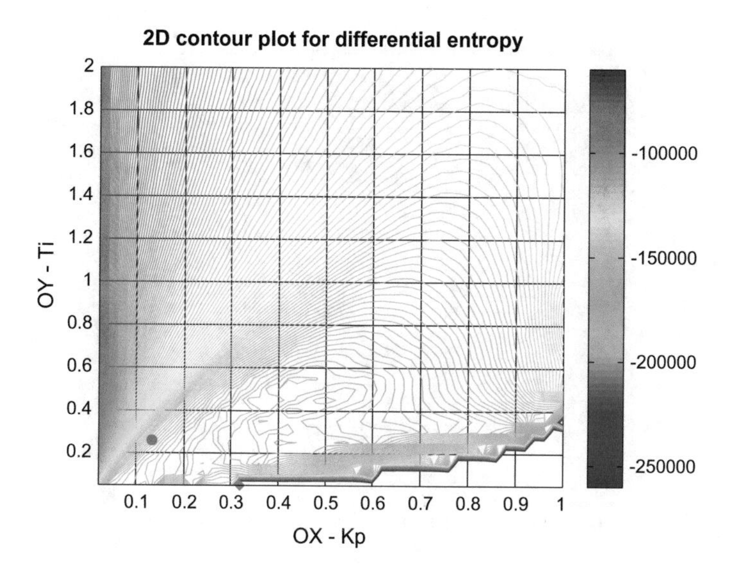

Fig. 11.128 Control quality surface for undisturbed $G_5(s)$: H^{DE}

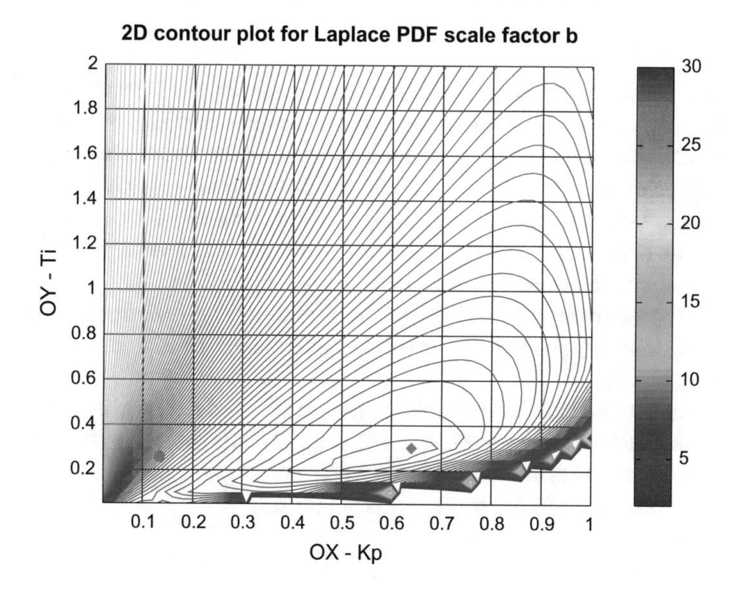

Fig. 11.129 Control quality surface for undisturbed $G_5(s)$: Laplace scaling b

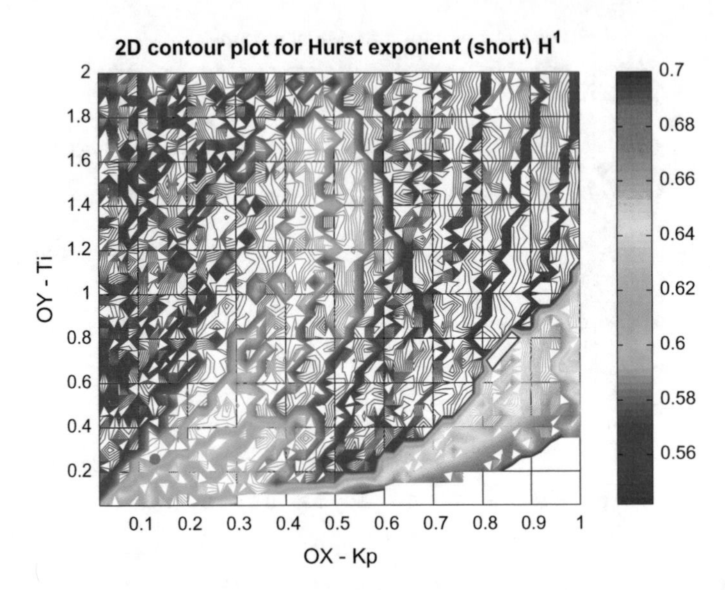

Fig. 11.130 Control quality surface for undisturbed $G_5(s)$: H^1 (the best value $H^1 = 0.5$)

the IAE. These shapes are almost identical. However, the detected results differ significantly. Square error and σ tend to the sluggish and non-overshoot control highly disliking aggressiveness in tuning. On the other hand absolute error and Laplace scale factor b are more realistic and closer to the expectations.

- Robust standard deviation (Fig. 11.125) performs similarly to the Cauchy and α-stable scale (Fig. 11.127) factors. The tuning seems to be relatively different from the optimal one, however obtained performance is quite similar.
- Both entropies share very similar shape. The detection (minimum value) tends towards extremely sluggish control and thus is rather useless for the assessment task (Fig. 11.128).
- All the Hurst exponents share similar shape of the surfaces, however on different levels. The short memory exponent H^1 (Fig. 11.130) should be the most informative. The value for the optimal control is not as expected, i.e. $H \approx 0.5$. The controller, according to the literature, is sluggish.

11.5.4 Control Performance Detection for Disturbed Loop

Now, the control loop is disturbed and the same set of the analyzes is performed. The way of the drawings preparation and surfaces analysis is exactly the same. Review of the attached disturbed control drawings brings up the following observations:

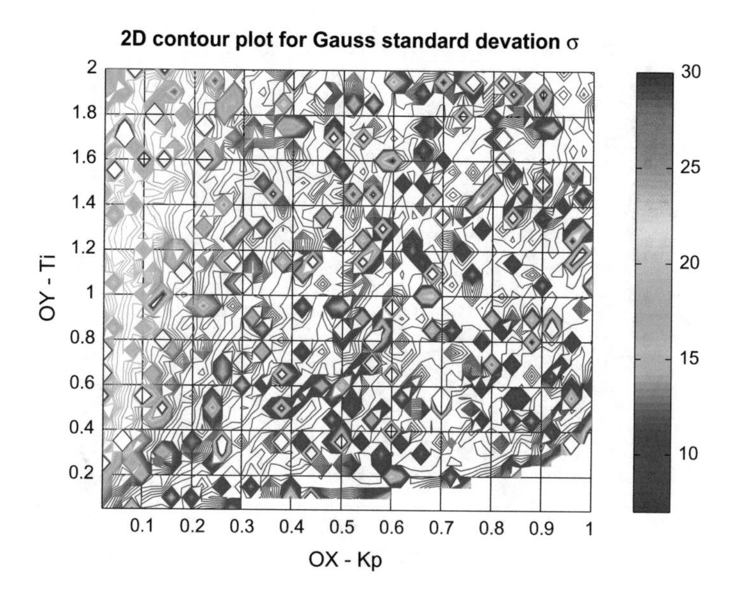

Fig. 11.131 Control quality surface for disturbed G_5(s): Gauss σ

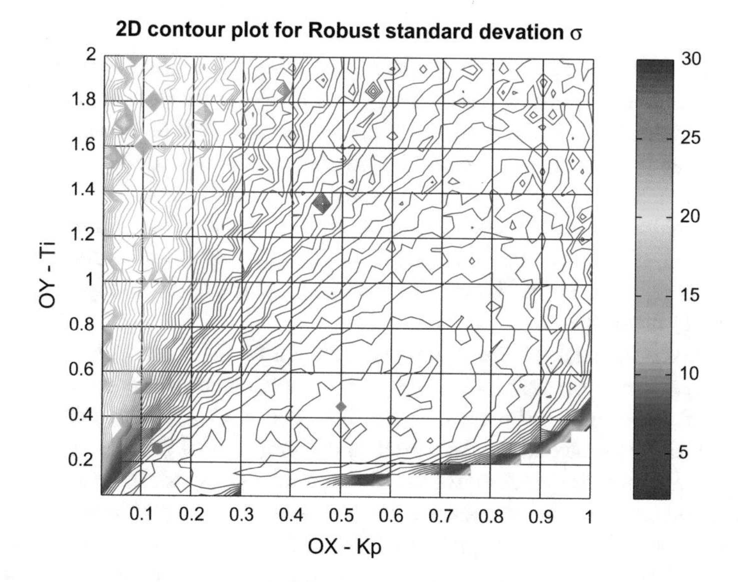

Fig. 11.132 Control quality surface for disturbed G_5(s): σ^{rob}

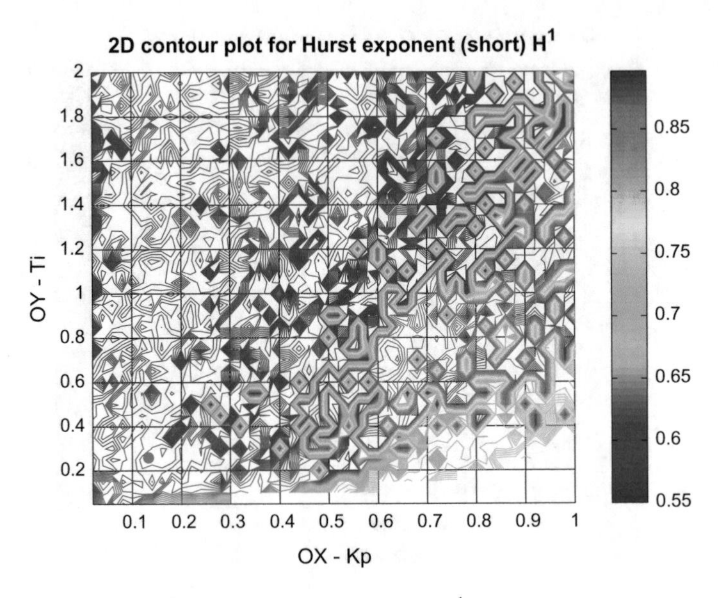

Fig. 11.133 Control quality surface for disturbed $G_5(s)$: H^1

- Similarly to the previous cases we observe the same properties. There are two types of the evaluated control surfaces. Part of them seems to be strongly affected by the disturbances showing "ragged" surface landscape, while the shape is smooth (unaffected by the disturbance *shadowing* effect) in the second group of the diagrams. The group of the disturbance sensitive measures consists of MSE, IAE, QE, normal standard deviation, Laplace scale factor and both entropies. On the other hand, the robust standard deviation, Cauchy and α-stable distribution factors together with all the Hurst exponents are robust against loop disturbances, especially the **outliers**.
- One surface for each of the groups is sketched as the representative. Gaussian standard deviation (Fig. 11.131) presents the existence of the *shadowing* effect for the above-mentioned measures. On the other hand, robust standard deviation (Fig. 11.132) is sketched for the visualization purposes of the robust surfaces.
- We also see that the detection for the disturbance robust indexes is slightly biased in comparison with the undisturbed case, however looks acceptable.
- Additionally, two Hurst exponents are presented, i.e. the short memory H^1 (Fig. 11.133) and the long memory H^3 (Fig. 11.134). Although the surfaces are smoother than for the classical measures the detection is no evident and indecisive. It is interesting to notice that the first one exhibits persistent (sluggish control) behavior, the longer memory one is opposite having anti-persistent properties.
- It is well seen that the introduction of the **outliers** (fat-tailed disturbance properties) into the control loop generates *shadowing* effect that makes the most of the CPA approaches questionable.

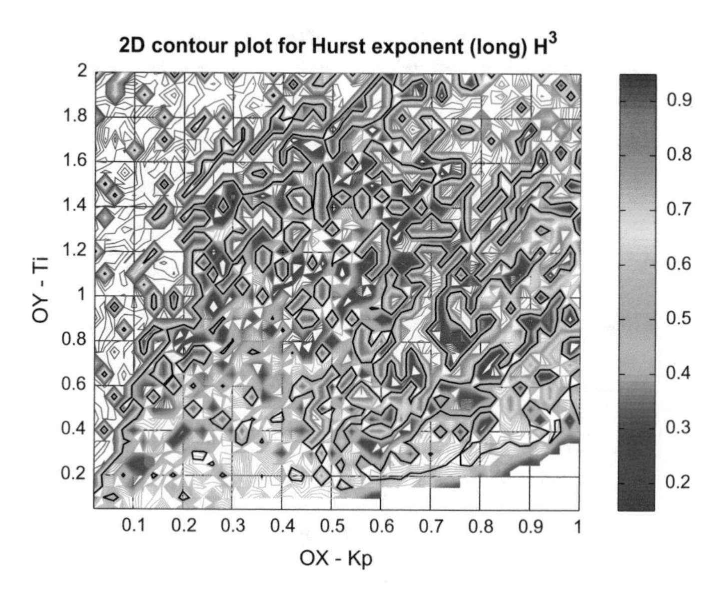

Fig. 11.134 Control quality surface for disturbed $G_5(s)$: H^3

11.6 Fast and Slow Modes Plant

The analysis starts with the evaluation of the optimal PID controller and is followed by the analysis of the tuning surfaces for all considered CPA indexes.

11.6.1 Optimal Controller Performance

The optimal controller for the $G_6(s)$ transfer function has been designed according to the predefined rules. The evaluated tuning parameters are presented below:

- gain: $k_p = 0.04$
- integration: $T_i = 20.03$
- derivation: $T_d = 5.03$

The graphical representation of the loop setpoint step response is sketched in Fig. 11.135, while Fig. 11.136 shows the disturbance step response. We see that the obtained tuning delivers the response with almost invisible overshoot. The controller output is smooth without any oscillations.

Next, the loop simulation has been run. The time trends (sample of first 100 s) for the selected loop profiles for setpoint, CV and MV are sketched in Fig. 11.137. The figure shows comparison of both undisturbed and disturbed behavior.

Respective control error histogram is shown in Fig. 11.138. It is clearly visible that its shape is not Gaussian with the significant tails. The Histogram can be modeled the

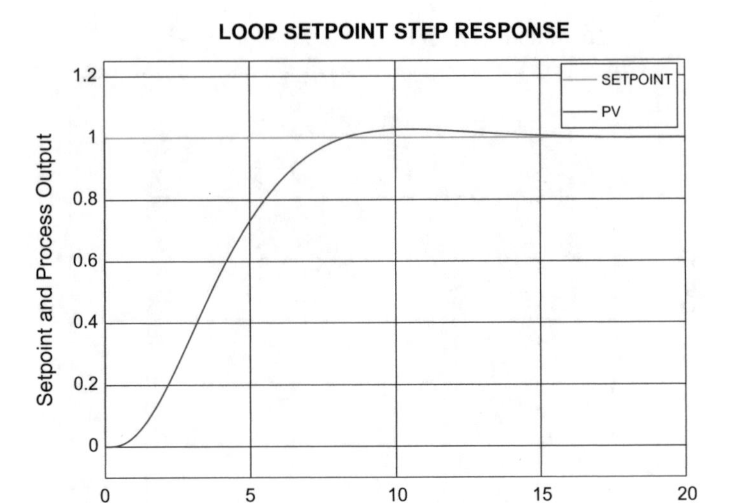

Fig. 11.135 Loop setpoint step response for $G_6(s)$

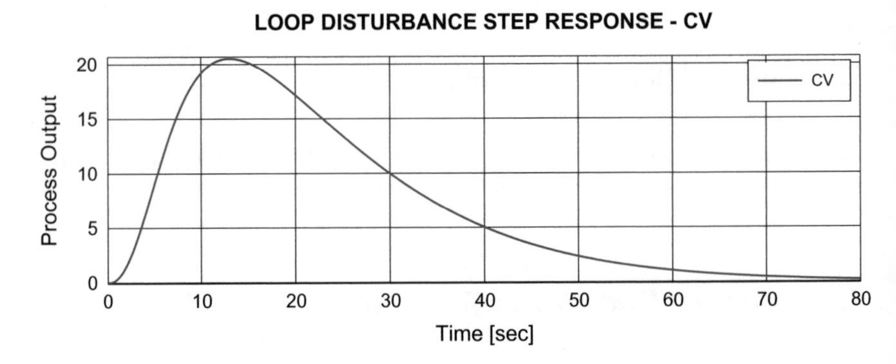

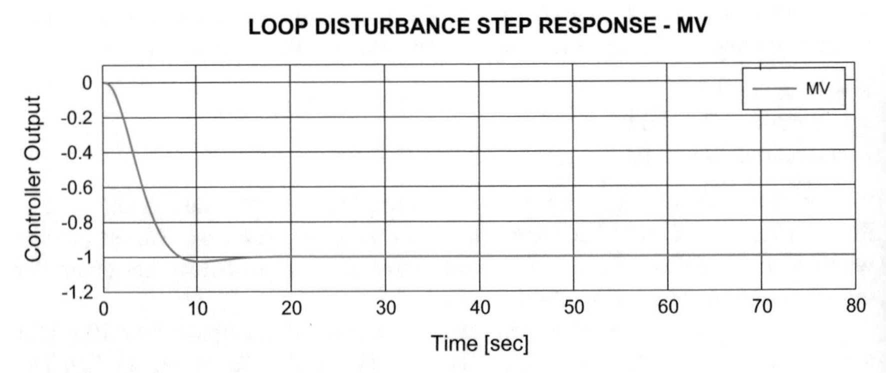

Fig. 11.136 Loop disturbance step response for $G_6(s)$

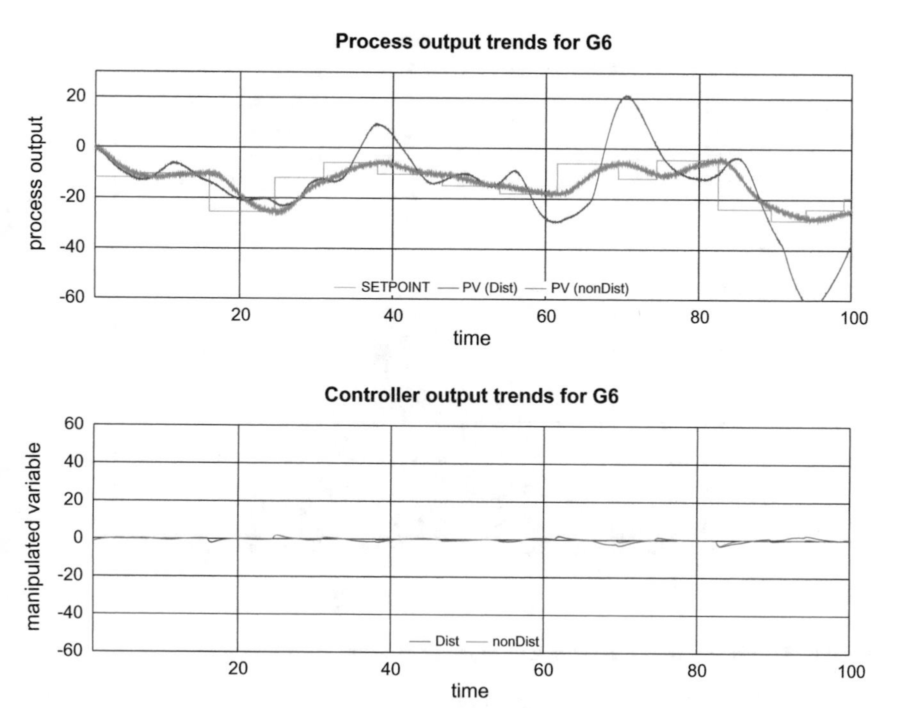

Fig. 11.137 Loop time trends in first 100 s for $G_6(s)$

most precisely by the α-stable distribution. The analog histogram for the disturbed simulation is sketched in Fig. 11.140.

Finally, the extended range R/S plot has been evaluated (Fig. 11.139). We clearly see that its character is not monofractal. The multi-persistent properties are well reflected with the two crossovers ($cross^1$ and $cross^2$) and three memory ranges described by the respective Hurst exponents (H^1, H^2 and H^3). The analog plot for the disturbed simulation is sketched in Fig. 11.141.

All the evaluated indexes for both undisturbed (nonDist) and disturbed (Dist) loop simulations are summarized in Table 11.11. It is also interested to see what tuning, i.e. controller parameters have been pointed out by each method. Table 11.12 summarizes these results. The table consists only of these measures that can detect the solution, i.e. point out the single result as the extreme index value (minimum or maximum). The settings highlighted show the results, which are close enough to the optimal solution, i.e. $k_p \in [0.03, 0.05]$ and $T_i \in [17.0, 25.0]$.

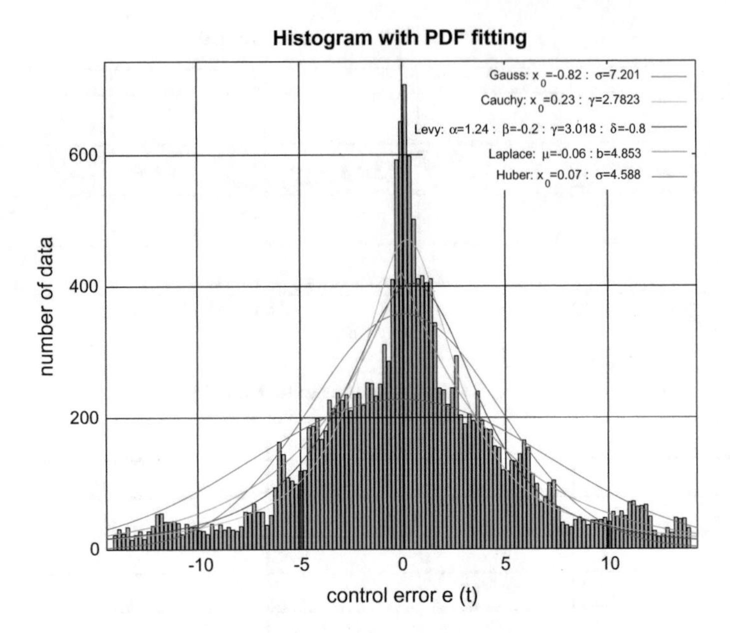

Fig. 11.138 Control error histogram with the fitted different PDFs for $G_6(s)$

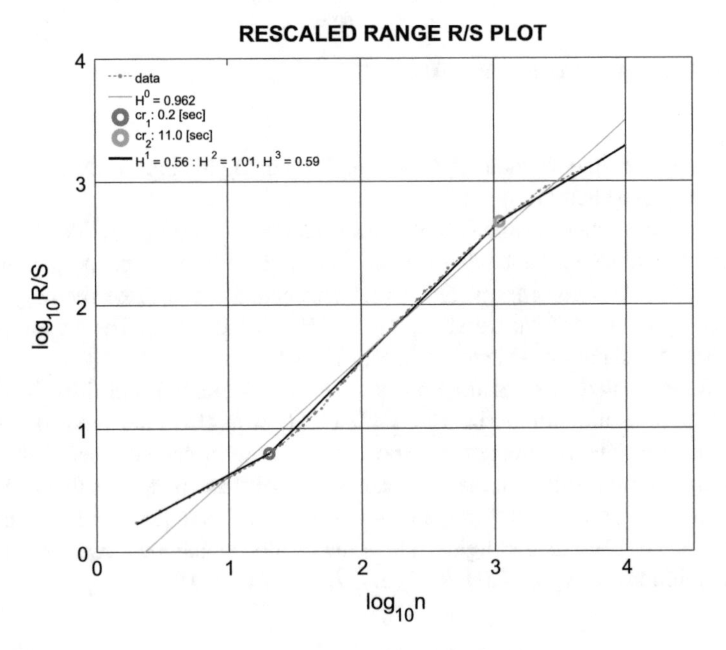

Fig. 11.139 Extended range R/S plot of the control error for $G_6(s)$

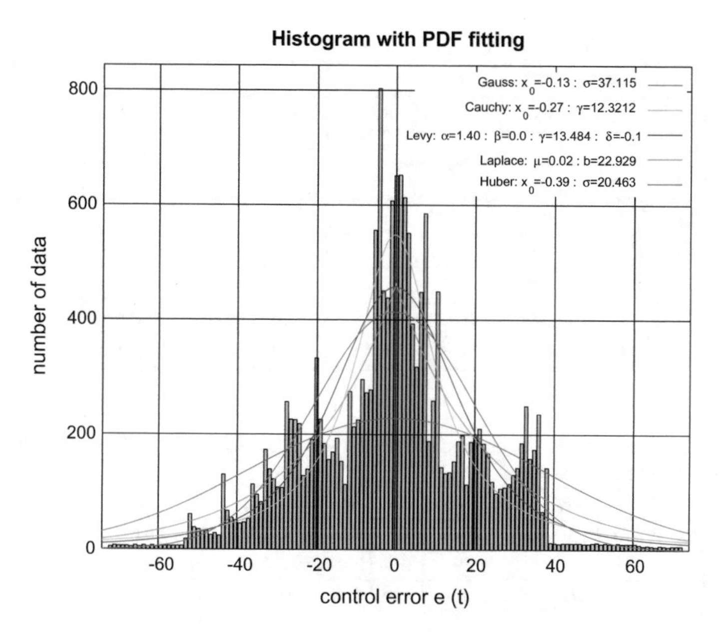

Fig. 11.140 Control error histogram with the fitted different PDFs in disturbed case for $G_6(s)$

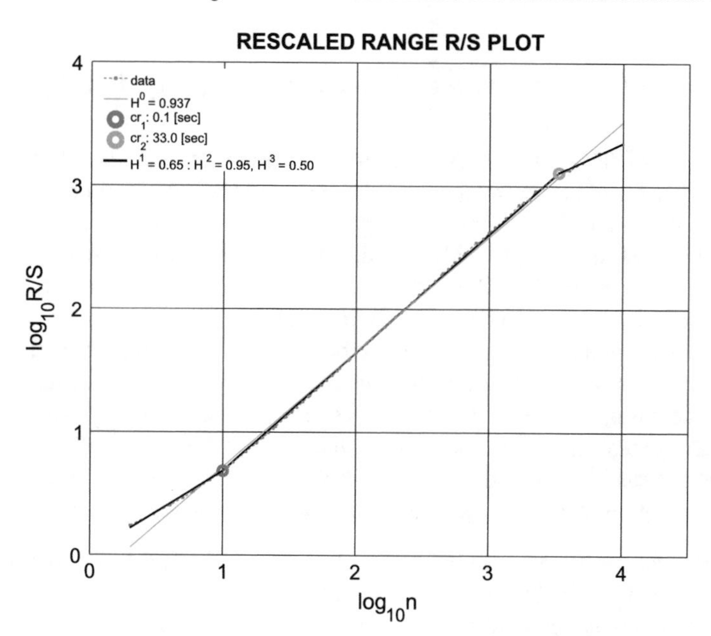

Fig. 11.141 Extended range R/S plot for the control error in disturbed case for $G_6(s)$

Table 11.11 Indexes for the optimal $G_6(s)$ loop controller

	T_s	$\kappa[\%]$	DR	II	AI	OI	RI	MSE	IAE	QE
nonDist	14.11	2.763	0.00	−0.732	1.0	0.000	NaN	195.601	8.976	196
Dist	–	–	–	–	–	–	–	2063.9	22.179	2065
	σ	γ^C	α	γ^L	b	σ^{rob}				
nonDist	13.985	4.005	0.447	2.820	8.992	6.900				
Dist	45.430	9.333	0.950	8.977	22.2	15.314				
	H^0	H^1	H^2	H^3	cr^1	cr^2	H^{RE}	H^{DE}		
nonDist	0.943	0.572	0.987	0.529	0.15	13.20	46.63	−129158		
Dist	0.941	0.671	0.626	0.956	0.10	49.50	46.66	−124044		

Table 11.12 Controller tuning pointed out by the selected measures for $G_6(s)$—the highlighted solutions are close to the optimal one

		T_s	$\kappa[\%]$	DR	MSE	IAE	QE	σ	γ^C	γ^L	b	σ^{rob}	H^{RE}	H^{DE}
nonDist	k_p	**0.03**	n/a	n/a	0.08	0.07	0.08	0.08	0.05	**0.04**	0.07	0.05	0.03	0.03
	T_i	**19.5**	n/a	n/a	59.5	59.5	59.5	59.5	28.5	**24.5**	59.5	30.5	58.5	58.5
Dist	k_p	n/a	n/a	n/a	0.08	0.09	0.08	0.08	0.18	0.04	0.09	0.08	0.03	0.03
	T_i	n/a	n/a	n/a	23.5	50.5	23.5	23.5	59.5	42.5	50.5	28.5	23.5	23.5

11.6.2 Control Performance Detection Using Step Indexes

The analysis of the performance assessment potential for the considered KPIs is presented in form of the control quality surfaces. The drawings are prepared according to the same scheme. The plots are in form of two dimensional surfaces presenting each KPI as the function of two PID coefficients: k_p and T_i. The third one, $T_d = 5.03$ is constant at its optimal, so the surface is a 2D cross-section. Other cross-sections are not presented as they do not bring additional value.

The position of the optimal controller is depicted with the magenta circle •. There are two variants for the diagram interpretation: the best index is at its minimum value or there is optimal set of the indexes. In the first case the red diamond ♦ depicts minimum index position. The black bold contour line for the best KPI values is used in the second case. Step based indexes are evaluated without the disturbance. Review of the drawings enables to bring up the following observations:

- Four the most representative indexes are presented. The surfaces of the basic settling time (Fig. 11.142) and the overshoot (Fig. 11.143) seems to work fine. Optimal solution obtained with the optimization has significant overshoot, especially in comparison with the previous cases. It might cause misplaced detection.
- Optimal controller lies close to the contour line depicting ideal II value (Fig. 11.144). It confirms reliability of the II. The character of the AI and OI surfaces is very similar. Ideal literature value $0.3 < AI < 0.7$ is relatively distant from the optimal controller (Fig. 11.145) and reflects rather sluggish control. Sim-

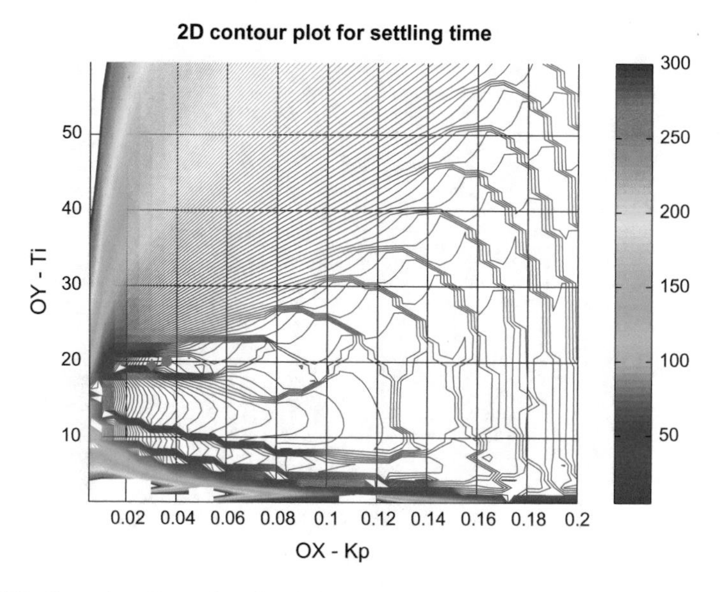

Fig. 11.142 Control quality surface for $G_6(s)$: T_s

ilar situation is visible for the Output Index, when optimal controller $OI \approx 0.0$ is distant from the literature $OI < 0.35$.
- R-Index and Decay Ratio are indecisive and very scattered and the respective plots are not included.

11.6.3 Control Performance Detection for Undisturbed Loop

The simulation of the loop using predefined setpoint profile is realized in two forms: undisturbed and disturbed one. Thus, the integral, statistical, fractal and entropy indexes are evaluated and analyzed. Diagrams sketched below present surfaces for the undisturbed simulation mode.

Review of the above undisturbed surfaces allows to bring about the following observations:

- The evaluation of the PID tuning for the fast and slow dynamics processes is a difficult task. It has been already reflected in the step test analysis. Similar observation can be concluded for the time series data analysis.
- Next, sketched plots present representative selection if the surfaces. First of all, they are all smooth. Secondly, the detected results are not close to the optimally identified tuning. Gaussian standard deviation σ (Fig. 11.146) is very close to the MSE (Fig. 11.148), IAE and Laplace scale factor. These shapes are almost

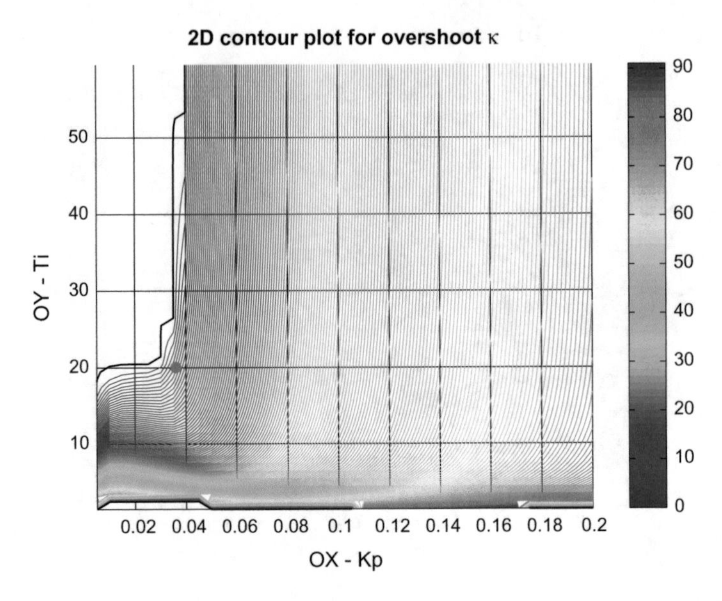

Fig. 11.143 Control quality surface for G_6(s): κ

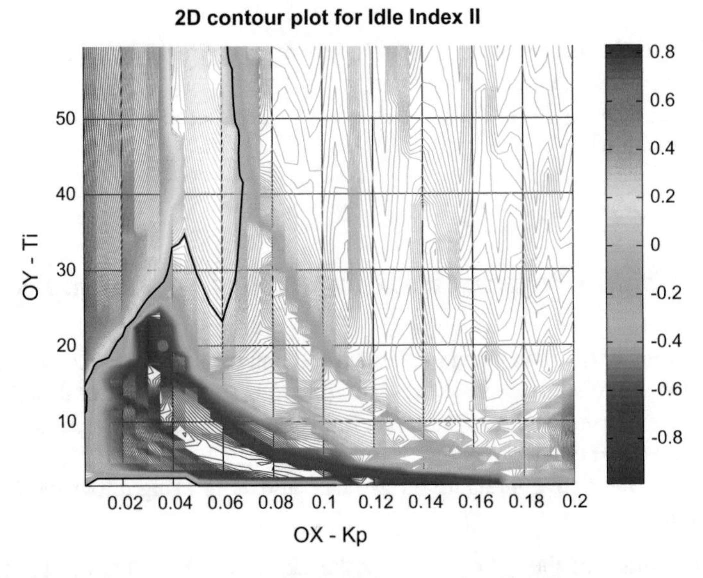

Fig. 11.144 Control quality surface for G_6(s): II (the best value $II = 0$)

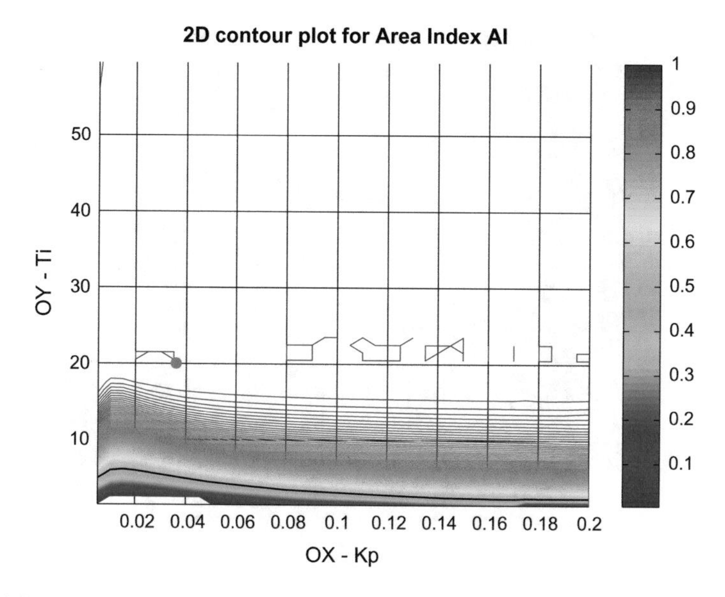

Fig. 11.145 Control quality surface for G_6(s): AI (the best value $AI = 0.5$)

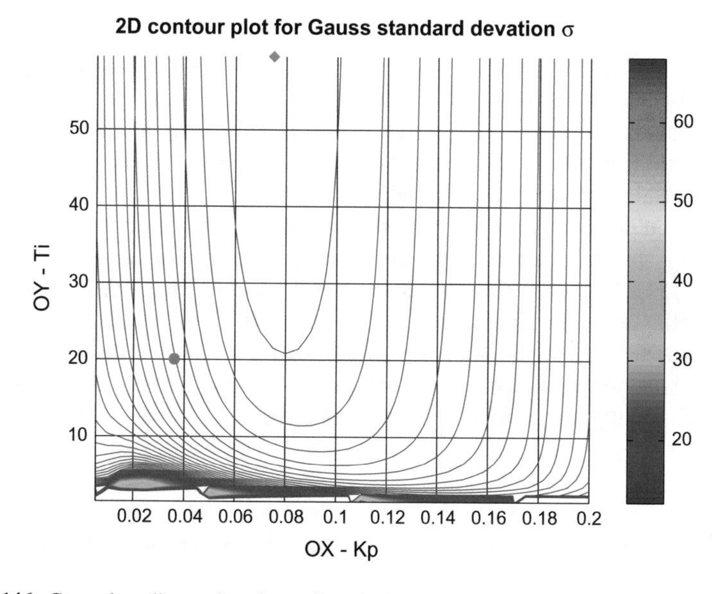

Fig. 11.146 Control quality surface for undisturbed G_6(s): Gauss σ

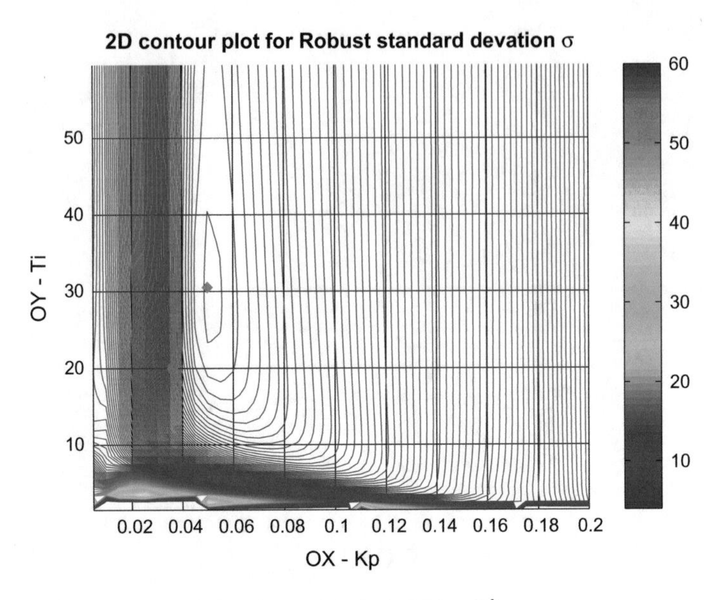

Fig. 11.147 Control quality surface for undisturbed $G_6(s)$: σ^{rob}

identical and the detected results differ significantly. They tend to the sluggish and non-overshoot control highly disliking aggressiveness in tuning.

- Robust standard deviation (Fig. 11.147) performs similarly to the Cauchy and α-stable scale (Fig. 11.149) factors. The tuning is relatively close the optimal one and the obtained performance is quite similar.
- Both entropies result with the very similar shapes. The detection tends towards extremely sluggish control and thus is rather useless for the assessment task (Fig. 11.151). It is interesting to observe that the shape of the entropies surfaces is very similar to the long memory Hurst exponent H^3 (Fig. 11.150).
- Hurst exponents share similar shape of the surfaces, however on different levels. The short memory exponents H^0 and H^1 tend towards sluggish control with $H^1 \rightarrow 1.0$. Additionally, the surface is flat in almost whole concerned domain.

11.6.4 Control Performance Detection for Disturbed Loop

Next, the control loop is disturbed and the same set of the analyzes is performed. The way of the drawings preparation and surfaces analysis is exactly the same. Review of the attached disturbed control drawings brings up the following observations:

- Similarly to the previous cases we observe the same properties. However, the process *difficulty* magnifies the effect. The disturbance *shadowing* is very significant and affects almost all measures, i.e. MSE (Fig. 11.152), IAE, QE, normal standard deviation, Laplace scale factor and both entropies. What is the most significant, it affects the long memory H^3 (Fig. 11.155) as well. In these cases the detection is clearly unreliable.

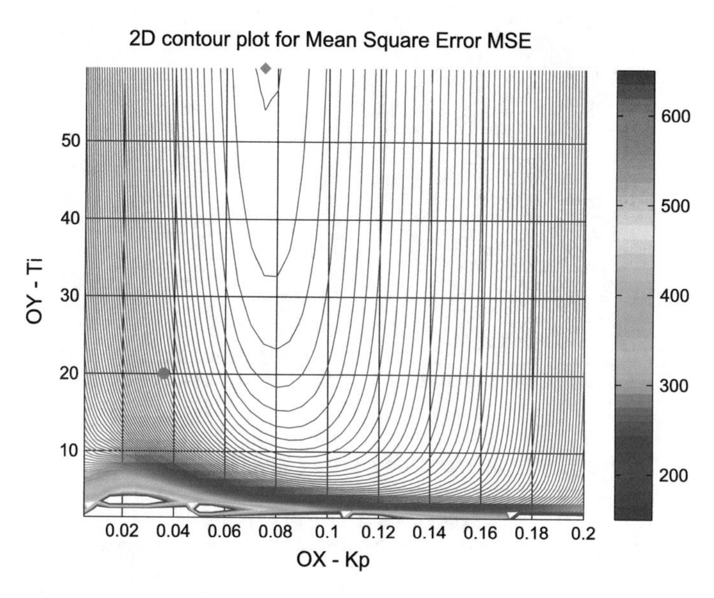

Fig. 11.148 Control quality surface for undisturbed $G_6(s)$: MSE

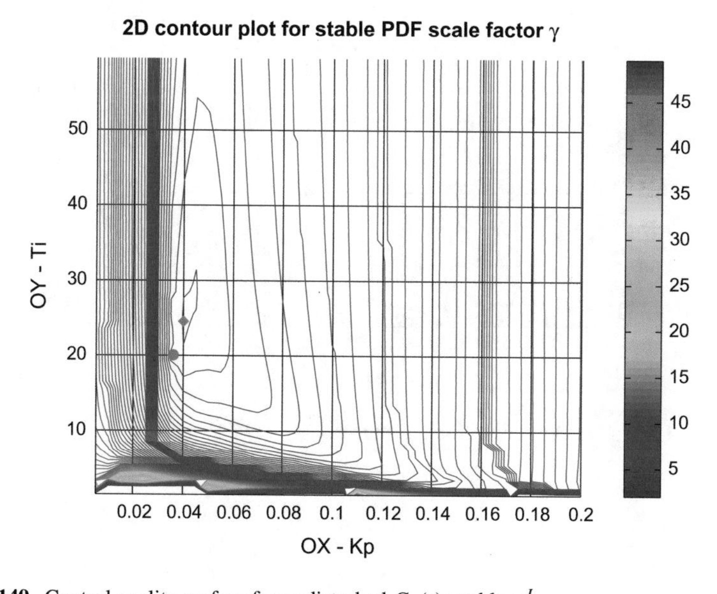

Fig. 11.149 Control quality surface for undisturbed $G_6(s)$: stable γ^L

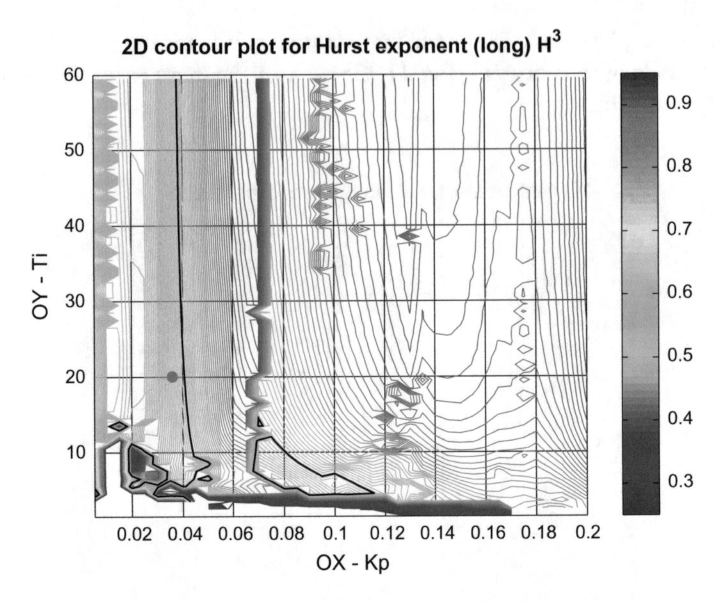

Fig. 11.150 Control quality surface for undisturbed $G_6(s)$: H^3

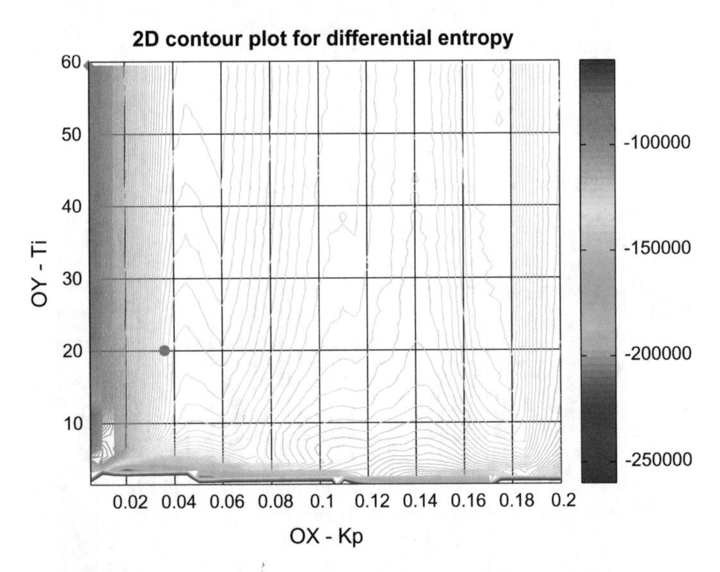

Fig. 11.151 Control quality surface for undisturbed $G_6(s)$: H^{DE}

- On the other hand, the robust standard deviation, Cauchy and α-stable distribution scale (Fig. 11.153) factors together with the Hurst exponents H^0, H^1 and H^2 are robust against loop disturbances, especially the **outliers**. The short memory Hurst exponent H^1 (Fig. 11.154) is the smoothest one, being the most robust against dis-

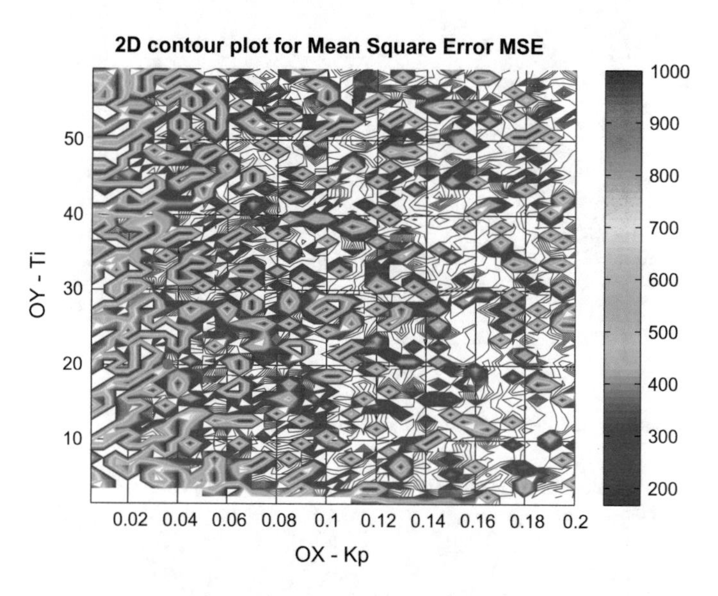

Fig. 11.152 Control quality surface for disturbed $G_6(s)$: MSE

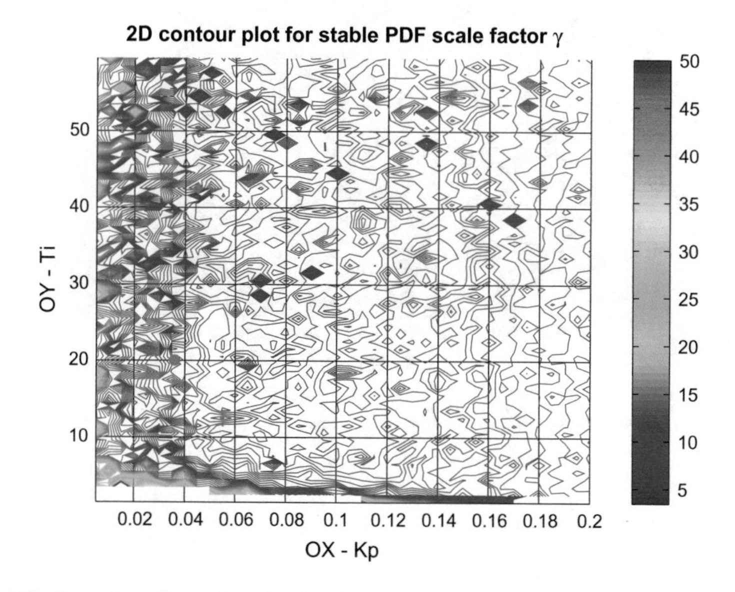

Fig. 11.153 Control quality surface for disturbed $G_6(s)$: stable γ

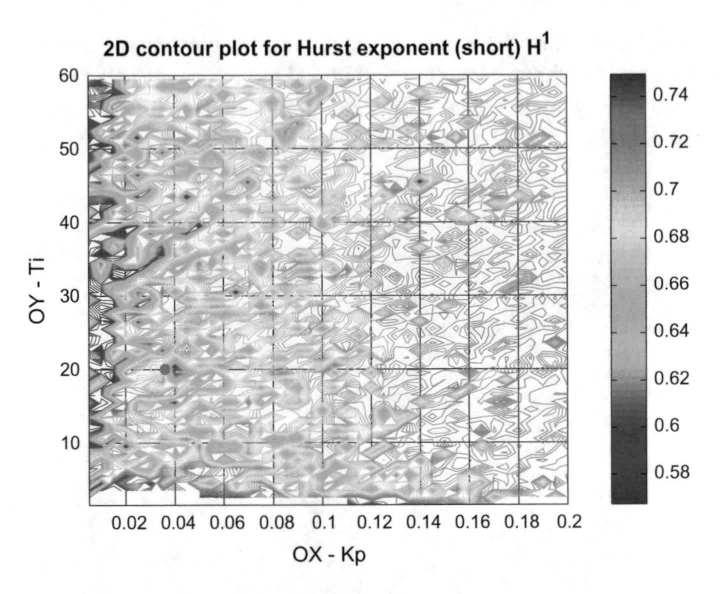

Fig. 11.154 Control quality surface for disturbed $G_6(s)$: H^1

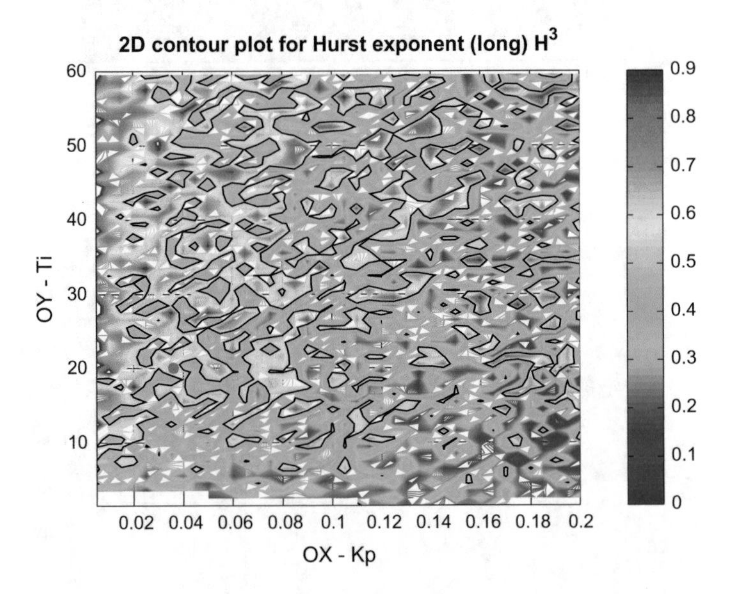

Fig. 11.155 Control quality surface for disturbed $G_6(s)$: H^3

turbances. The detection in all cases is not very easy, with only α-stable distribution scale delivering some evident result.

- It is well seen that the introduction of the **outliers** (fat-tailed disturbance properties) into the control loop generates *shadowing* effect that makes the most of the CPA approaches questionable.

11.7 Summary of the Results

Several simulation scenarios showing efficiency of the considered CPA measures have been executed, presented and commented. The analysis focused on the main aspects of successful assessment, i.e. proper detectability, consistency of the results, robustness to the possible process disturbances and finally clearness in the results interpretation. The analysis has covered the most popular control philosophy, which is used in process control, i.e. PID-based univariate control. Six different processes have been simulated in two scenarios, i.e. undisturbed and disturbed. The fat-tailed disturbances have been used to reflect industrial complexity.

The analysis has started with the step response based measures. These simulations have been analyzed in the undisturbed environment. Although, it looked simple not all the measures have behaved as assumed. Two basic ones, i.e. settling time T_{set} and the overshoot κ have exhibited exactly as expected. The settling time has been showing almost in all the cases predesigned tuning optimized for the low overshoot and the minimal time of settlement. This optimal setting has been also positioned very close to the zero-overshoot contour plot of the overshoot surfaces.

In contrary, the Decay Ratio and R-Index have been failing for all the cases in proper and reliable support to indicate control quality. The obtained surfaces have been very scattered, not intuitive and indecisive. Such a performance has been obtained for all the considered transfer function, what makes very questionable their practical industrial applicability as the CPA measure. Three other step response indexes (AI, OI and II) exhibit relatively in the similar way, though some differences related to different kind of the processes exist. It is well visible that Idle Index (II) delivers the most reliable results consistent with the settling time and the overshoot. Concluding above results summery the analysis of the control loop using the step response can be limited to the generally acceptable and widely used two main indexes, i.e. settling time T_{set} and the overshoot κ. Some additional information might be obtained with the use of the Idle Index as it may distinguish between aggressive and sluggish control within one index.

The assessment using normal operation data reveals impact of the disturbances on the achieved assessment quality. The influence is significant and non-negligible. It is well visible in all the simulated scenarios. It is expected and demanded that the reliable and robust assessment measure should exhibit in the explicit and conclusive way, which ought to result in smooth control surface clearly indicating loop quality and the optimal solution. Control surface smoothness should be the must in the undisturbed situations, as there is no hindering reason.

The assessment reliability and clearness should be also featured in the presence of the loop disturbances. This aspect has been investigated for all the considered process examples. The main observation is that for some measures (Gaussian standard deviation, MSE or entropies) the disturbances cause extremely ragged landscapes of the control surfaces. This effect has been previously noticed in the literature [6, 7] and has been called the *shadowing* effect. The fat-tailed loop disturbances (properly reflecting the signal properties common in process industry) cause significant distortions into the control surface. They introduce multiple sharp "valleys" and "hills" into the smooth plot causing the detection ambiguous, random and unrepeatable. Such highly and randomly ragged behavior *shadows* real underlying relationship of the smooth control surface.

Review of the obtained surfaces enables to cluster the measures into groups resembling in behavior and properties. This affinity manifests itself in the similar control surface shapes and similar sensitivity to the disturbances:

(1) Mean Square Error (MSE) and Gaussian standard deviation (σ),
(2) Integral Absolute Error (IAE) and Laplace distribution scaling factor (b),
(3) robust standard deviation σ^{rob} and scaling factors for Cauchy distribution and especially with the α-stable probabilistic density function,
(4) differential H^{DE} and rational H^{RE} entropy,
(5) all the persistence measures of the Hurst exponent H^0, H^1, H^2 and H^3.

The behavior may be explained. Statistical theory [12] enables to explain the similarities within first three groups. The classical and canonical approach to the regression estimator dates back to Gauss and Legendre (see [11]) and is associated with the minimization of the square error (11.1).

$$\min_{\hat{\theta}} \sum_{i=1}^{n} \varepsilon_i^2 \tag{11.1}$$

Afterwards, Gauss has introduced the normal PDF $N(x_0, \sigma)$ (4.1) as the error distribution, for which the least square estimator (3.2) is optimal. This relation causes similar properties of the MSE and normal standard deviation σ. The combination of Gaussian assumptions and mean square error has become a standard method in estimation and data analysis.

However, classical Gaussian regression mechanism has drawbacks. It is not robust against regression outliers, i.e. *single points deviating from the linear pattern followed by the majority of the data*. Thus the outlier is according to [9] '*an observation which deviates so much from other observations as to arouse suspicions that it was generated by a different mechanism*'. Such points can totally destroy the estimation.

The researchers have started to work on other approach, called robust regression and have tried to design other estimators being not affected by outliers. To measure the sensitivity of the regression mechanism against the outliers, the notion of the breakdown point has been proposed [12], i.e. the measure of how many outliers the regressor can cope with. It appears that the least square estimator has a breakdown point of 0%.

In 1887 Edgeworth [8] suggested that outliers may significantly influence the estimation, because the residuals ε_i are squared. Therefore, he has proposed the new regression index in form of the absolute error (IAE) (11.2).

$$\min_{\hat{\theta}} \sum_{i=1}^{n} |\varepsilon_i| \tag{11.2}$$

Such a formulation corresponds to the earlier Laplace research on one dimensional observations and his error law, now called the double exponential or Laplace PDF (4.15). Although the breakdown point for the regression using IAE is still 0%, it protects the results against some types of the outliers and thus is sometimes preferred over the MSE.

Above statistical interpretations explain two significant observations: the similarities between the indexes and the way they are affected by the shadowing effect. As the MSE/σ selection is highly affected by the outliers introduced with the disturbance, the IAE/b exhibits better, however still affected with the *shadowing*.

However, further explanation is required about robust ML-estimator of the standard deviation and its relationship with the scale factors of the fat-tail distributions. First of all, the analysis shows that if a monotone, bounded ψ-function is used, the breakdown point of the corresponding M-estimator is approximately 50% [13]. This property might explain high robustness of robust standard deviation σ^{rob} measure against fat-tail disturbances. On the other hand α-stable distribution is fitted to the data using ML-estimator, very similar in the methodology as the one used in the robust scale M-estimator. This fact may be the reason of the similar behavior of the considered measures.

The above argumentation explains the similarities of the first three groups of the measures and their ability to cope with the disturbance shadowing effect. The similarities in performance of both entropies seem to be quite natural. However it is unfortunate that both of them are fully unreliable both in the detection of the good controller tuning in the undisturbed case and in the robustness against disturbances.

The last group consists of the considered persistence measures exemplified with the Hurst exponents. Four different exponent values have been evaluated. The single (global) exponent presents the power law for all the history ranges neglecting the exhibiting crossover points and persistence changes in different scales. To reflect this possibility, the double crossover extended range R/S plots have been prepared and according to them the respective values of three Hurst exponents (H^1, H^2 and H^3) have been evaluated. As the crossover points are not reflecting loop control quality as such, they are not further considered in the analysis.

It is very interesting that in general the short memory Hurst exponent (the one, which is of the interest in case of dynamic loop behavior) has been always slightly biased toward higher values $0.5 \rightarrow 0.6$, i.e towards sluggish operation. The respective exponents are collected in Table 11.13. The bias is well visible. Actually this effect has been previously observed in the industrial fractal assessment and the research [7].

According to the previous research [14] persistent properties reflect sluggish controller performance $H \in (0.5; 1)$, while anti-persistent ones $H \in (0; 0.5)$ indicate aggressive tuning often characterized by oscillatory operation. Thus independent control error realizations described by $H \approx 0.5$ should characterize good loop performance. The observed shift means that the control error for the optimal should not be considered as the independent variable, but some persistence should be expected. It is probably brought about by the feedback.

It is interesting to notice that in some situations the unexpected values for the Hurst exponent $H > 1.0$ also appear. The meaning of the Hurst exponent has been considered by several researchers in different contexts [2, 3, 10]. Table 11.14 summarizes the observations. Originally, it has been assumed that $H \in (0, 1)$. Further research has started to consider and analyze situations with $H > 1.0$. There might be several reasons for such behavior, like non-stationarity or unsuccessful detrending.

Once the properties and the similarities between the measures have been discussed, their efficiency, i.e. the ability to indicate optimal tuning needs some discussion. First of all we have to distinguish the methods according to their requirements:

- step response needed:
  overshoot, settling time, Decay Ratio
- disturbance response needed:
  Idle Index, Area Index, Output Index, R-Index
- historical data needed:
  statistical factors, fractal and persistence measures, entropies

The results observed within the SISO simulations are summarized in Table 11.15. The results confirm the common popularity of the step-based methods. They are good, once there are no disturbances and the response is available (collected or

Table 11.13 Hurst exponents H^1 for all transfer functions

	G1	G2	G3	G4	G5	G6
nonDist	0.591	0.614	0.614	0.622	0.592	0.572
Dist	0.565	0.638	0.622	0.645	0.606	0.671

Table 11.14 Meaning of the Hurst exponent—assumed finite variance

	Time series properties
$H = 1.5$	Brownian noise (random walks)
$H \in (1.0, 2.0)$	Fractional Brownian noise
$H \in (0.5, 1.0)$	Persistent fractional Gaussian noise—long range dependency
$H = 0.5$	White noise
$H \in (0.0, 0.5)$	Anti-persistent fractional Gaussian noise—intermediate range dependency
$H \leqslant 0$	Over-differencing

Table 11.15 Summary of the SISO loop assessment observations ("++" strongly recommended, "+" recommended, "?" questionable, "–" not suggested)

	Real poles		Nonminimumphase		Time delay		Oscillatory		Fast and slow	
	nonDist	Dist	nonDist	Dist	nonDist	Dist	nonDist	Dist	nonDist	Dist
Overshoot	++	–	++	–	++	–	++	–	++	?
Settling time	++	–	++	–	++	–	++	–	++	?
Decay Ratio	–	–	–	–	–	–	–	–	–	–
Idle Index	++	–	++	–	++	–	++	–	+	–
Area Index	?	–	?	–	?	–	?	–	?	–
Output Index	+	–	+	–	+	–	+	–	+	–
R-Index	–	–	–	–	–	–	–	–	–	–
MSE	+	–	+	–	+	–	+	–	+	–
IAE	++	?	++	?	++	?	++	–	++	–
QE	+	–	+	–	+	–	+	–	+	–
Normal σ	+	–	+	–	+	?	+	?	+	?
Cauchy PDF γ	++	+	++	+	++	+	++	+	++	+
Stable PDF α	–	–	–	–	–	–	–	–	–	–
Stable PDF γ	++	++	++	++	++	++	++	++	++	+
Laplace PDF b	++	?	++	?	++	?	++	?	++	?
Robust σ	++	+	++	++	++	++	++	+	++	+
Hurst exponent	Requires further attention									
Entropy H^{DE}	–	–	–	–	–	–	–	–	–	–
Entropy H^{DE}	–	–	–	–	–	–	–	–	–	–

evaluated). Apart from the overshoot and settling time, the Idle Index is reliable, especially being supported by Output Index. The simulations have shown that Decay Ratio, Area Index and R-Index might be misleading and as such are not suggested.

Integral indexes seem to be sensitive to the disturbances. It is mostly visible with the square error based measures of the MSE and QR. Absolute integral measure IAE gives more reasonable results. However, they are also subject to the disturbance shadowing effect. Minimum variance benchmark results are discouraging, while entropies are not suggested. Gaussian measures should be restricted to the non disturbed cases. The fat tailed distribution factors together with the robust standard deviation give the most reliable results and are strongly suggested.

References

1. Åström, K.J., Hägglund, T.: New tuning methods for PID controllers. In: Proceedings 3rd European Control Conference, pp. 2456–2462 (1995)
2. Bryce, R.M., Sprague, K.B.: Revisiting detrended fluctuation analysis. Sci. Rep. **2**, 315 (2012)
3. Ceballos, R.F., Largo, F.F.: On the estimation of the hurst exponent using adjusted rescaled range analysis, detrended fluctuation analysis and variance time plot: a case of exponential distribution. Imp. J. Interdiscip. Res. **3**(8), 424–434 (2017)

4. Domański, P.D.: Fractal measures in control performance assessment. In: Proceedings of IEEE International Conference on Methods and Models in Automation and Robotics MMAR, Miedzyzdroje, Poland, pp. 448–453 (2016)
5. Domański, P.D.: Non-Gaussian and persistence measures for control loop quality assessment. Chaos Interdiscip. J. Nonlinear Sci. **26**(4), 043,105 (2016)
6. Domański, P.D.: Non-Gaussian statistical measures of control performance. Control Cybern. **46**(3), 259–290 (2017)
7. Domański, P.D.: Statistical measures for proportional-integral-derivative control quality: simulations and industrial data. Proc. Inst. Mech. Eng. Part I J. Syst. Control Eng. **232**(4), 428–441 (2018)
8. Edgeworth, F.Y.: On observations relating to several quantities. Hermathena **6**, 279–285 (1887)
9. Hawkins, D.M.: Identification of Outliers. Chapman and Hall, London (1980)
10. Løvsletten, O.: Consistency of detrended fluctuation analysis (2018). arXiv:1609.09331v2, [math.ST]
11. Plackett, R.L.: Studies in the history of probability and statistics. XXIX the discovery of the method of least squares. Biometrika **59**(2), 239–251 (1972)
12. Rousseeuw, P.J., Leroy, A.M.: Robust Regression and Outlier Detection. Wiley, New York (1987)
13. Ruckstuhl, A.: Robust fitting of parametric models based on M-estimation, iDP Institute of Data Analysis and Process Design, ZHAW Zurich University of Applied Sciences in Winterthur (2016)
14. Spinner, T., Srinivasan, B., Rengaswamy, R.: Data-based automated diagnosis and iterative retuning of proportional-integral (PI) controllers. Control Eng. Pract. **29**, 23–41 (2014)

Chapter 12
MIMO PID Based Control

You always admire what you really don't understand.

– Blaise Pascal

MIMO linear PID based control examples are reviewed in this chapter. Two main cases are considered: cascaded control and disturbance decoupling feedforward. The effects of poor tuning of the upstream and downstream controller in the cascaded configuration are assessed together with the impact of the feedforward settings.

12.1 Cascaded Control

Cascaded control configuration is sketched in Fig. 10.1. Two described benchmark cases (10.20) and (10.22) are evaluated and compared. Cascaded control structure is a multi loop configuration with the upstream (master, outer, top) and downstream loops (slave, inner, bottom). Thus, the assessment differs from the SISO configuration with single controller. Loop quality assessment has to be modified accordingly, as not all indexes are relevant or useful.

The upstream loop is the most important from the loop quality perspective and as such it is mostly considered in the assessment. First off all the step indexes are limited only to the most common ones, i.e. the overshoot and the settling time. It follows the observation that these indexes happened to be the most informative in all previous investigations. They are evaluated in the no disturbance environment.

All further evaluations are done in the loop configuration embedding two measurement noises at the output of each process transfer function (denoted $d_1(t)$ and

P. D. Domański, *Control Performance Assessment: Theoretical Analyses and Industrial Practice*, Studies in Systems, Decision and Control 245,
https://doi.org/10.1007/978-3-030-23593-2_12

Table 12.1 Selected CPA indexes used in cascaded configuration

No	Method domain	Index	Acronym
1	Step response	Overshoot	κ
2		Settling time	t_s
3	Time-domain methods	Mean Square Errors	MSE
4		Integral Absolute Error	IAE
5		Quadratic Error	QE
6	Normal PDF (Gauss)	Standard deviation	σ
7	α-stable PDF factors	Stability	α
8		Scale	γ^L
9	Laplace PDF factors	Scale	b
10	Robust statistics	Standard deviation	σ^{rob}
11	Alternative	Multiple Hurst exponents	H^0, H^1, H^2, H^3
12		Differential entropy	H^{DE}
13		Rational entropy	H^{RE}

$d_2(t)$, respectively) and the process disturbance $z(t)$ realized through filtered fat-tailed Cauchy noise.

The loop quality assessment indexes are evaluated for both loops, i.e. for the inner (downstream) loop using inner control error signal and for the outer (upstream) loop using the main control error signal. The list of considered factors is limited, due to the multi loop configuration, the results of the previous SISO loops analysis and the assessment practicability (see Table 12.1).

Visualization of the assessment is done for two scenarios. The optimal controllers tunings are evaluated first. Next, the poor tuning of the upper controller is evaluated, i.e. three stages of the sluggishness are assessed and three stages of the tuning aggressiveness. The calculated indexes are limited to the considered loop, as the modifications in the upstream controller affect only outer loop performance. Similar approach is further performed for the downstream controller. In that case the indexes are calculated for both loops. Simulated scenarios are presented and specified in Table 12.2.

12.1.1 Cascaded Control Example No 1

In the first example cascaded control with the upstream PID controller $R_1(s)$ and downstream proportional PID controller $R_2(s)$ is assessed. The process comprises of the respective two transfer functions:

Table 12.2 Simulation scenarios for cascaded control

ScenNum	Description
1	Very sluggish upstream control
2	More sluggish upstream control
3	Sluggish upstream control
4	Optimal control
5	Aggressive upstream control
6	More aggressive upstream control
7	Very aggressive upstream control
8	Very sluggish downstream control
9	More sluggish downstream control
10	Sluggish downstream control
11	Optimal control
12	Aggressive downstream control
13	More aggressive downstream control
14	Very aggressive downstream control

$G_1^{C1}(s) = \frac{1}{5s+1}e^{-2s},$

$G_2^{C1}(s) = \frac{1}{s+1}e^{-s}.$

At first assumed ideal controllers has been tuned:

- upstream controller R_1: $k_p = 1.201$; $T_i = 7.243$; $T_d = 1.879$;
- downstream controller R_2: $k_p = 0.423$; $T_i = 1.065$; $T_d = 0.413$.

The main (upstream) loop is described by settling time $T_{set} = 20.31$ [s] and overshoot $\kappa = 0.98$ [%] and the downstream loop by settling time $T_{set} = 9.77$ [s] and overshoot $\kappa = 1.34$ [%]. The general step response for the main cascaded loop is presented in Fig. 12.1. The trends for the optimal loop are sketched in Fig. 12.2. Similarly to the previously defined assessment procedure the statistical analysis using histogram and PDF fitting is presented in Fig. 12.3 (upstream) and Fig. 12.4 (downstream), while rescaled range R/S plots are sketched in Figs. 12.5 and 12.6, respectively.

Observation of the rescaled range R/S plots shows that upstream loop has clear multi-persistence character with three scales separated by two crossover points. In contrary the downstream loop plot shows mono-persistence character with single Hurst exponent. Thus H^0 index is the most representative for the inner loop, while short scale measure of H^1 is more appropriate for the main loop analysis.

The first group of scenarios, i.e. $ScenNum = 1, \ldots, 7$ compares different settings of the main loop, so there is no need to compare the performance of the inner loop as its quality is not affected. The effect of the poor tuning is well visible on the comparison plots for the overshoot and the settling time. Respective diagrams for

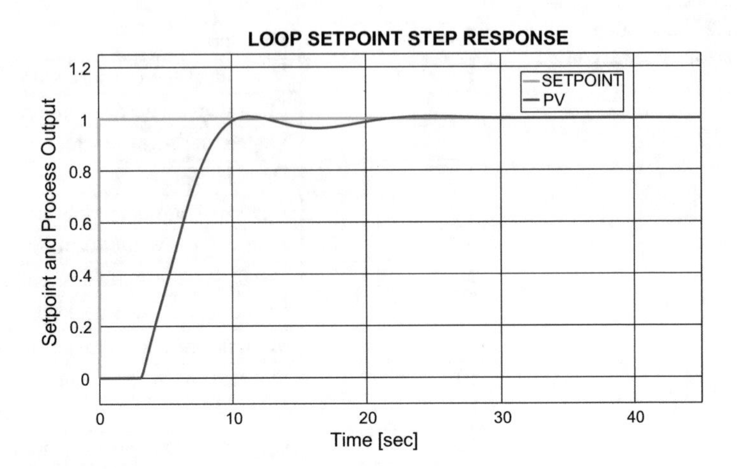

Fig. 12.1 Optimal main controller step response for case 1

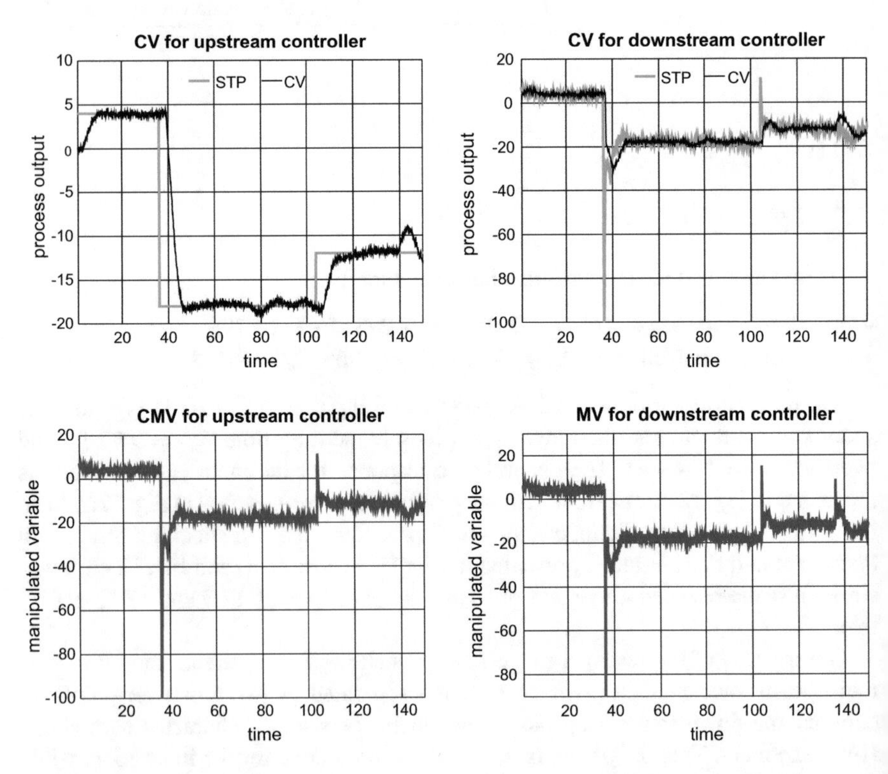

Fig. 12.2 Optimal controller time trends for case 1 (left—upstream, right—downstream)

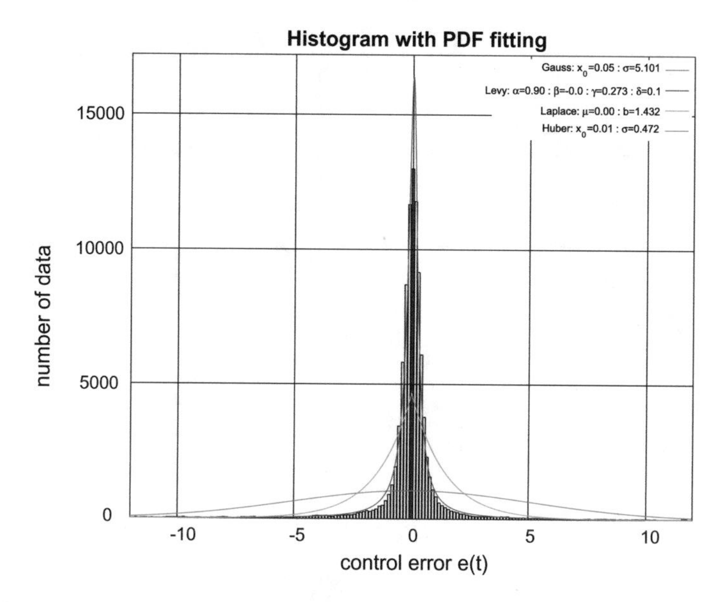

Fig. 12.3 Histogram for optimal main loop (case 1)

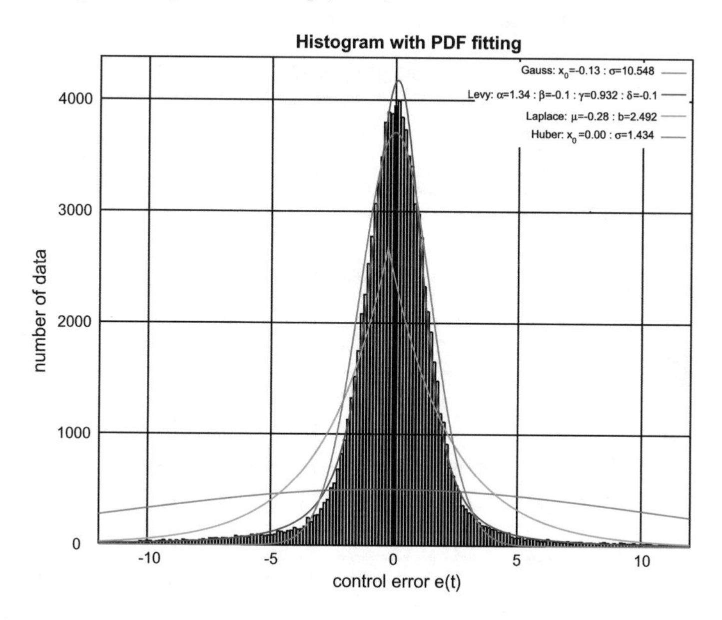

Fig. 12.4 Histogram for optimal inner loop (case 1)

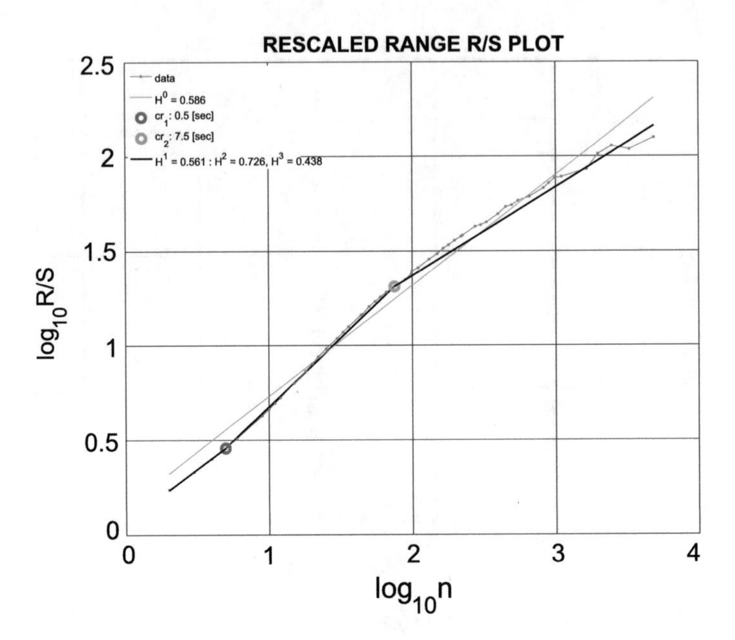

Fig. 12.5 R/S plot for optimal main loop (case 1)

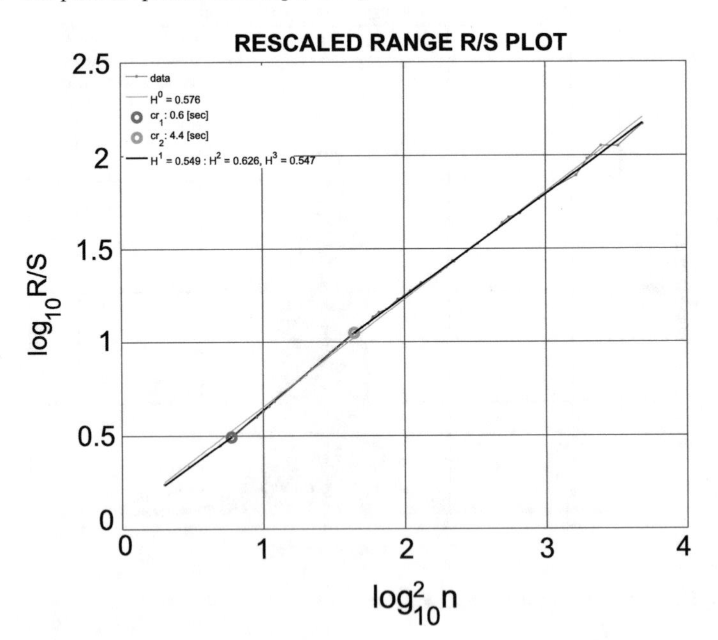

Fig. 12.6 R/S plot for optimal inner loop (case 1)

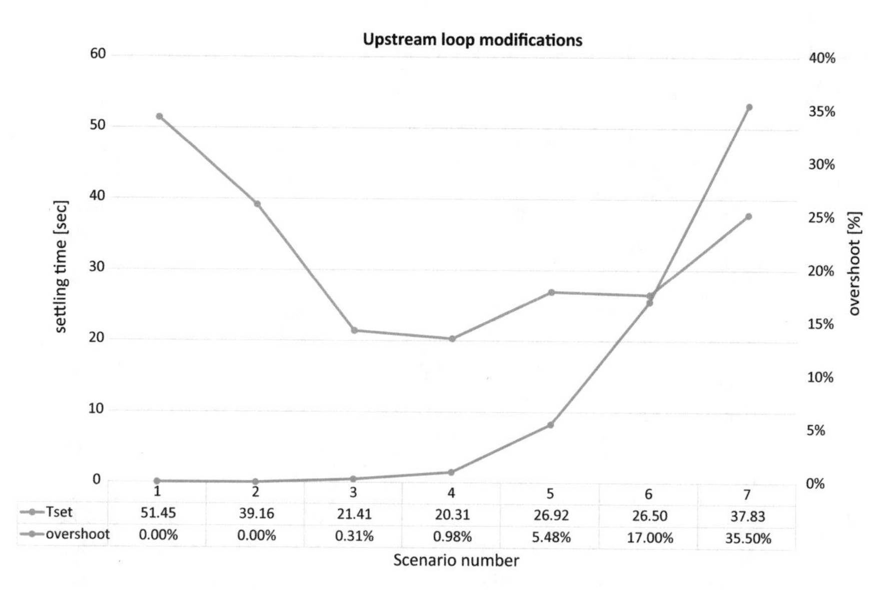

Fig. 12.7 Upstream loop tuning effect (case 1): settling time and overshoot

the upstream loop tunings is presented in Fig. 12.7. The effect of the upstream loop tuning is straightforward. The aggressive control causes diminishing of the settling time and increases the overshoot.

All the evaluated performance indexes are summarized in Table 12.3. We notice that practically all the indexes are monotonously increasing or decreasing in case of the Hurst exponents, except robust standard deviation and stable scale factor. The monotonous character of the relationship disables proper detection with most of the factors. Only two enlisted ones tend to indicate minimum associated with the best control quality. Representative relationships are presented in Fig. 12.8. Majority of the indexes behaves like sketched two indexes with the monotonic increase, i.e. IAE and scale factor b of the Laplace PDF. Additionally we clearly see the equivalence between IAE and Laplace distribution [2].

The second group of the relationships is exemplified with the two the most robust indexes, i.e. stable scale factor and robust standard deviation. Although the minimum is very flat it exists, thus these factors show the possibility to point out the good tuning and differentiate the poor one. Hurst exponents do not vary a lot and the differences between extreme scenarios is less than 10%.

Further analysis is conducted for the effect of the modifications done for the downstream controller. Downstream loop tuning effect is well visible in diagrams sketched in Fig. 12.9. The effect of the downstream PID parameters tuning is quite opposite. While the impact of the downstream controller on the indexes for the inner loop is as expected, the respective effect in the main loop is more indicative. It points out the good control more evidently.

Table 12.3 Case 1 upstream controller tuning effect on performance indexes (green color indicates optimal settings)

ScenNum	1	2	3	4	5	6	7
MSE	84.1	93.1	109.3	111.3	120.9	149.2	197.8
IAE	2.18	2.25	2.44	2.47	2.62	3.11	3.95
QE	130.6	140.5	158.3	160.8	171.6	203.3	259.1
σ	9.17	9.65	10.45	10.55	10.99	12.21	14.06
α	1.43	1.41	1.35	1.34	1.31	1.26	1.19
γ^L	0.912	0.905	0.923	0.932	0.973	1.143	1.435
b	2.205	2.280	2.462	2.492	2.644	3.132	3.961
σ^{rob}	1.387	1.384	1.421	1.434	1.507	1.785	2.274
H^0	0.590	0.590	0.577	0.576	0.570	0.553	0.539
H^1	0.547	0.545	0.548	0.549	0.552	0.554	0.565
H^2	0.643	0.632	0.629	0.626	0.625	0.620	0.623
H^3	0.550	0.560	0.548	0.547	0.543	0.528	0.511
H^{RE}	101.87	110.95	124.04	126.97	134.07	150.64	169.75
H^{DE}	−913500	−969862	−1058689	−1071061	−1116569	−1230496	−1360074

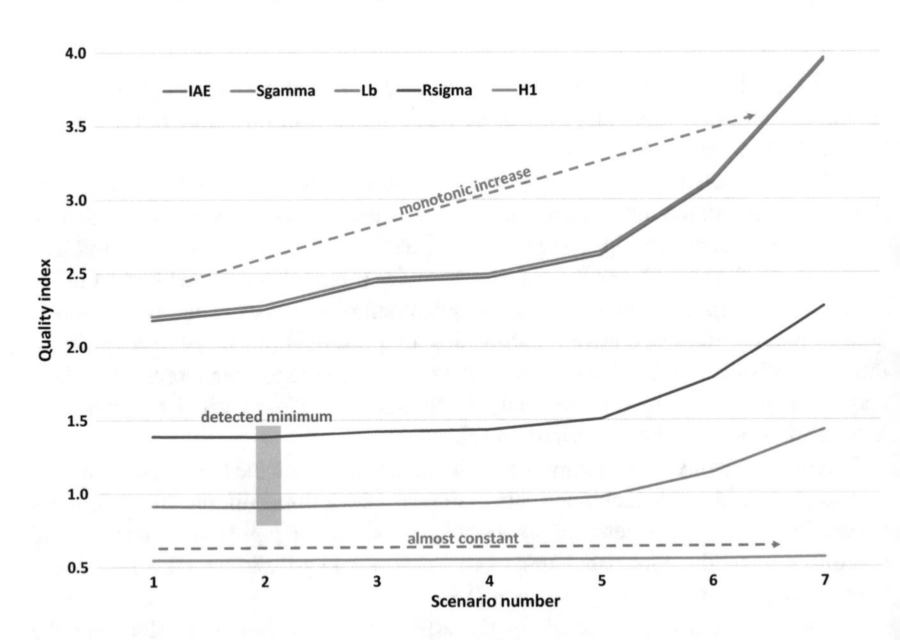

Fig. 12.8 Upstream loop tuning effect (case 1): selected indexes

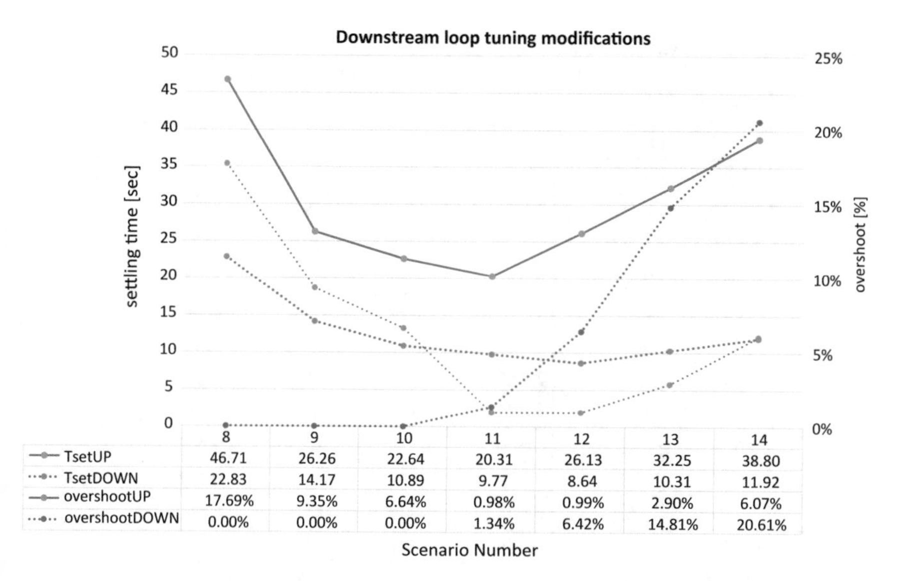

Fig. 12.9 Downstream loop tuning effect (case 1): settling time and overshoot

All the evaluated indexes for both loops are summarized in Table 12.4. The results are more constructive than for the previous case. First of all, more indexes have the ability to point out some minimum value of the index, which is associated with the best loop performance. For the upstream loop behavior this is achievable with the robust indexes γ^L and σ^{rob} together with Laplace scaling b and IAE and moreover with Hurst exponents H^1 and H^2. We see that integral and statistical indexes are pointing out slightly more sluggish settings denoted by the scenario Sc10, while persistence measures of the rescaled range tend to indicate more aggressive tuning of the scenario **Sc12**.

Downstream loop is more sensitive to the detection. Means square error MSE, quadratic error QE and normal standard deviation indicate sluggish scenario **Sc9**. IAE and scaling b point out slightly aggressive scenario **Sc12**, while two robust indicators γ^L and σ^{rob} detect exactly optimal tuning of the scenario Sc11.

The observations for the cascaded control detection may be summarized in few words. Detection of the optimal control is not se easy. The main loop is slow and in that case the poor tuning is not well visible. In such case only robust measures based on the α-stable distribution (γ^L) or factors using robust statistics (σ^{rob}) give some indications, although biased towards sluggish control.

In case of the mis-tuning for the inner loop the detections seems to be easier, though similar indexes give reliable and appropriate indications. Finally, the equivalences of mean square error and normal standard deviation together with Laplace scale and mean absolute error have been observed.

Table 12.4 Case 1 downstream controller tuning effect on performance indexes for both loops (the best indexes are highlighted in bold fonts, green color indicates optimal settings)

ScenNum	Upstream						
	8	9	10	11	12	13	14
MSE	87.7	94.6	98.8	111.3	130.5	157.2	187.1
IAE	2.69	2.46	**2.44**	2.47	2.63	3.01	3.45
QE	125.5	135.9	142.2	160.8	191.9	238.5	293.8
σ	9.36	9.72	9.93	10.55	11.42	12.54	13.68
α	1.24	1.35	1.35	1.34	1.34	1.21	1.14
γ^L	1.000	0.940	**0.924**	0.932	0.969	1.046	1.147
b	2.708	2.486	**2.464**	2.492	2.650	3.033	3.469
σ^{rob}	1.575	1.449	**1.423**	1.434	1.496	1.659	1.848
H^0	0.614	0.595	0.590	0.576	0.551	0.530	0.508
H^1	0.557	0.557	0.550	0.549	**0.537**	0.546	0.551
H^2	0.720	0.670	0.653	0.626	**0.598**	0.601	0.617
H^3	0.521	0.541	0.546	0.547	0.535	0.517	0.497
H^{RE}	106.30	114.73	118.71	126.97	133.31	141.63	155.64
H^{DE}	−812199	−923458	−966820	−1071061	−1195839	−1310894	−1463004
ScenNum	Downstream						
	8	9	10	11	12	13	14
MSE	25.7	**25.5**	25.6	26.0	26.7	28.0	29.7
IAE	1.74	1.55	1.50	1.43	**1.42**	1.43	1.48
QE	67.5	**66.5**	**66.5**	67.0	68.3	71.0	74.5
σ	5.07	**5.05**	5.06	5.10	5.17	5.29	5.45
α	0.83	0.86	0.87	0.90	0.91	0.89	0.90
γ^L	0.343	0.289	0.281	**0.273**	0.285	0.288	0.305
b	1.742	1.546	1.497	1.432	**1.424**	1.432	1.475
σ^{rob}	0.634	0.519	0.495	**0.472**	0.493	0.503	0.532
H^0	0.591	0.586	0.586	0.586	0.587	0.585	0.582
H^1	0.740	0.546	0.548	0.561	0.691	0.682	0.670
H^2	0.589	0.732	0.725	0.726	0.601	0.604	0.626
H^3	0.571	0.398	0.404	0.438	0.572	0.571	0.568
H^{RE}	42.55	44.47	45.93	47.40	48.99	51.52	54.92
H^{DE}	−252521	−298618	−317765	−360185	−405567	−458057	−493920

12.1.2 *Cascaded Control Example No 2*

In the second example cascaded control with two PID controllers $R_1(s)$ and $R_2(s)$ for both upstream and downstream loops is assessed. The process consists of two transfer functions:

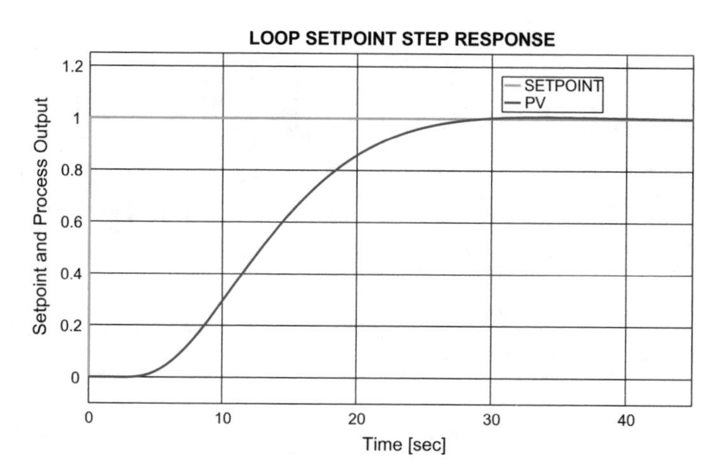

Fig. 12.10 Optimal main controller step response for case 2

$$G_1^{C2}(s) = \frac{10(-0.5s+1)}{(3s+1)^3(s+1)^2} e^{-0.5s},$$

$$G_2^{C2}(s) = \frac{3}{(1.33s+1)} e^{-0.3s}.$$

We see that the main loop is much slower, with larger delay and nonminimumphase zero. At first, the ideal controllers has been tuned:

- upstream controller R_1: $k_p = 0.058$; $T_i = 7.864$; $T_d = 1.965$;
- downstream controller R_2: $k_p = 0.367$; $T_i = 1.350$; $T_d = 0.091$.

The main (upstream) loop is described by settling time $T_{set} = 27.41$ [s] and overshoot $\kappa = 0.76$ [%] and the downstream loop by settling time $T_{set} = 4.28$ [s] and overshoot $\kappa = 0.11$ [%]. The general step response for the main cascaded loop is presented in Fig. 12.10. The trends for the optimal loop are sketched in Fig. 12.11. Similarly to the previously defined assessment procedure the statistical analysis using histogram and PDF fitting is presented in Fig. 12.12 (upstream) and Fig. 12.13 (downstream), while rescaled range R/S plots are sketched in Figs. 12.14 and 12.15, respectively.

Observation of the rescaled range R/S plots shows that upstream loop has clear multi-persistence character with three scales separated by two crossover points. In contrary the downstream loop plot shows mono-persistence character with single Hurst exponent. Thus H^0 index is the most representative for the inner loop, while short scale measure of H^1 is more appropriate for the main loop analysis.

The first group of scenarios, i.e. $ScenNum = 1, \ldots, 7$ compares different settings of the main loop, so there is no need to compare the performance of the inner loop as its quality is not affected. The effect of poor tuning is well visible on the comparison plots for the overshoot and the settling time. The diagrams for the upstream loop tunings is presented in Fig. 12.16. The effect of the upstream loop tuning is straightforward.

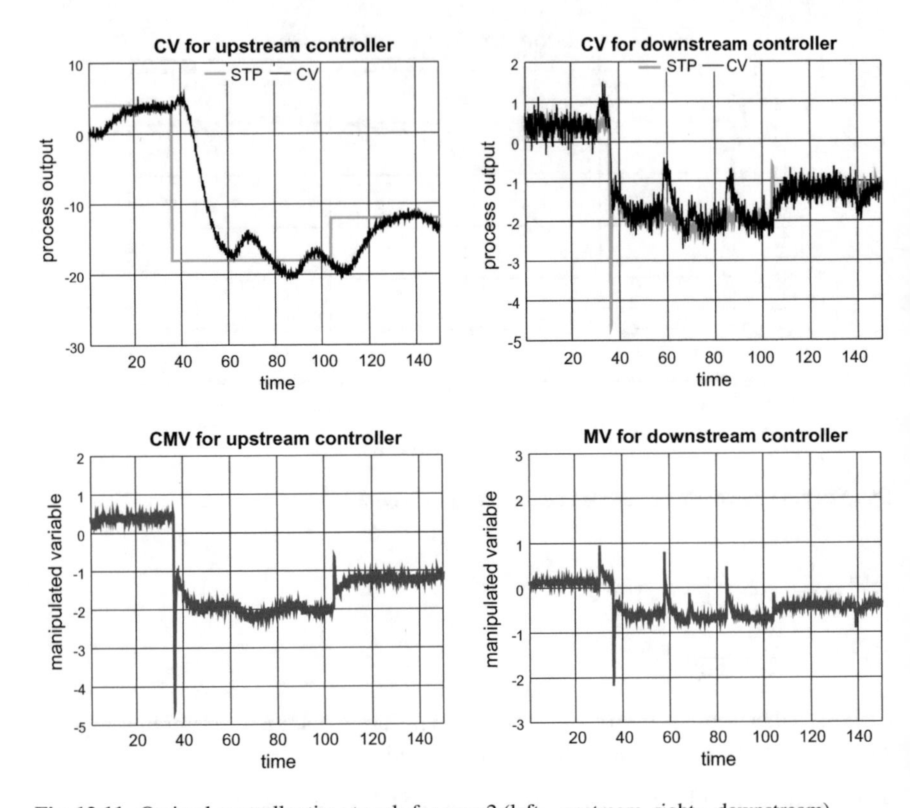

Fig. 12.11 Optimal controller time trends for case 2 (left—upstream, right—downstream)

The aggressive control causes diminishing of the settling time and increases the overshoot.

All the evaluated performance indexes are summarized in Table 12.5. We notice that in the considered case all the indexes are monotonously changing. They do not show any settings to be optimal. The monotonous character of the relationship disables proper detection with most of the factors. Representative curves are presented in Fig. 12.17. The equivalence between IAE and Laplace distribution is also visible as previously. Analogously, the curves for MSE and normal standard deviation exhibit similarly.

In the considered case there are no indexes pointing out any solution. Hurst exponents, as in previous case, do not vary a lot and the differences between extreme scenarios is not significant.

Further analysis is conducted to evaluate the effect of the modifications performed for the downstream controller. Inner loop tuning effect is well visible in diagrams sketched in Fig. 12.18. The effect of the downstream PID parameters tuning is quite opposite. While the impact of the downstream controller on the indexes for the inner loop is as expected, the respective effect in the main loop is more indicative. It points out the good control more evidently.

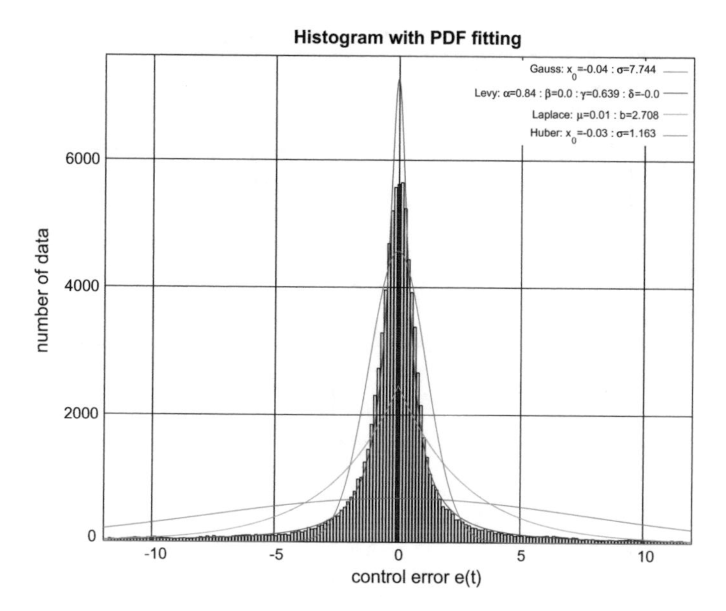

Fig. 12.12 Histogram for optimal main loop (case 2)

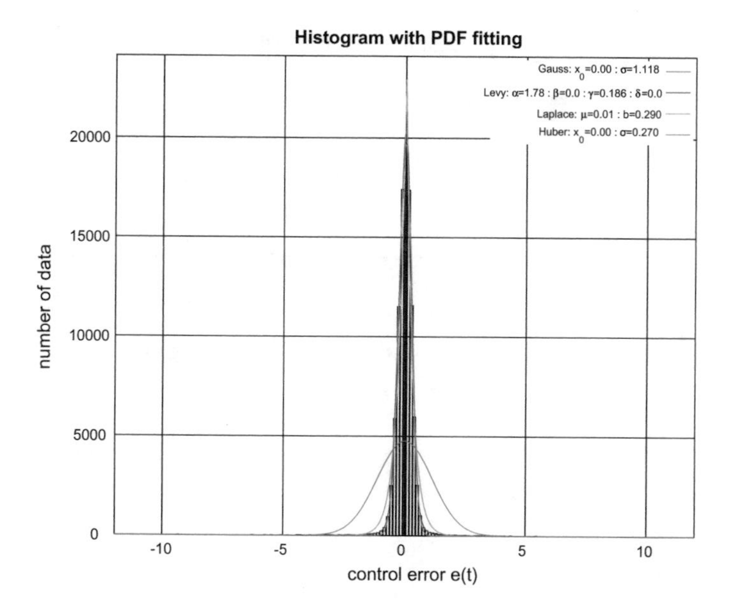

Fig. 12.13 Histogram for optimal inner loop (case 2)

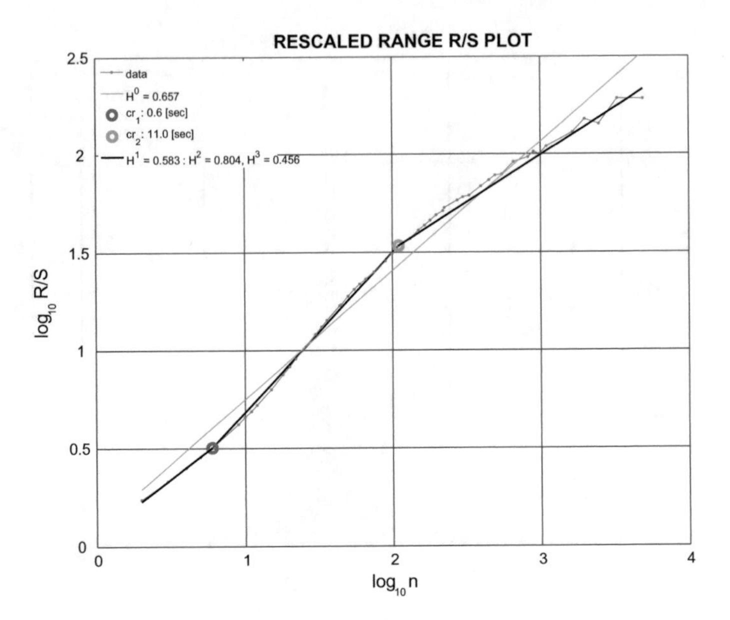

Fig. 12.14 R/S plot for optimal main loop (case 2)

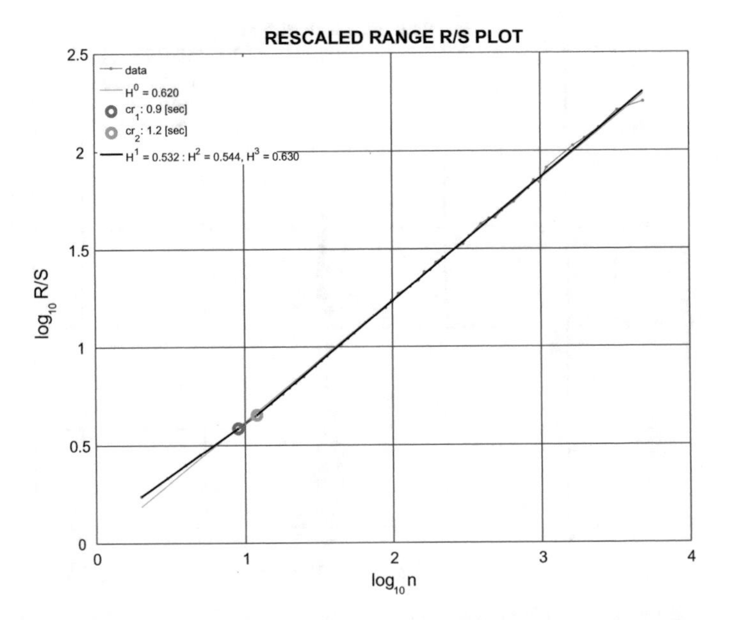

Fig. 12.15 R/S plot for optimal inner loop (case 2)

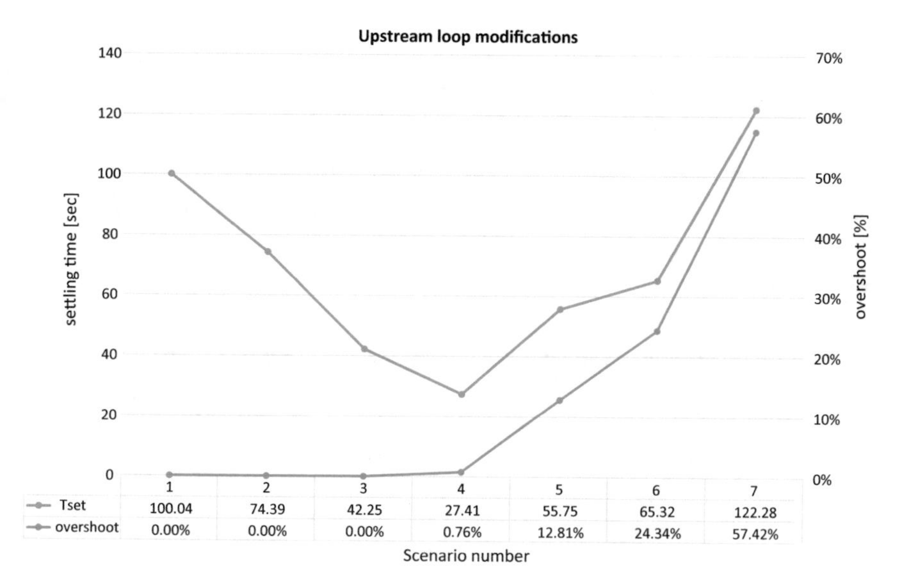

Fig. 12.16 Upstream loop tuning effect: settling time and overshoot

Table 12.5 Case 2 upstream controller tuning effect on performance indexes (green color indicates optimal settings)

ScenNum	1	2	3	4	5	6	7
MSE	1.205	1.219	1.244	1.251	1.276	1.337	1.519
IAE	0.275	0.279	0.286	0.288	0.298	0.315	0.364
QE	1.289	1.305	1.334	1.342	1.372	1.440	1.650
σ	1.098	1.104	1.115	1.119	1.130	1.156	1.232
α	1.806	1.775	1.777	1.768	1.770	1.782	1.728
γ^L	0.177	0.178	0.182	0.183	0.189	0.199	0.225
b	0.275	0.279	0.286	0.289	0.298	0.315	0.364
σ^{rob}	0.259	0.260	0.266	0.267	0.276	0.291	0.331
H^0	0.613	0.610	0.613	0.615	0.618	0.609	0.611
H^1	0.538	0.530	0.532	0.538	0.532	0.532	0.539
H^2	0.515	0.541	0.620	0.622	0.539	0.175	0.602
H^3	0.619	0.617	0.418	0.419	0.626	0.613	0.603
H^{RE}	13.20	13.41	13.31	13.09	13.63	14.28	15.35
H^{DE}	−180108	−179905	−180877	−179358	−178363	−177596	−174930

All the evaluated indexes for both loops are summarized in Table 12.6. The results are more constructive than for the previous case. First of all, some indexes have the ability to point out some minimum value of the index, which is associated with the best loop performance. For the upstream loop behavior detection is achievable with

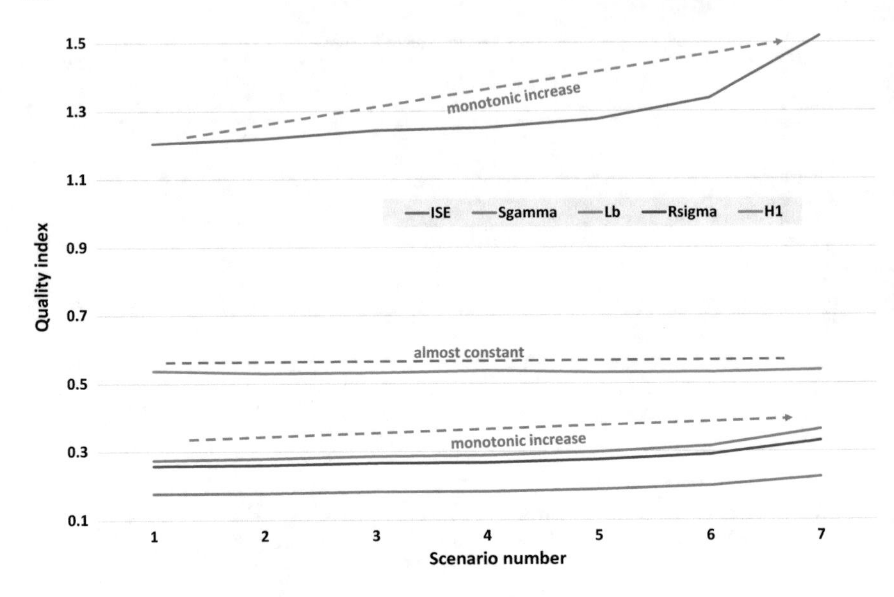

Fig. 12.17 Upstream loop tuning effect for case 2: selected indexes

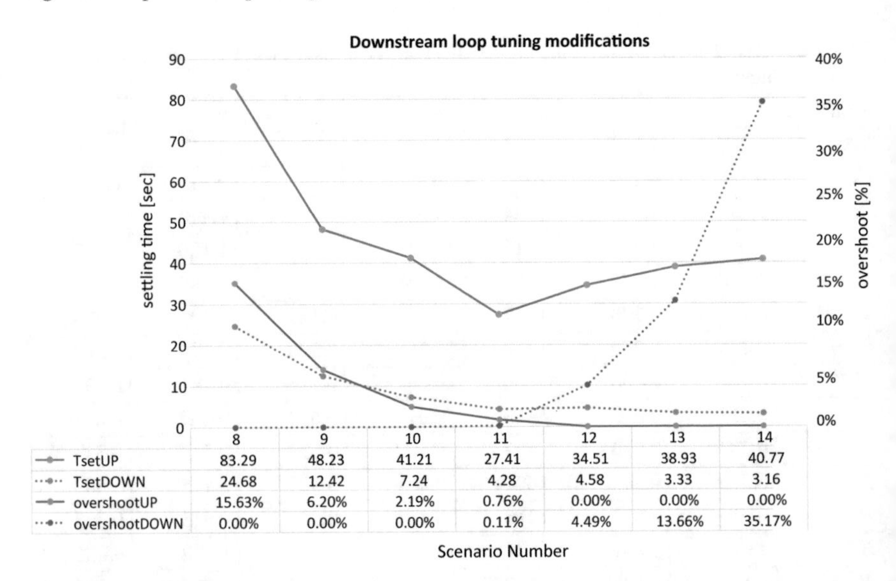

	8	9	10	11	12	13	14
TsetUP	83.29	48.23	41.21	27.41	34.51	38.93	40.77
TsetDOWN	24.68	12.42	7.24	4.28	4.58	3.33	3.16
overshootUP	15.63%	6.20%	2.19%	0.76%	0.00%	0.00%	0.00%
overshootDOWN	0.00%	0.00%	0.00%	0.11%	4.49%	13.66%	35.17%

Fig. 12.18 Downstream loop tuning effect for case 2: settling time and overshoot

the robust indexes γ^L and σ^{rob} together with Laplace scaling b and IAE and moreover with Hurst exponents H^2 and H^3. This result is exactly the same as in previous case. The integral and statistical indexes are pointing out exact optimal settings denoted by scenario **Sc11**, while persistence measures of the rescaled range tend to indicate other solutions.

Downstream loop is allows only two robust indicators γ^L and σ^{rob} to detect exactly optimal tuning of the scenario Sc11, while Hurst exponents H^1 suggests more aggressive performance.

The observations for the second cascaded control example may be summarized in few words. Detection of the optimal control is not easy as well. The main loop is slow and in that case the poor tuning is not well visible. In case of the mis-tuning for the inner loop the detections seems to work. Robust measures based on the α-stable distribution (γ^L) or factors using robust statistics (σ^{rob}) give reliable and accurate indications. As previously, the equivalences of mean square error and normal standard deviation together with Laplace scale and mean absolute error have been observed.

12.1.3 Summary of the Observations

The results of control performance assessment in case of the MIMO cascaded control seems to be more challenging, than univariate PID control. It is especially visible with the upstream loop performance. None of the considered indexes has been able to detect any solution. In contrary, the tuning of the downstream controller has been causing measurable effects in both control errors (main and inner loop).

The measures exhibit consistently in all cases. The index robustness against outliers (fat tails) is required for the proper and reliable loop quality detection. Only two robust measures have been indicating solutions optimal or close to the optimal one. This is scale factor γ^L of the α-stable distribution and robust standard deviation σ^{rob} evaluated with the logistic ψ scale M-estimator.

The second group of the reliable loop quality estimators consists of two equivalent measures: absolute error integral IAE and Laplace distribution scaling factor b. Ads we know from the statistical research they are less robust than the previous ones. Nonetheless they are also able to point out some solution being relatively close to the optimal one. In contrary other measures (like popular mean square error od normal standard deviation) do not behave properly in the situations with disturbances exhibiting outliers.

Concluding, the cascaded control tests confirm observations for the SISO loops. A good CPA measure has to be robust against outliers. Indicators originating from the robust or fat tail domain enable such an ability.

12.2 Feedforward Disturbance Decoupling

Multivariate control configuration including feedforward disturbance decoupling is analyzed in this section. The loop configuration is presented in Fig. 10.2. The practical industrial feedforward configuration is considered in form presented in Fig. 10.3. Two benchmark cases (10.24) and (10.24) are analyzed and compared. The considered MIMO configuration consists of the two element: feedback controller and

Table 12.6 Case 2 downstream controller tuning effect on performance indexes for both loops (the best indexes are highlighted in bold fonts, green color indicates optimal settings)

ScenNum	Upstream						
	8	9	10	11	12	13	14
MSE	0.763	0.961	1.139	1.251	1.536	1.777	2.121
IAE	0.325	0.296	0.290	**0.288**	0.289	0.292	0.302
QE	0.814	1.026	1.220	1.342	1.678	1.996	2.584
σ	0.873	0.980	1.067	1.119	1.239	1.333	1.456
α	1.622	1.736	1.764	1.768	1.772	1.788	1.801
γ^L	0.204	0.188	0.184	**0.183**	**0.183**	0.184	0.191
b	0.325	0.297	0.290	**0.289**	0.290	0.292	0.302
σ^{rob}	0.303	0.276	0.269	**0.267**	0.268	0.269	0.279
H^0	0.679	0.650	0.624	0.615	0.593	0.588	0.564
H^1	0.545	0.530	0.537	0.538	0.530	0.532	0.531
H^2	0.722	0.658	0.631	0.622	0.598	**0.529**	0.531
H^3	0.586	0.420	**0.367**	0.419	0.443	0.593	0.570
H^{RE}	9.42	11.42	12.48	13.09	15.48	16.38	20.00
H^{DE}	−76358	−122438	−158557	−179358	−239581	−290825	−439282
ScenNum	Downstream						
	8	9	10	11	12	13	14
MSE	61.504	61.407	60.459	59.946	57.923	56.272	54.989
IAE	3.178	2.845	2.738	2.703	2.646	2.610	2.592
QE	61.897	61.787	60.832	60.318	58.290	56.636	55.352
σ	7.842	7.836	7.776	7.743	7.611	7.501	7.416
α	0.872	0.838	0.824	0.834	0.848	0.849	0.859
γ^L	0.855	0.686	**0.629**	**0.629**	0.637	0.633	0.643
b	3.178	2.845	2.738	2.703	2.646	2.610	2.592
σ^{rob}	1.550	1.242	1.151	**1.149**	1.153	1.146	1.157
H^0	0.655	0.656	0.656	0.657	0.660	0.661	0.662
H^1	0.605	0.583	0.585	0.583	**0.582**	**0.582**	0.586
H^2	0.833	0.813	0.804	0.804	0.800	0.801	0.796
H^3	0.424	0.443	0.454	0.455	0.461	0.461	0.454
H^{RE}	66.39	71.30	72.00	72.35	70.72	69.62	68.49
H^{DE}	−449561	−539796	−569097	−579513	−579759	−571656	−562295

feedforward disturbance decoupler. The performance of the feedback controller is independent on the feedforward tuning and as such exhibits the same as the SISO control case. Thus there is no need to consider CPA analysis of the feedback controller. The analysis focuses on the settings and performance of the feedforward block.

Analogously to the cascaded control (Sect. 12.1), two simulation examples are presented: (10.24) and (10.25). The loop is additionally impacted with the measurement noise generated as the Gaussian random signal with some standard deviation σ. The disturbance that has to be decoupled is in form of the rectangle step wave with added fat-tail noise generated as the stochastic variable with the α-stable symmetric distribution. The disturbance parameters are given in details for each of the simulation cases. The step indexes are limited only to the most common ones, i.e. the overshoot and the settling time. It follows the observation that these indexes happened to be the most informative in all previous investigations. They are evaluated only for the feedback controller in the undisturbed simulations.

The loop quality assessment indexes are evaluated only for the feedback control error for the varying parameters of the decoupler. The list of considered CPA factors is limited (Table 12.1), due to the multi loop configuration, the results of the previous SISO loops analysis and the assessment practicability.

12.2.1 Feedforward-Feedback Example No 1

The first presented example uses two tank process with the transfer functions [1] shown below. The ideal feedback PID controller has been tuned first with its parameters: $k_p = 1.540$; $T_i = 1.920$; $T_d = 0.440$.

$$G^{FF1}(s) = \frac{1}{(s+1)^2}e^{-0.2s}, \ G_D^{FF1}(s) = \frac{1}{s+1}e^{-0.1s}.$$

The main feedback loop (without the impact of the disturbance) is described by settling time $T_{set} = 3.64$ [s] and overshoot $\kappa = 0.00$ [%]. The general step response for the main cascaded loop is presented in Fig. 12.19. The industrial feedforward (as in Fig. 10.3) has been designed. The obtained parameters are: $K = 30.0$; $T_1 = 0.11$; $T_2 = 0.17$. The step response of the feedforward unit is sketched in Fig. 12.20.

Setpoint which is being tracked in the simulations is in the shape of the rectangular wave with varying height and width of the steps (upper plot in Fig. 12.24). The simulations are disturbed. Disturbance variable, which is being decoupled consists of the varying step wave with added α-stable symmetrical, zero-mean noise and $\alpha = 1.8$ and $\gamma = 1.6$ (middle plot in Fig. 12.24). The loop output is disturbed with zero-mean Gaussian noise and $\sigma = 0.3$.

The trends for the optimal loop are sketched in Fig. 12.21. Similarly to the previously defined assessment procedure the statistical analysis using histogram and PDF fitting is presented in Fig. 12.22, while rescaled range R/S plots are sketched in Fig. 12.23. The histogram clearly shows the fat-tailed character of the loop. The R/S plot depicts multi-persistent properties of the loop with the tuning exhibiting slightly sluggish performance, which is justified by the simulations.

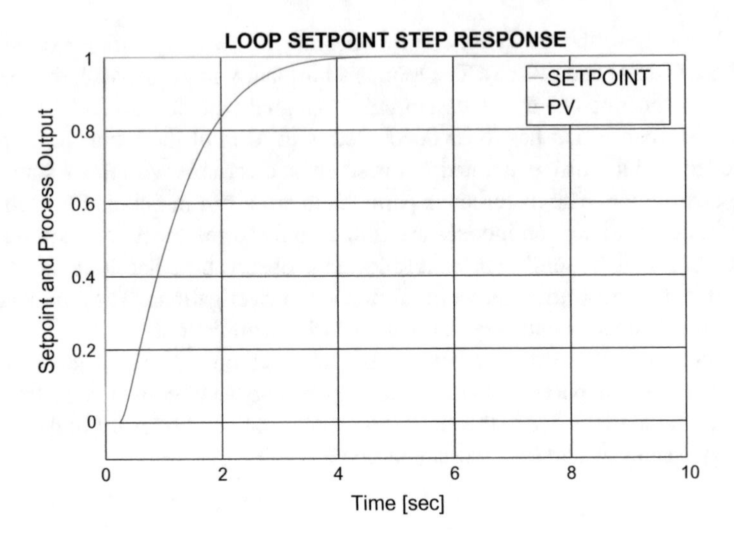

Fig. 12.19 Optimal feedback response

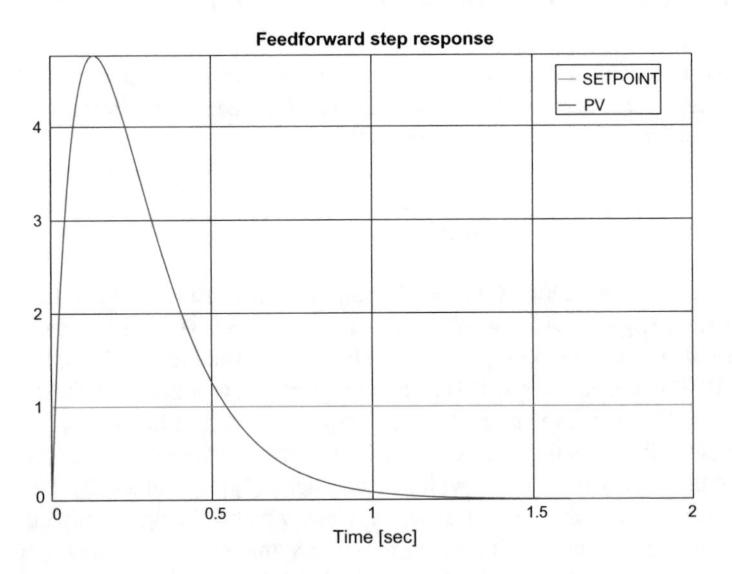

Fig. 12.20 Optimal feedforward response

Seven different feedforward tuning scenarios has been analyzed, starting for no feedforward scenario Sc1 and than increasing the importance of the feedforward through the optimal tuning Sc4 and finishing at the most intensive one Sc7. The parameters for the selected feedforward settings are presented in Table 12.7. The step responses for all of them is sketched in Fig. 12.24.

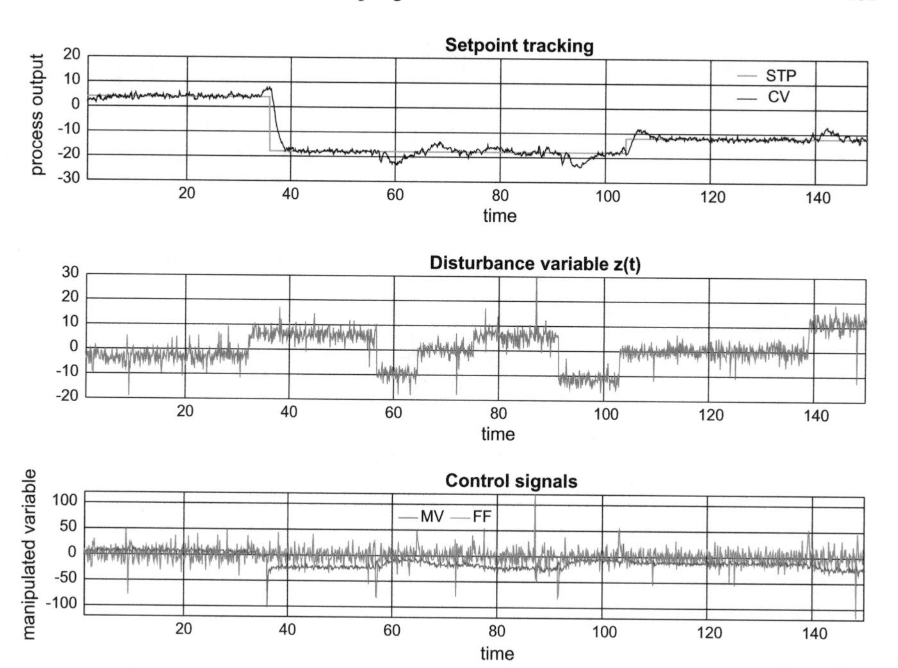

Fig. 12.21 Optimal feedback-feedforward loop time trends for case 1

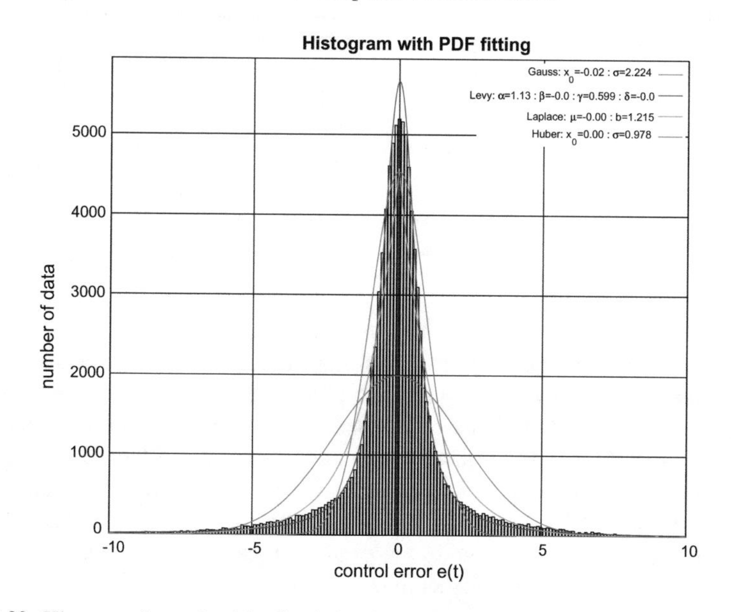

Fig. 12.22 Histogram for optimal feedback-feedforward loop (case 1)

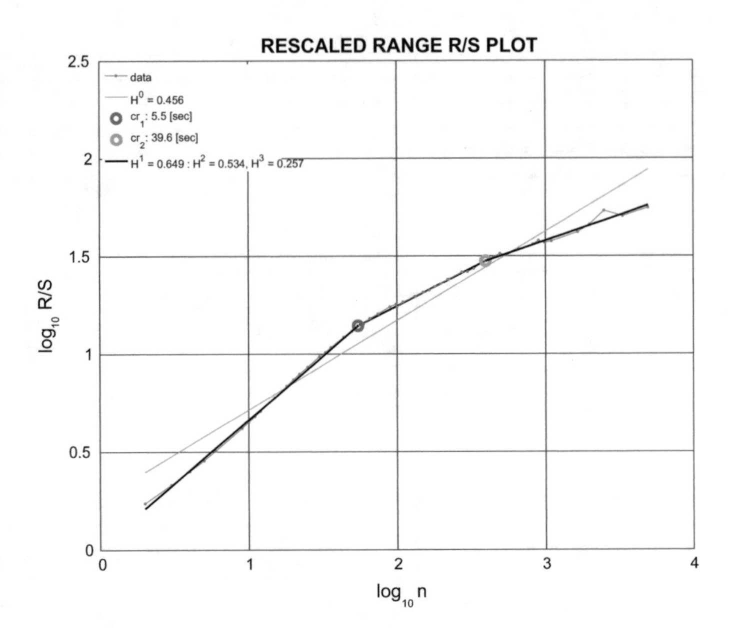

Fig. 12.23 R/S plot for optimal feedback-feedforward loop (case 1)

Table 12.7 Feedforward parameters settings for the simulation example 1

	K	T_1	T_2
Sc1	No feedforward		
Sc2	7.0	0.11	0.15
Sc3	15.0	0.11	0.15
Sc4	30.0	0.11	0.17
Sc5	30.0	0.11	0.19
Sc6	30.0	0.10	0.20
Sc7	30.0	0.10	0.23

The simulations have been run and all the considered indexes have been calculated for the selected scenarios. Their values are presented in Table 12.8. The results are constructive and consistent. Some of the indexes have the ability to point out some minimum value of the index, which is associated with the best loop performance or the one close to the optimal. Detection is feasible with all the majority of the indexes. Additionally, the Hurst exponents H^0 (single persistence) and H^1 (the shortest memory) tend towards optimal independent stochastic process $H = 0.5$. Entropies are not able to detect loop control quality as in the previous cases.

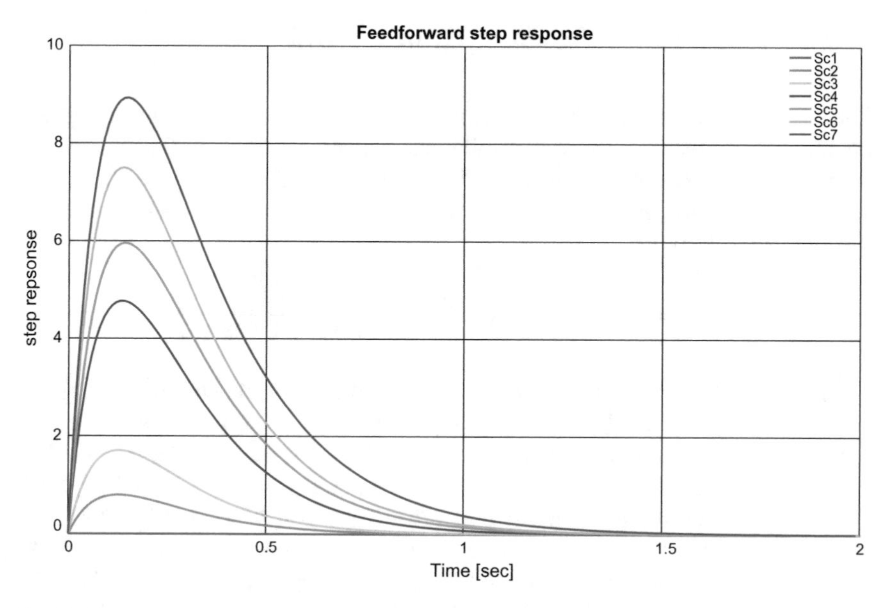

Fig. 12.24 Optimal feedback-feedforward loop time trends for case 1

Table 12.8 Case 1 feedback-feedforward loop tuning effect on performance indexes (the best indexes are highlighted in bold fonts, green color indicates optimal feedforward)

	Sc1	Sc2	Sc3	Sc4	Sc5	Sc6	Sc7
MSE	6.37	5.62	4.99	**4.94**	6.20	8.37	12.95
IAE	1.29	1.23	**1.18**	1.22	1.44	1.74	2.23
QE	52.21	50.90	**49.84**	50.08	52.46	56.54	64.92
σ	2.524	2.370	**2.233**	**2.223**	2.491	2.892	3.598
α	1.058	1.043	**1.035**	1.131	1.188	1.193	1.189
γ^L	0.558	0.528	**0.511**	0.604	0.760	0.945	1.221
b	1.290	1.227	**1.177**	1.216	1.436	1.741	2.231
σ^{rob}	0.915	0.878	**0.856**	0.984	1.226	1.521	1.971
H^0	0.437	0.445	0.449	**0.455**	0.442	0.434	0.420
H^1	0.606	0.626	0.647	**0.545**	0.622	0.602	0.588
H^2	0.494	0.506	0.527	0.654	0.515	0.488	0.471
H^3	0.270	0.241	0.259	0.317	0.256	0.284	0.277
H^{RE}	20.22	20.24	20.06	20.19	20.62	21.67	24.22
H^{DE}	−137134	−137566	−138246	−142939	−144144	−142706	−159158

12.2.2 Feedforward-Feedback Example No 2

The second presented example uses is more challenging with the plant witnessing nonminimumphase response with oscillations (damping ration $\zeta = 0.23$). The ideal feedback PID controller has been tuned first with its parameters: $k_p = 1.540$; $T_i = 1.920$; $T_d = 0.440$. It is the same as in previous case, as the feedback process transfer function is the same.

$$G^{FF2}(s) = \frac{1}{(s+1)^2} e^{-0.2s}, \; G_D^{FF2}(s) = \frac{-2s+1}{4s^2 + 0.92s + 1} e^{-1.5s}.$$

The main feedback loop (without the impact of the disturbance) is described by settling time $T_{set} = 3.64$ [s] and overshoot $\kappa = 0.00$ [%]. The general step response for the main cascaded loop is presented in Fig. 12.25. The industrial feedforward (as in Fig. 10.3) has been designed. The obtained parameters are: $K = 17.0$; $T_1 = 24.0$; $T_2 = 26.6$. The step response of the feedforward unit is sketched in Fig. 12.26.

Setpoint whicj is being tracked in the simulations is in the shape of the rectangular wave with varying height and width of the steps (upper plot in Fig. 12.30). The simulations are disturbed. Disturbance variable, which is being decoupled consists of the varying step wave with added α-stable symmetrical, zero-mean noise and $\alpha = 1.8$ and $\gamma = 1.6$ (middle plot in Fig. 12.30). The loop output is disturbed with zero-mean Gaussian noise and $\sigma = 0.3$.

The trends for the optimal loop are sketched in Fig. 12.27. Similarly to the previously defined assessment procedure the statistical analysis using histogram and PDF fitting is presented in Fig. 12.28, while rescaled range R/S plots are sketched in

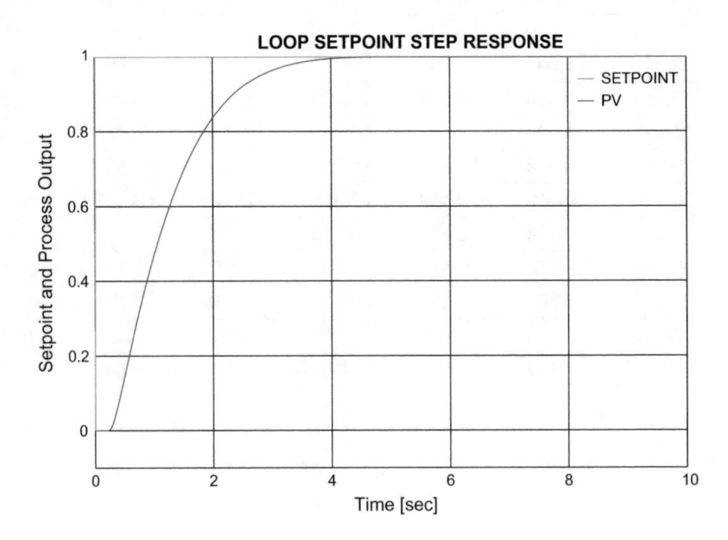

Fig. 12.25 Optimal feedback response

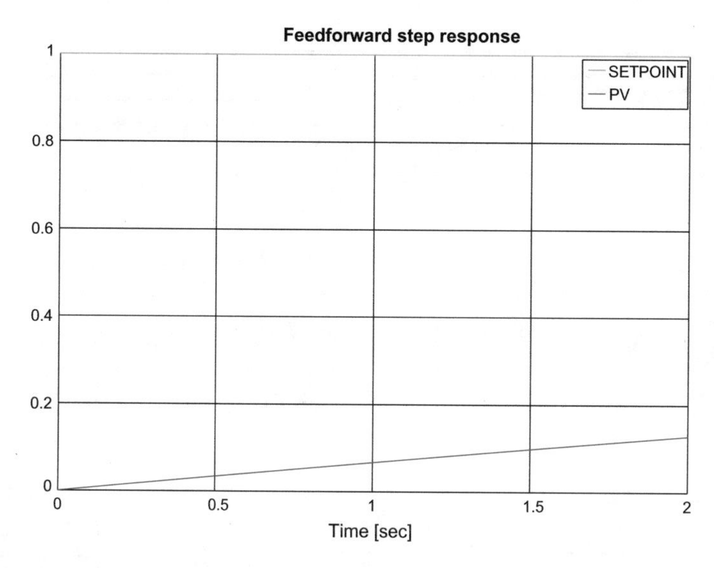

Fig. 12.26 Optimal feedforward response

Table 12.9 Feedforward parameters settings for the simulation example 2

	K	T_1	T_2
Sc1	No feedforward		
Sc2	5	10	11
Sc3	12	20	22
Sc4	17	24	26.6
Sc5	17	24	32.6
Sc6	17	24	39.6
Sc7	17	24	48.6

Fig. 12.29. The histogram clearly shows the fat-tailed character of the loop. The R/S plot depicts multi-persistent properties of the loop with the tuning exhibiting slightly sluggish performance, which is proven in the simulations.

Seven different feedforward tuning scenarios has been analyzed, starting for no feedforward scenario **Sc1** and than increasing the importance of the feedforward through the optimal tuning **Sc4** and finishing at the most intensive one **Sc7**. The parameters for the selected feedforward settings are presented in Table 12.9. The step responses for all of them is sketched in Fig. 12.30.

The simulations have been run and all the considered indexes have been calculated for the selected scenarios. Their values are presented in Table 12.10. The results are less consistent than in the previous case. It is so, due to the challenging character of the disturbance impact character (nonminimumphase + delay). Feedforward parameters have been tuned according to the MSE index, which is clearly visible. Some of the indexes have the ability to point out the minimum value of the index, which is

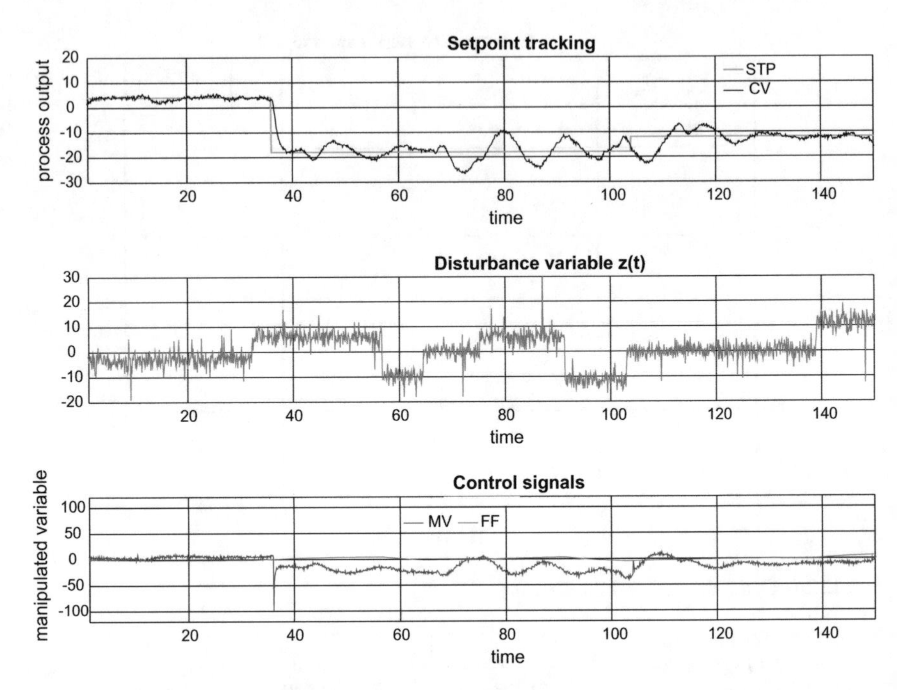

Fig. 12.27 Optimal feedback-feedforward loop time trends for case 2

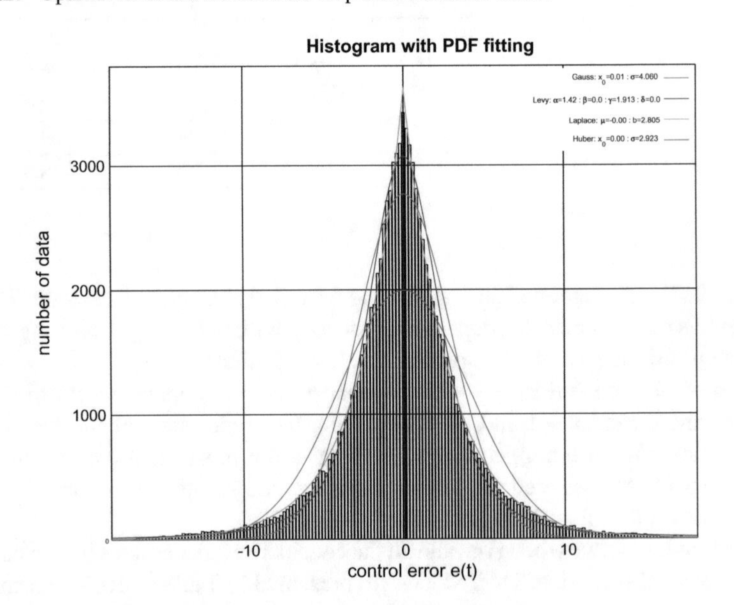

Fig. 12.28 Histogram for optimal feedback-feedforward loop (case 2)

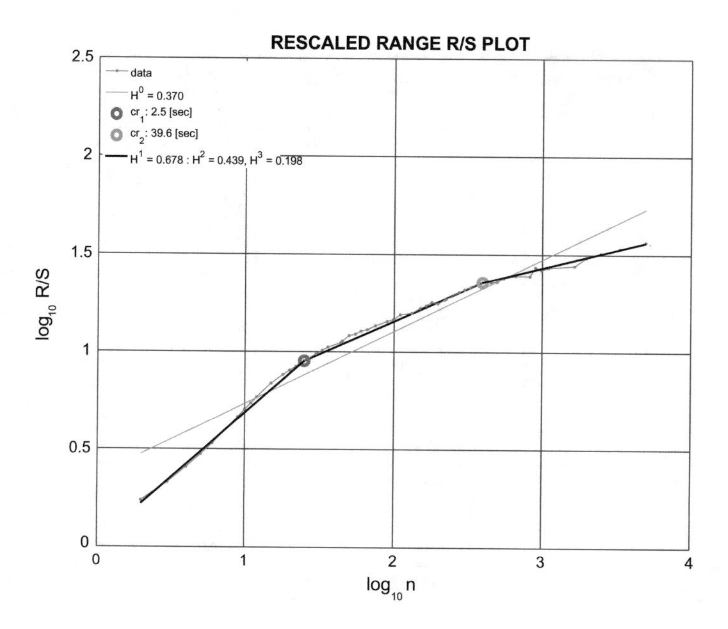

Fig. 12.29 R/S plot for optimal feedback-feedforward loop (case 2)

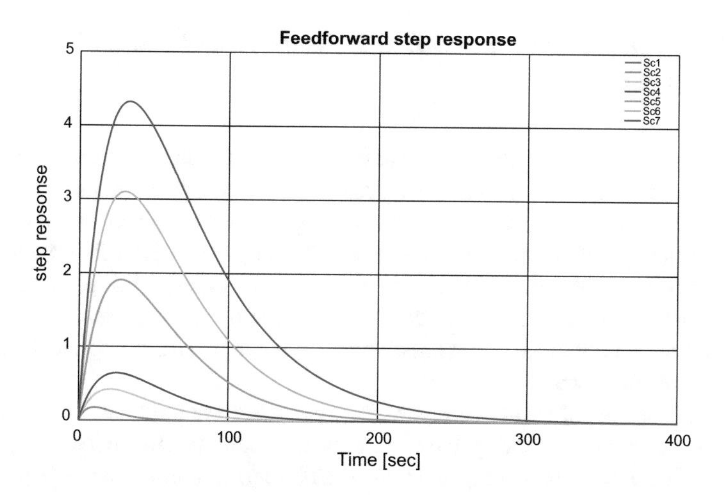

Fig. 12.30 Optimal feedback-feedforward loop time trends for case 2

associated with the no feedforward performance (scenario Sc1). This effect shows the limitation of the industrial formulation of the feedforward, which may not be efficient in case of the nonminimumphase disturbance impact character. Entropies are not able to detect loop control quality similarly to the previous cases.

Table 12.10 Case 1 feedback-feedforward loop tuning effect on performance indexes (the best indexes are highlighted in bold fonts, green color indicates optimal feedforward)

	Sc1	Sc2	Sc3	Sc4	Sc5	Sc6	Sc7
MSE	16.70	16.84	16.56	**16.48**	17.31	19.14	21.62
IAE	**2.78**	2.79	2.80	2.81	3.01	3.26	3.51
QE	73.05	72.32	67.09	**64.67**	74.33	111.74	174.65
σ	4.087	4.103	4.069	**4.060**	4.160	4.375	4.650
α	**1.359**	1.368	1.397	1.423	1.604	1.695	1.729
γ^L	**1.802**	1.823	1.873	1.914	2.274	2.577	2.841
b	**2.777**	2.793	2.796	2.806	3.008	3.255	3.515
σ^{rob}	**2.796**	2.825	2.881	2.925	3.374	3.788	4.157
H^0	0.376	0.372	0.371	0.371	0.381	0.404	0.432
H^1	**0.676**	0.679	0.677	0.678	0.686	0.696	0.705
H^2	0.478	0.458	0.457	0.454	0.502	0.548	0.591
H^3	0.175	0.171	0.187	0.199	0.208	0.182	0.166
H^{RE}	23.03	22.91	22.69	22.57	22.46	22.07	21.89
H^{DE}	−143431	−142118	−141436	−141026	−135722	−131555	−129231

12.2.3 *Summary of the Feedback-Feedforward Control Observations*

Feedback-feedforward control may be assessed in the similar way as the SISO loops or the cascaded control, however the approach is even more challenging. One has to consider two separate impacts: setpoint tracking and measurable disturbance decoupling. It is clearly visible that the character and the properties of the disturbance affect the control loop. It has to be noted that the decoupling element has been considered in the industrial form consisting of two lags with the opposite signs. Although such a formulation is very limited in its performance, it can be easily implemented using the DCS/PLC blockware.

The feedback-feedforward configuration assessment should focus on both aspects of control, i.e. setpoint tracking and disturbance decoupling. As the feedback control has been considered extensively in the SISO PID loop configurations (Chap. 11), the feedforward disturbance decoupling has bee analyzed in details.

Performance of the measures is consistent with previous results and expectations. The index robustness against outliers (fat tails) is required for the proper and reliable loop quality detection. It is optimally achievable with scale factor γ^L of the α-stable distribution and robust standard deviation σ^{rob} evaluated with the logistic ψ scale M-estimator. Unfortunately, other measures (like popular mean square error od normal standard deviation) do not behave properly in the situations with disturbances exhibiting outliers.

Feedback-feedforward control assessment examples confirm observations for the SISO and cascaded loops. Reliable CPA approach has to be robust against outliers. Indicators originating from the robust or fat tail domain have such an ability.

References

1. Adam, E.J., Marchetti, J.L.: Designing and tuning robust feedforward controllers. Comput. Chem. Eng. **28**(9), 1899–1911 (2004)
2. Rousseeuw, P.J., Leroy, A.M.: Robust Regression and Outlier Detection. Wiley, New York, NY, USA (1987)

Chapter 13
Industrial Simulated Examples

Science is the great antidote to the poison of enthusiasm and superstition.

– Adam Smith

13.1 Drum Level Control

Drum level control is used as the first example originating from the industry. The level control process is crucial for the overall performance of the general boiler controls, as its improper design or poor tuning affect other boiler controls often generating oscillations in other variables and loops, like in steam temperature and pressure or load demand control. The example has been already considered by the researchers [6, 8].

The process consists of three main elements: steam-water separation in a drum, feedwater control valve and the varying steam consumption disturbing the operation. The so called three element control, i.s. cascaded control configuration with the feedforward disturbance decoupling (see Fig. 10.9) constitutes the industrial drum level control standard [3]. There might be found several formulations of the process equations. The simplified ones (10.26) proposed in [2] forms the good compromise between the simplicity and the process nonminimumphase properties (Figs. 13.1, 13.2 and 13.3).

The tracked setpoint used in the simulations is in the shape of the rectangular wave with varying height and width of the steps (upper plot in Fig. 13.4). The simulations are disturbed. Disturbance variable, which is being decoupled consists of the varying step wave with added α-stable symmetrical, zero-mean noise and $\alpha = 1.8$ and $\gamma = 1.6$ (middle plot in Fig. 13.4). The inner loop output is disturbed with zero-mean Gaussian noise and $\sigma_2 = 0.1$, while the noise in the main loop is more significant with $\sigma_2 = 0.5$.

P. D. Domański, *Control Performance Assessment: Theoretical Analyses and Industrial Practice*, Studies in Systems, Decision and Control 245,
https://doi.org/10.1007/978-3-030-23593-2_13

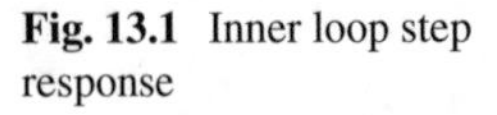

Fig. 13.1 Inner loop step response

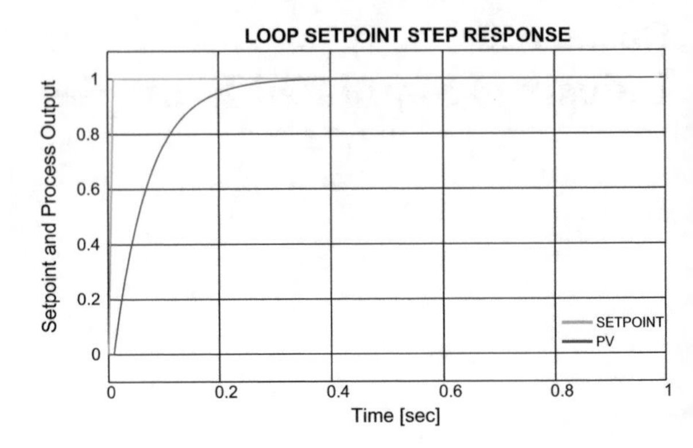

Fig. 13.2 Main loop step response

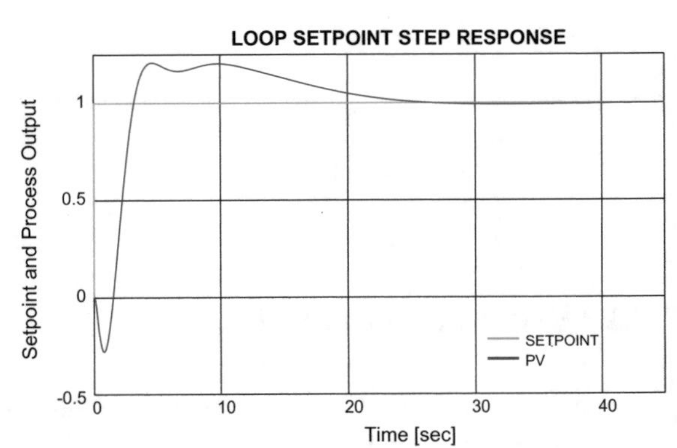

At first the so called optimal control configuration has been tuned. It consists of three elements:

1. inner loop PI controller $R_2(s)$ with parameters: $k_p = 2.35$ and $T_i = 0.15$,
2. main loop PID controller $R_1(s)$ with parameters: $k_p = 1.5$, $T_i = 40.5$ and $T_d = 2.2$,
3. and feedforward element $R_{FF}(s)$ with settings: $K = 2.5$; $T_1 = 0.001$; $T_2 = 0.2$.

The designed control loops are described by the basic performance indexes, i.e. the settling time and the overshoot. The inner valve control loop consisting of the valve transfer function $G_v(s)$ and PI controller $R_2(s)$ exhibits settling time $T_{set} = 0.24$ [s] and overshoot $\kappa = 0.00$ [%]. The respective loop step response is presented in Fig. 13.1. Tuning of the main loop is challenging. The nonminimumphase behavior deteriorates the control performance. Secondly the compromise between the control time and overshoot is needed. The step response (Fig. 13.2) features settling time $T_{set} = 24.46$ [s] and relatively large overshoot $\kappa = 20.8$ [%]. The feedforward element has been tuned for the fast reaction. The designed parameters are: $K = 16.9$; $T_1 = 0.005$; $T_2 = 0.025$. The respective feedforward unit step response is sketched in Fig. 13.3.

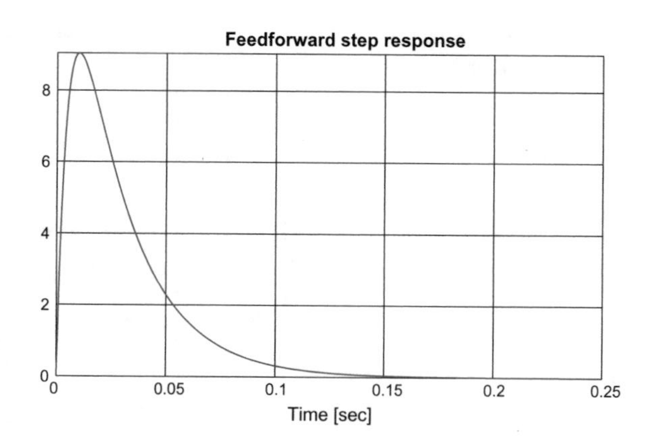

Fig. 13.3 Feedforward step response

13.1.1 Simulations

All the parts of the three element control has been put together and the simulation impacted by the disturbances has been run. Time trends of the main process variables are shown in Fig. 13.4. The deteriorating effect of the disturbances is well seen together with the noisy and persistent character of the disturbance. Similarly to the previously defined assessment procedure the statistical analysis using histogram and PDF fitting is presented in Fig. 13.5, while rescaled range R/S plots are sketched in Figs. 13.6. The histogram shows the fat-tailed loop character with the best fitting by

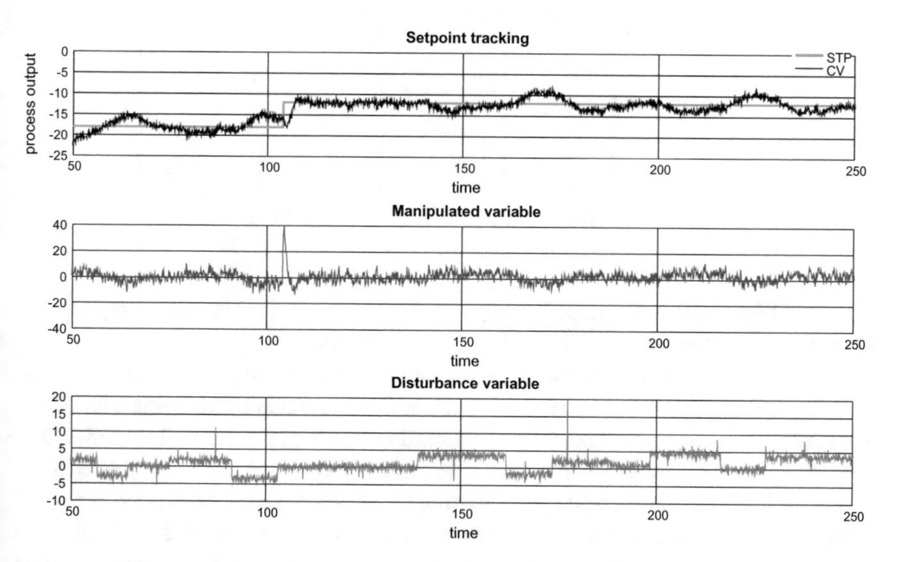

Fig. 13.4 Drum level control performance

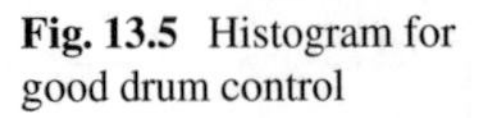

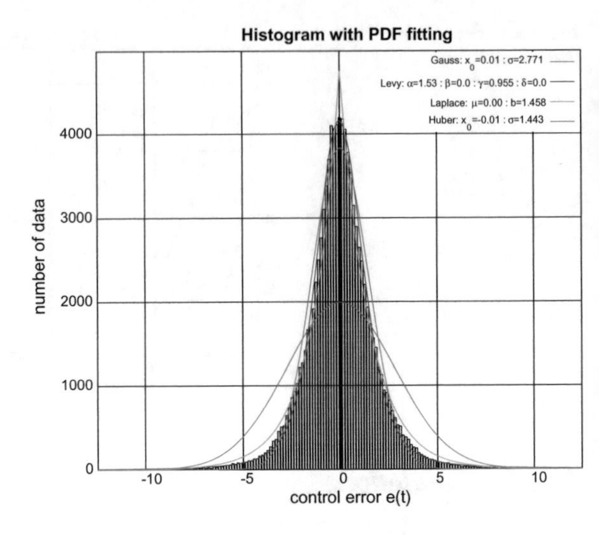

Fig. 13.5 Histogram for good drum control

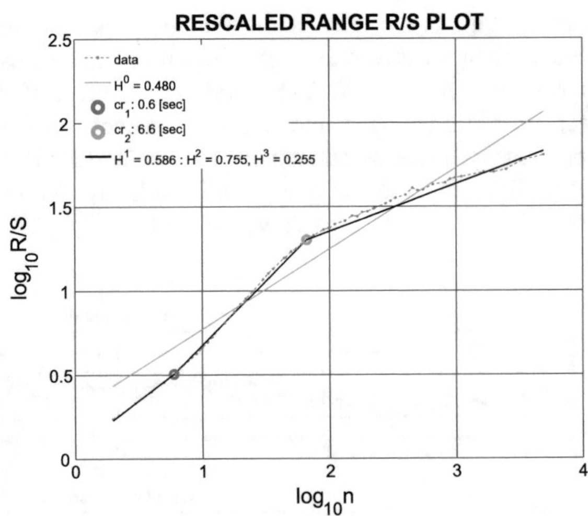

Fig. 13.6 R/S plot for good drum control

robust PDFs. The R/S plot depicts multi-persistent loop properties exhibiting slightly sluggish performance ($H = 0.59$), visible in simulations.

Each of the control configuration elements has been prepared in three forms. The controllers apart from the designed ones, has sluggish and aggressive form. Parameters of the controllers and feedforward units are presented in Table 13.1. The feedforward element can be from one side turned off or tuned to be very intensive.

As each of the control units participating in the three element drum level control exhibits three different parameters setting the overall number of the simulations is equal to 27. The analysis starts with the evaluation of the basic loops dynamic properties, i.e the setting time and the overshoot, which are presented in Table 13.2. The fast inner loop dynamics is visible with short settling time. The main loop does

Table 13.1 Settings of the drum level control units

Control unit	Label	No	k_p	T_i	T_d	Description
R2	R2num	1	0.9	0.25	–	Sluggish
		2	2.35	0.15	–	Good
		3	10	0.04	–	Aggressive
R1	R1num	1	1.05	15.5	1.3	Sluggish
		2	1.65	9.5	2.3	Good
		3	3.2	75.5	1.7	Aggressive
			K	T_1	T_2	
R_{FF}	FFnum	1	–	–	–	No feedforward
		2	16.9	0.005	0.025	Good
		3	16.9	0.1	0.4	Intensive

not perform very well, with significant oscillations. Optimal control performance is highlighted in green color.

Table 13.3 presents the summary of all calculated indexes for each simulated scenario. The table is divided into three sections according to the tuning of the inner control loop. Unfortunately, the reading of such table results representation is difficult. Tthe radar plots have been prepared for clear visualization of the results.

Previous analysis have shown several important aspects that might improve radar visualization. There are expected equivalences between the indexes, i.e. MSE is equivalent with normal standard deviation σ and IAE with Laplace scaling b. Thus there is no need to represent them all. The MSE and IAE will be used in the radar plots. Robust indexes of scale factor γ^L for the α stable distribution and robust standard deviation σ^{rob} evaluated with the logistic ψ scale M-estimator have been always proving their efficiency so they are also represented in the radar plots. Additionally, the short memory Hurst exponent H^1 is added, while entropies are excluded from the plots.

Three radar plots has been prepared reflecting different tuning of the inner control loop (controller R_2). The real values of the measures Ind are scaled Ind^{sc} against the best found value of each considered index Ind^{opt} (13.1) to be presented in the plot. The unitary index value $Ind^{sc} = 1$ is the best one. Only Hurst exponent (Fig. 13.10) remains unscaled, what enables to show its variations from the value $H = 0.5$.

$$Ind^{sc} = \frac{Ind^{opt}}{Ind} \tag{13.1}$$

The first three plots visualize behavior of the integral (MSE and IAE) and statistical indexes (γ^L and σ^{rob}) for three situations. Figure 13.7 presents the indexes for the situation, where the inner controller is tunned to be very slow, i.e. $R2num = 1$. Figure 13.8 compares the indexes in case of the optimal inner controller settings ($R2num = 2$), and Fig. 13.9 shows according relationships for the aggressive inner

Table 13.2 Drum level loops performance indexes (green color indicates optimal tuning)

SimN	R2num	Ffnum	R1num	Inner loop		Main loop	
				t_s	κ	t_s	κ (%)
1	1	1	1	0.920	0%	40.96	29.5
2	1	1	2			23.80	33.4
3	1	1	3			41.50	104.4
4	1	2	1			40.96	29.5
5	1	2	2			23.80	33.4
6	1	2	3			41.50	104.4
7	1	3	1			40.96	29.5
8	1	3	2			23.80	33.4
9	1	3	3			41.50	104.4
10	2	1	1	0.290	0%	41.15	26.4
11	2	1	2			24.46	20.8
12	2	1	3			24.10	93.4
13	2	2	1			41.15	26.4
14	2	2	2			24.46	20.8
15	2	2	3			24.10	93.4
16	2	3	1			41.15	26.4
17	2	3	2			24.46	20.8
18	2	3	3			24.10	93.4
19	3	1	1	0.140	11.1%	41.23	25.7
20	3	1	2			24.63	20.0
21	3	1	3			18.21	83.0
22	3	2	1			41.23	25.7
23	3	2	2			24.63	20.0
24	3	2	3			18.21	83.0
25	3	3	1			41.23	25.7
26	3	3	2			24.63	20.0
27	3	3	3			18.21	83.0

loop performance with $R2num = 3$. The last presented radar plot (Fig. 13.10) shows only one index, i.e. Hurst exponent H^1 for all considered tuning configurations of the three element control.

At first case, i.e. sluggish inner control, the MSE index indicates optimal control ($SimN = 2, 5, 8$). Optimal solution is also indicated by other measures in case of intensive feedforward ($FFnum = 3$). In two other cases robust measures suggest more aggressive setup. Sluggish tuning is always rejected.

The optimal tuning of the inner loop reveals very similar performance. The MSE measure exhibits towards selected optimal control, while other, robust, indexes tends towards more aggressive configurations. Exactly the same performance is observed in

Drum level all performance indexes (green color indicates optimal tuning)

SimN	MSE	IAE	QE	σ	α	γ^L	L	σ^{rob}	H^0	H^1	H^2	H^3	H^{RE}	H^{DE}
1	13.10	2.33	19.0	3.619	1.622	1.675	2.325	2.50	0.521	0.566	0.794	0.270	28.2	−156452
2	8.42	1.53	25.0	2.902	1.483	0.965	1.530	1.47	0.488	0.743	0.599	0.231	29.5	−183825
3	12.55	1.54	117.5	3.542	1.574	0.827	1.535	1.23	0.482	0.567	0.667	0.271	34.3	−214984
4	12.99	2.31	18.9	3.605	1.622	1.661	2.311	2.48	0.521	0.784	0.665	0.207	28.2	−158016
5	8.37	1.52	24.9	2.894	1.487	0.958	1.521	1.46	0.487	0.594	0.755	0.272	29.3	−183063
6	12.42	1.53	116.5	3.524	1.580	0.821	1.528	1.22	0.486	0.571	0.666	0.288	34.8	−215228
7	16.82	2.73	23.8	4.101	1.466	1.882	2.734	2.86	0.503	0.578	0.769	0.269	28.6	−157722
8	10.86	1.85	31.1	3.295	1.236	1.012	1.854	1.62	0.463	0.701	0.544	0.228	29.5	−179382
9	15.60	2.11	144.4	3.950	1.356	1.201	2.108	1.86	0.461	0.562	0.533	0.241	35.1	−212031
10	12.31	2.26	18.1	3.508	1.654	1.653	2.258	2.45	0.521	0.566	0.797	0.261	28.1	−161064
11	7.70	1.47	24.9	2.775	1.524	0.959	1.466	1.45	0.481	0.586	0.755	0.264	29.8	−192850
12	9.38	1.26	108.2	3.063	1.876	0.776	1.256	1.12	0.499	0.547	0.686	0.288	34.2	−242810
13	12.20	2.24	18.0	3.494	1.654	1.643	2.244	2.43	0.524	0.600	0.798	0.269	28.2	−161639
14	7.67	1.46	24.8	2.769	1.536	0.956	1.457	1.44	0.481	0.587	0.754	0.259	29.1	−192942
15	9.40	1.25	108.8	3.066	1.854	0.768	1.250	1.11	0.498	0.545	0.682	0.285	34.8	−243231
16	15.71	2.66	22.6	3.963	1.508	1.871	2.657	2.83	0.505	0.572	0.775	0.265	28.6	−160789
17	9.89	1.78	30.7	3.144	1.299	1.039	1.784	1.65	0.462	0.710	0.545	0.236	29.8	−189649
18	11.61	1.68	132.2	3.407	1.437	0.996	1.684	1.51	0.479	0.595	0.558	0.224	35.4	−238128
19	12.07	2.24	17.9	3.474	1.668	1.650	2.240	2.44	0.521	0.786	0.709	0.216	27.9	−160305
20	7.48	1.45	24.8	2.735	1.535	0.953	1.446	1.44	0.482	0.585	0.755	0.263	29.0	−193181
21	8.32	1.17	103.9	2.885	1.959	0.758	1.168	1.09	0.503	0.548	0.693	0.292	34.7	−251549
22	12.01	2.23	17.8	3.465	1.667	1.639	2.228	2.42	0.523	0.605	0.798	0.262	28.1	−163508
23	7.45	1.44	24.7	2.729	1.542	0.951	1.438	1.43	0.477	0.584	0.755	0.254	29.0	−192949
24	8.28	1.16	103.6	2.878	1.911	0.745	1.161	1.08	0.502	0.539	0.692	0.280	34.4	−250125
25	15.44	2.64	22.3	3.929	1.526	1.875	2.640	2.83	0.503	0.570	0.776	0.256	28.8	−163131
26	9.61	1.77	30.7	3.100	1.319	1.049	1.768	1.66	0.460	0.710	0.543	0.229	30.1	−192443
27	10.27	1.55	126.1	3.205	1.464	0.929	1.552	1.41	0.481	0.609	0.571	0.220	34.9	−244452

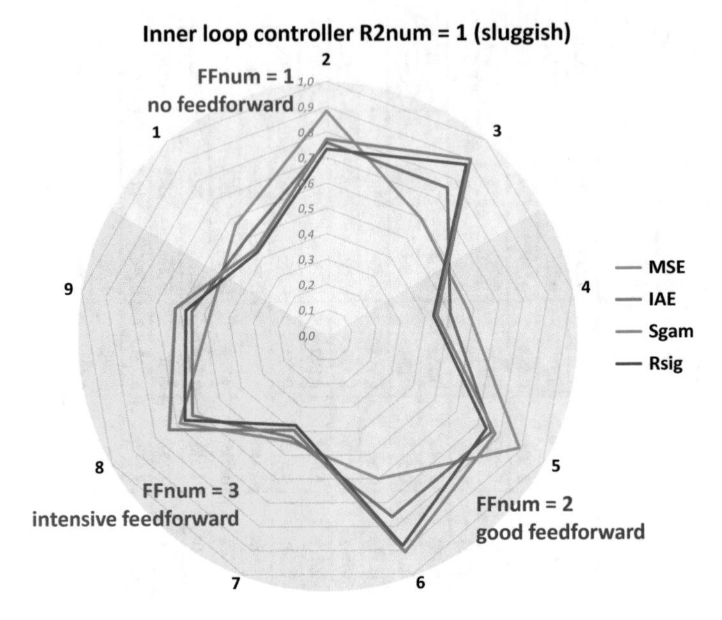

Fig. 13.7 Radar plot for drum control performance—R2num = 1

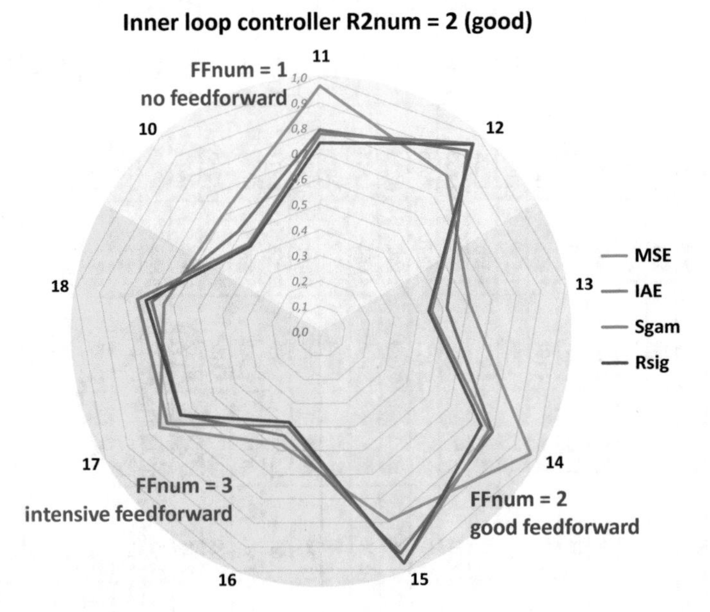

Fig. 13.8 Radar plot for drum control performance—R2num = 2

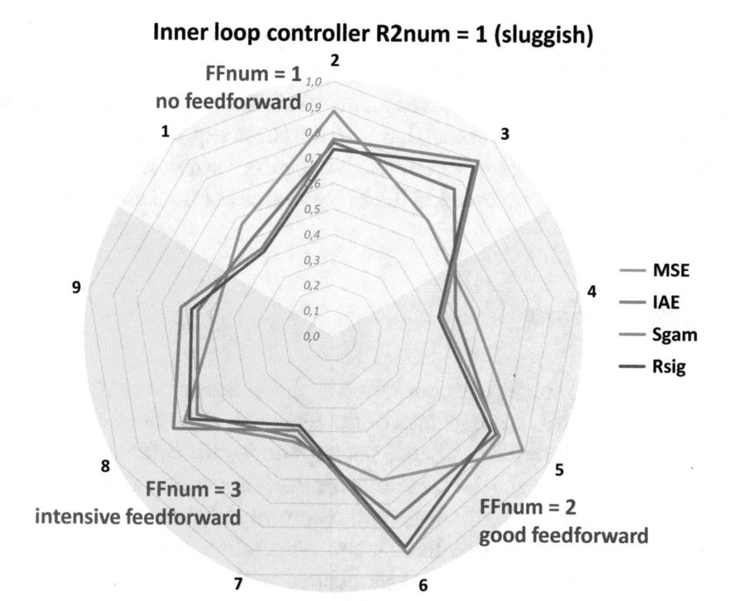

Fig. 13.9 Radar plot for drum control performance—R2num = 3

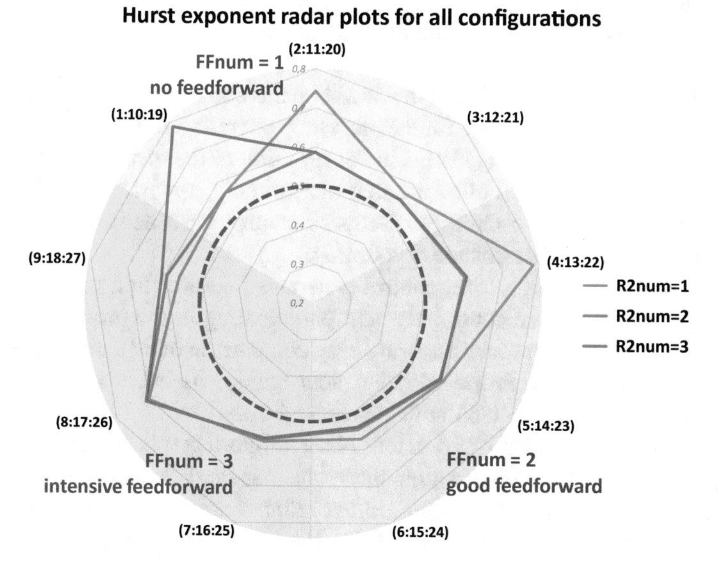

Fig. 13.10 Radar plot for drum control Hurst exponent

the third, inner loop aggressive tuning setup. Additionally, the very similar behavior of γ^L and σ^{rob} in all situations is clearly observed.

Short memory Hurst exponent gives further insight into the control performance. In general goof feedforward is associated with the Hurst exponent values closer to the independent realization of $H = 0.5$. Intensive feedforward in fact generates slower control reaction, which is reflected in the higher values of the Hurst exponent.

13.1.2 Observations

Three element control strategy is considered as the industrial standard for the steam boiler drum level control. It constitutes of the feedback cascaded two elements supported with the feedforward disturbance decoupling feedforward. Actually, the control is challenging, as it forms the backbone for the other steam boiler controls. If the drum level is not controlled satisfactory, the control deficiencies, like oscillations, propagate to other process parts, like steam temperature, pressure or flow performance. The control challenge is also associated with the swell and shrink effect resulting in the nonminimumphase behavior.

The inner loop in the cascaded structure is an easy part. It controls the water flow through the feedwater valve. Its design and tuning does not generate problems. The upstream main control loop and the feedforward has to cope with the nonminimumphase process properties.

The designed control consists of the downstream loop PI controller $R_2(s)$ and the PID algorithm being responsible for the main loop operation $R_1(s)$. The feedforward disturbance decoupling $R_{FF}(s)$ helps in the rejection of the disturbances generated by varying steam demand. Following that the example comprises effects featured in all previous examples of the SISO feedback control, cascaded configuration and industrial feedforward disturbance decoupling.

Assessment of the drum level control generates interpretation problems as one has to cope with the three control element participating in the process. There is a need for multi-criteria approach allowing easy comparison of different measures for different control element. Proposed radar plots enable possible visualization much easier in interpretation than tabular data presentation.

The assessment also confirms previous observations that robust measures of scale factor γ^L for the α stable distribution and robust standard deviation σ^{rob} evaluated with the logistic ψ scale M-estimator allow reliable assessment. We also observe that they tend to indicate more aggressive loop settings comparing to the MSE and equivalent normal standard deviation σ. Apart from previously observed index equivalences, i.e. $MSE \Leftrightarrow \sigma$ and $IAE \Leftrightarrow b$ there is also noticeable equivalence $\gamma^L \Leftrightarrow \sigma^{rob}$. These equivalences might help to minimize the number of used measures in the assessment process and to protect the engineer from using different measures bringing to the picture the same information. Additionally, short memory Hurst exponent H^1 can be also used in the assessment process, making possible differentiation between the type of poor tuning, i.e. the diversity between aggressive and sluggish control.

The considered drum level three element control example enable testing of the methods and measures in the challenging industrial example simulated together with the fat-tailed disturbances. The complexity of the control strategy demands fusion of different indexes measured for different control variables. Complexity also demands better visualization tools merging different information. Radar plot is proposed showing its ability to combine various information in one picture.

13.2 Linear Binary Distillation Column

Linear binary distillation column is often considered as an important and reasonable example of the multivariate control. It consists of the TITO (Two Input Two Output) model heaving internal cross-coupling and external disturbance. The disturbance variable is considered as the unmeasured one in the considered case. The industrial feedforward is used to decouple internal cross dependencies simulated through transfer functions $G_{12}(s)$ and $G_{21}(s)$ of the model (10.27).

Apart from the selected linear distillation column model of Wood and Berry [7], there are many other formulations of the distillation column TITO model [1, 4, 5]. The selection has been done due to the model popularity as the control benchmark, mostly for the MPC. Proposed linear control structure is sketched in Fig. 10.11.

The setpoint applied in the simulations is in the shape of the rectangular wave with varying height and width of the steps (upper plot in Fig. 13.11). The simulations are disturbed. Disturbance variable $X_f(t)$ consists of the varying step wave with added α-stable symmetrical, zero-mean noise and $\alpha = 1.8$ and $\gamma = 1.6$. Both loop outputs are disturbed with zero-mean Gaussian noise and $\sigma_2 = 0.02$. At first, the so called optimal control configuration has been tuned. It consists of four elements:

1. $R_1(s)$ PID controller with parameters: $k_p = 0.49$, $T_i = 9.0$ and $T_d = 1.10$,
2. $R_2(s)$ PID controller with parameters: $k_p = -0.25$ and $T_i = 1.15$,
3. feedforward element $F_1(s)$ with settings: $K = 1.4$; $T_1 = 0.4$; $T_2 = 0.45$,
4. and feedforward element $F_2(s)$ with settings: $K = -2.5$; $T_1 = 1.8$; $T_2 = 1.9$.

The designed control loops are described with basic performance indexes, i.e. the settling time and the overshoot. The $X_d(t)$ control loop with main $R_1(s)$ controller and $F_1(s)$ decoupler exhibits settling time $T_{set} = 28.25$ [s] and overshoot $\kappa = 6.72$ [%]. The second, $X_b(t)$ control loop with main $R_2(s)$ controller and $F_2(s)$ feedforward exhibits settling time $T_{set} = 27.81$ [s] and overshoot $\kappa = 7.69$ [%]. Tuning of the loops is very challenging, mainly due to the cross-dependencies between the loops through transfer functions $G_{12}(s)$ and $G_{21}(s)$.

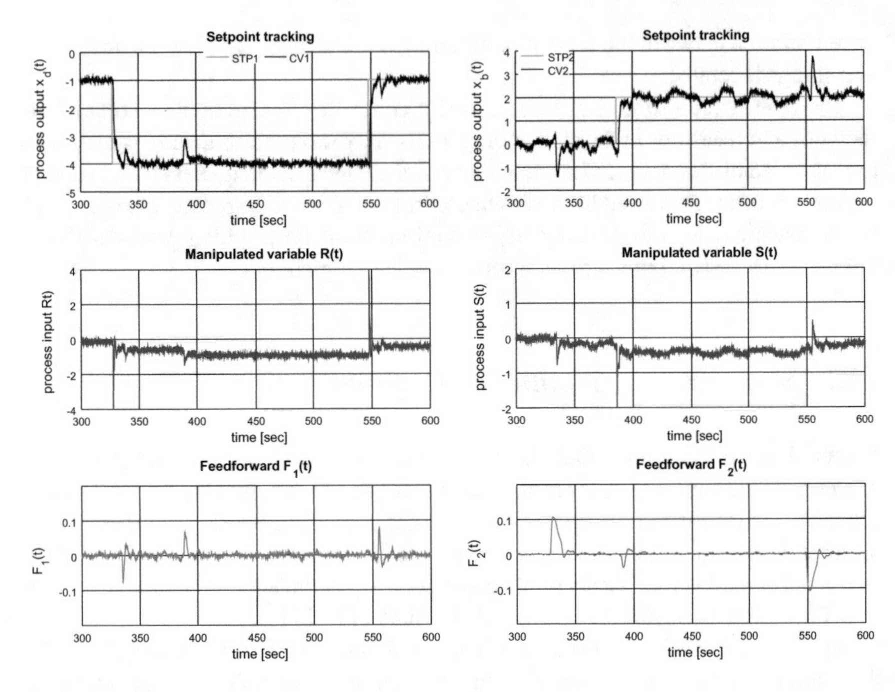

Fig. 13.11 Drum level control performance

13.2.1 Simulations

Simulation trends of the main process variables for the whole TITO system with embedded disturbance and measurement noises are shown in Fig. 13.4. The deteriorating effect of the disturbances is well seen, together with noisy and persistent character of the disturbance. Similarly to the previously defined assessment procedure the statistical analysis using histogram and PDF fitting is presented in Fig. 13.12, while rescaled range R/S plots are in Fig. 13.13. Histograms show the fat-tailed loop character with the best fitting by robust PDFs, i.e. α-stable and to some extend the robust one. The R/S plot depicts slightly multi-persistent loop properties exhibiting slightly sluggish performance ($H = 0.59$), visible in time trends as well.

Each of the controllers $R_1(s)$ and $R_2(s)$ has been tuned in three forms (associated with different performance). The controllers, apart from the designed ones, has sluggish or aggressive form. It is assumed that the parameters of feedforward decouplers do not change within the simulations. Selected parameters of the controllers and feedforward units are presented in Table 13.4.

Overall number of the simulations is equal to 9, as each of the controllers participating in the column control exhibits three different parameters settings. The analysis

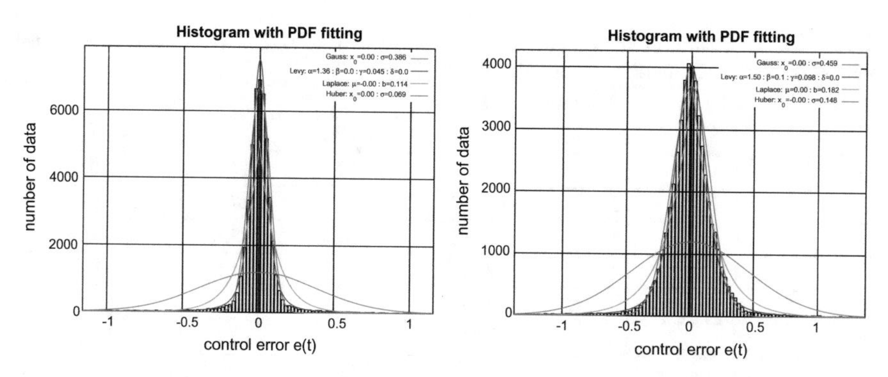

Fig. 13.12 Histograms for good column control (R_1 loop—left, R_2 loop—right)

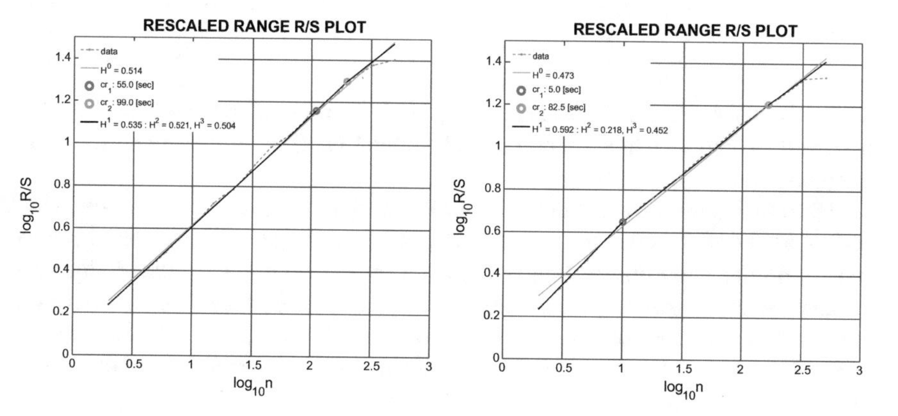

Fig. 13.13 RS plots for good column control (R_1 loop—left, R_2 loop—right)

Table 13.4 Settings of the elements participating on the distillation column TITO control

Control unit	Label	No	k_p	T_i	T_d	Description
R1	R1num	1	0.09	12.0	0.20	Sluggish
		2	0.49	9.0	1.10	Good
		3	0.69	6.0	1.30	Aggressive
R2	R2num	1	−0.05	12.5	0.15	Sluggish
		2	−0.25	9.5	1.15	Good
		3	−0.35	8.5	1.25	Aggressive

Table 13.5 Distillation column control loops performance indexes (green color indicates optimal tuning) for different simulation setups (denoted as SimN)

SimN	R1num	R2num	R1 loop		R2 loop	
			T_{set}	κ (%)	T_{set}	κ (%)
1	1	1	71.71	2.73	41.51	0.62
2	1	2	71.71	2.73	27.81	7.69
3	1	3	71.71	2.73	22.81	44.70
4	2	1	28.25	6.72	41.51	0.62
5	2	2	28.25	6.72	27.81	7.69
6	2	3	28.25	6.72	22.81	44.70
7	3	1	19.84	13.61	41.51	0.62
8	3	2	19.84	13.61	27.81	7.69
9	3	3	19.84	13.61	22.81	44.70

starts with the evaluation of basic loops dynamic properties, i.e the settling time T_{set} and the overshoot κ, which are presented in Table 13.5. Optimal control performance is highlighted in green color.

Table 13.6 presents the summary of all calculated indexes for each simulated scenario. The table is divided into two sections according to the considered control loop. An interesting observation is visible. The best solutions for the R_1 loop are obtained with sluggish control of the other loop and the opposite. Such a conclusion is clear. Sluggish control in one loop generates smaller disturbance to the opposite one and vice versa. There is a need for the joint measures describing the overall TITO system performance.

Following, Table 13.7 compares unified values of the indexes. The unification is done through averaging, as both controlled variables vary in the similar regimes. It is well seen that the selected optimal control is properly detected with the majority of the indexes. Only equivalent indexes of IAE and Laplace scaling b have a tendency towards aggressive tuning. The feature of the Hurst exponent to differentiate between sluggish $H > 0.5$ and aggressive control $H < 0.5$ is also observed.

Unfortunately, the readability of such large tabular representation is difficult. The radar plots have been prepared to improve visualization of the results. Previous analysis have used different ways for radar visualization. Distillation column configuration includes two coupled loops with the indexes separately calculated for each control error. Thus the first radar representation (Fig. 13.14) includes behavior of the main indexes, i.e. MSE, IAE, scale factor γ^L for the α stable distribution (denoted `Sgam` on the plot), robust standard deviation σ^{rob} evaluated with the logistic ψ scale M-estimator (denoted `Rsig` on the plot) and the single Hurst exponent H^0 (denoted `H0` on the plot). The real values of the indexes (except Hurst exponent) are scaled as in the previous example (13.1). Such a selection uses previously acquired knowledge about the properties and applicability of the CPA indexes. The selection of the H^0

Table 13.6 Distillation column all performance indexes with highlighted in bold the best indexes (green color indicates optimal tuning)

SimN	Loop R1													
	MSE	IAE	QE	σ	α	γ^L	L	σ^{rob}	H^0	H^1	H^2	H^3	H^{RE}	H^{DE}
1	0.717	0.485	0.775	0.854	1.105	0.225	0.482	0.375	0.691	0.583	0.700	0.687	5.67	−26864
2	0.926	0.512	0.990	0.969	0.807	0.130	0.508	0.256	0.686	0.601	0.750	0.454	5.76	−28214
3	0.966	0.519	1.031	0.990	0.812	0.135	0.515	0.261	0.680	0.603	0.744	0.446	5.86	−27860
4	0.118	0.112	0.197	0.353	1.497	0.060	0.112	0.090	0.523	0.559	0.609	0.507	3.76	−33791
5	0.143	0.114	0.231	0.386	1.369	0.045	0.114	0.069	0.502	0.554	0.553	0.396	3.89	−34321
6	0.181	0.147	0.274	0.432	1.225	0.058	0.146	0.092	0.503	0.572	0.602	0.480	4.38	−33019
7	0.099	0.090	0.207	0.325	1.630	0.052	0.091	0.076	0.493	0.570	0.572	0.343	3.30	−34905
8	0.119	0.100	0.240	0.352	1.528	0.046	0.101	0.069	0.462	0.526	0.493	0.435	3.49	−34494
9	0.194	0.213	0.339	0.448	1.280	0.115	0.215	0.182	0.445	0.572	0.624	0.350	4.36	−30952
SimN	Loop R2													
	MSE	IAE	QE	σ	α	γ^L	L	σ^{rob}	H^0	H^1	H^2	H^3	H^{RE}	H^{DE}
1	0.824	0.555	0.840	0.889	1.379	0.343	0.541	0.514	0.671	0.610	0.690	0.666	5.69	−25471
2	0.175	0.161	0.198	0.411	1.569	0.095	0.157	0.143	0.477	0.602	0.622	0.449	3.81	−28167
3	0.170	0.143	0.199	0.403	1.503	0.076	0.140	0.116	0.466	0.591	0.497	0.363	3.69	−28477
4	0.961	0.595	0.978	0.964	1.177	0.311	0.579	0.496	0.668	0.606	0.683	0.661	5.77	−25683
5	0.215	0.183	0.240	0.459	1.504	0.098	0.182	0.148	0.474	0.584	0.615	0.446	4.05	−27459
6	0.260	0.225	0.295	0.507	1.226	0.105	0.222	0.165	0.465	0.577	0.581	0.414	4.16	−27573
7	0.991	0.600	1.009	0.981	1.154	0.306	0.584	0.491	0.666	0.598	0.749	0.658	5.80	−25633
8	0.246	0.194	0.272	0.492	1.521	0.101	0.193	0.153	0.471	0.566	0.561	0.431	4.70	−27449
9	0.490	0.419	0.541	0.701	1.275	0.240	0.418	0.382	0.441	0.599	0.663	0.355	5.14	−26235

Table 13.7 Distillation column all performance indexes in the unified form (averaged values) with highlighted in bold the best indexes (green color indicates optimal tuning)

SimN	Loop R1													
	MSE	IAE	QE	σ	α	γ^L	L	σ^{rob}	H^0	H^1	H^2	H^3	H^{RE}	H^{DE}
1	0.770	0.520	0.808	0.871	1.242	0.284	0.512	0.444	0.681	0.596	0.695	0.676	5.68	−26167
2	0.550	0.336	0.594	0.690	1.188	0.113	0.333	0.199	0.581	0.602	0.686	0.451	4.78	−28190
3	0.568	0.331	0.615	0.697	1.158	0.105	0.328	0.188	0.573	0.597	0.620	0.404	4.78	−28169
4	0.539	0.354	0.587	0.659	1.337	0.186	0.345	0.293	0.596	0.582	0.646	0.584	4.77	−29737
5	0.179	0.149	0.236	0.422	1.436	0.072	0.148	0.109	0.488	0.569	0.584	0.421	3.97	−30890
6	0.220	0.186	0.285	0.469	1.226	0.082	0.184	0.128	0.484	0.574	0.592	0.447	4.27	−30296
7	0.545	0.345	0.608	0.653	1.392	0.179	0.338	0.284	0.580	0.584	0.661	0.501	4.55	−30269
8	0.182	0.147	0.256	0.422	1.525	0.074	0.147	0.111	0.466	0.546	0.527	0.433	4.10	−30971
9	0.342	0.316	0.440	0.575	1.277	0.177	0.316	0.282	0.443	0.585	0.644	0.353	4.75	−28594

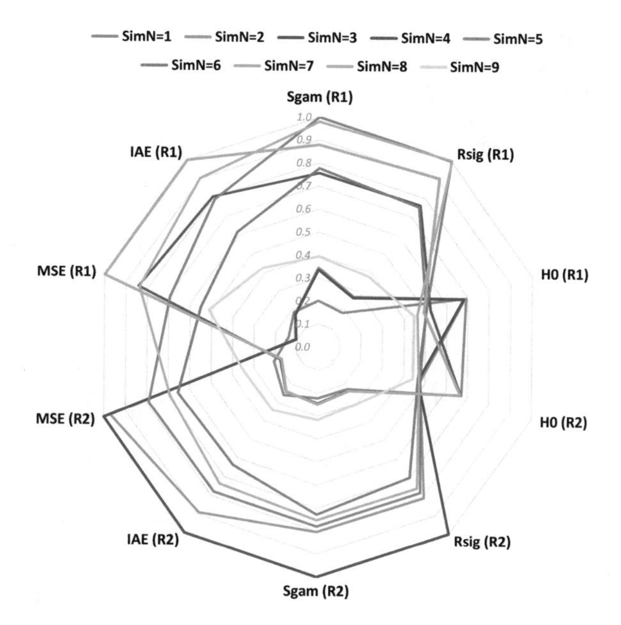

Fig. 13.14 Radar plot for distillation column control performance—comparison of both loops (red—optimal tuning)

Hurst exponent is justified with close to mono-persistent character of the R/S plots. Upper half of the plot presents indexes for the R_1 loop while the lower one for the loop R_2.

Radar visualization facilitates observation of the properties already discussed above. The effect of the selection as optimal loop scenario exhibiting sluggish performance for the opposite loop is well seen, so optimal system tuning is not selected. The second radar plot (Fig. 13.15) shows unified system performance measured with the averaging of the loops indexes. As only unified indexes are shown, the number of shown measures is extended. They are grouped in a way to visualize equivalence between some of them.

It is now well seen that the pairs: MSE and normal standard deviation σ and IAE and Laplace scaling b tend to behave similarly. Additionally, the very similar behavior of γ^L and σ^{rob} is also noticeable. Hurst exponent gives further insight into the control performance. In general goof feedforward is associated with the Hurst exponent values closer to the independent realization of $H = 0.5$. The entropies are not so good. The optimal tuning is pointed out by all the indexes visualizing compromise between the performance of each of two loops.

13.2.2 Observations

Multivariate process control poses specific challenges due to the significant cross-coupling between the manipulated and controlled variables. The selection of the

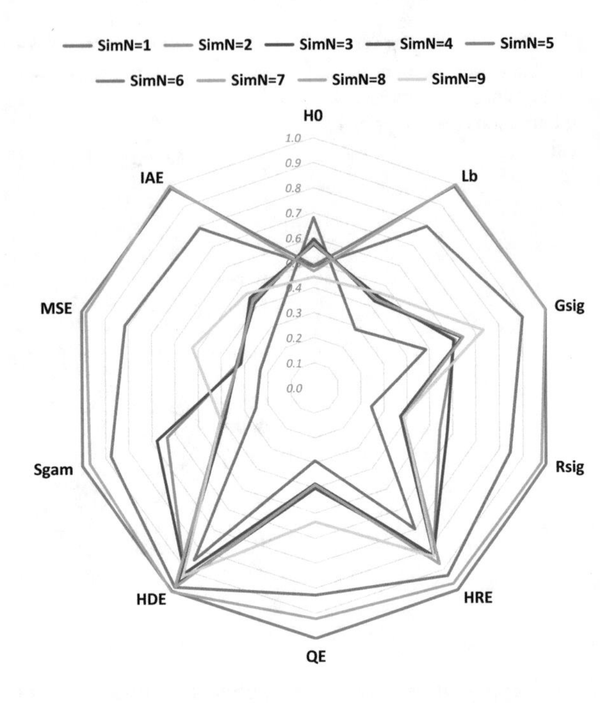

Fig. 13.15 Radar plot for distillation column control performance—unified system measures (red—optimal tuning)

multi-loop control system enhanced with the disturbance decoupling enables satisfactory process control, but at some cost. The system tuning has to be done with cautiousness, as the improvement in one loop may deteriorate the performance of the other loop. This effect is well seen in Table 13.6 and mostly in radar plot (Fig. 13.14).

Compromise is required and the loop costs has to be considered. Unification of the indexes has been done. It has been performed in the simplest way through the averaging of the loop indexes. Unified system approach enables proper assessment of the multivariate system. The observation that in the multivariate case the overall system approach is required not the loop-by-loop assessment is the main outcome of the considered case.

The simulations has additionally confirmed previous observations about the equivalence between some indexes, so there is no need to consider all of them in the assessment. The analysis of the loop statistical properties has confirmed for the next time that the outliers are common and the fat-tailed approach is required. Thus the indexes of the α-stable distribution or robust statistics explain loop performance (reflected in the histogram) in an appropriate way.

Finally, some comments should be devoted to the results visualization aspect. The multi-criteria approach is required as the system performance is often presented with different numbers (various indexes for many loops). Table representation does not enable transparent visualization. Proposed radar plots allow to reflect and show various multi-criteria properties in a single diagram.

References

1. Al-Shammari, A., Faqir, N., Binous, H.: Model predictive control of a binary distillation column. In: Petrova, V.M. (ed.) Advances in Engineering Research, vol. 8, pp. 255–270. Nova Science Publishers (2014)
2. Bequette, B.: Process Control: Modeling, Design, and Simulation, 1st edn. Prentice Hall Press, Upper Saddle River, NJ, USA (2002)
3. Gilman, G.F.: Boiler Control Systems Engineering, 2nd edn. ISA International Society for Automation, Research Triangle Park, NC, USA (2010)
4. Marquardt, W., Amrhein, M.: Development of a linear distillation model from design data for process control. Comput. Chem. Eng. **18**, S349–S353 (1994). European Symposium on Computer Aided Process Engineering–3
5. Minh, V.T., Rani, A.M.A.: Modeling and control of distillation column in a petroleum process. Math. Probl. Eng. (2009). Article ID 404,702
6. Pawlak, M.: Performance analysis of power boiler drum water level control systems. Acta Energ. **29**(4), 81–89 (2016)
7. Wood, R.K., Berry, M.W.: Terminal composition control of a binary distillation column. Chem. Eng. Sci. **28**(9), 1707–1717 (1973)
8. Zhang, Z.: Performance assessment for the feedforward-cascade drum water level control system. Int. J. Control. Autom. **9**(12), 96–98 (2016)

Chapter 14
MPC Control Examples

Our patience will achieve more than our force.

–Edmund Burke

The assessment covers the respective aspects of the MPC design, i.e. embedded model quality and its fitness, like gain delay or the dynamics, control horizon designation an improper selection of the performance index tuning coefficients. They are validated with univariate and multivariate processes.

Model Predictive Control assessment generates other challenges than single or multi-loop PID based control. The difference is fundamental as the control philosophy differ. In case of the multi-loop PID-based configurations the loops and all the feedforward units are tuned separately. It makes the tuning process difficult, tedious and without explicit indications. Multi-criteria considerations and cross-coupling is hard to be formalized. Actually, such a tuning is mostly based on the control engineer experience, rather than on the strict procedures.

MPC algorithm forms single entity. Multi-criteria aspects are expressed within the performance index, while internal cross-coupling is hidden inside of the models. Although, the MIMO MPC tuning is still a complex problem, it is easier and unambiguous, than complex multi-loop feedback structures with several feedforward units.

The control performance assessment of the MPC shifts the operational perspective comparing to the multi-loop PID configurations. Although we still have separate control errors connected with each controlled variable, the controller is a single entity. Thus the integration and coordination between the loops is already solved. The assessed control quality of one loop internally incorporates a compromise with the performance of other loops. The assessment tasks migrates from the consec-

P. D. Domański, *Control Performance Assessment: Theoretical Analyses and Industrial Practice*, Studies in Systems, Decision and Control 245,
https://doi.org/10.1007/978-3-030-23593-2_14

utive single loop controllers and their parameters (SISO loop level) to the single multivariate APC system level.

The assessment for the MIMO predictive control system should cover the respective aspects of the MPC design:

- Embedded model quality and its fitness, like gain delay or the dynamics
- Control horizon designation,
- Improper selection of the performance index tuning coefficients μ_n and λ_n for the consecutive controlled variables ($n = 1, \ldots, n_{\mathrm{y}}$).

The presentation of the CPA task for the MPC control is divided into two examples. At first (Sect. 14.1), the SISO case will be considered, which enables to present basic aspects of the MPC control quality assessment and properties of different approaches. Next (Sect. 14.2), the MIMO distillation column example is addressed to highlight the multivariate features of the MPC control performance assessment.

14.1 SISO Linear Process Example

This section presents summary and extension of the results that can be found in [5, 6]. The following continuous-time dynamic SISO process is considered to demonstrate various aspects of the univariate MPC control quality assessment:

$$G(s) = \frac{K}{(T_1 s + 1)(T_2 s + 1)} e^{-T_0 s}, \tag{14.1}$$

with assumed nominal parameters: $K = 2$, $T_0 = 4$, $T_1 = 3$, $T_2 = 10$.

In the considered case the GPC version of the Model Predictive Control is considered. It has been introduced by Clarke in 1987 [2, 3] with further extensions and further specific considerations in 1989 [1]. The algorithm is well established in scientific community and industrial practice with many reported successful implementations. The selection of this algorithm is brought about by the fact that the internal model parameters, i.e. delay, gain or time constants (model dynamics) have direct engineering interpretation. They are clearly understood and normally used by control engineer during design, tuning and daily maintenance.

As the process model in a digital ARMA form is required to evaluate the control rule, the exact discrete form formulation of model (14.1) has been calculated. Assuming the sampling period of 0.5 s, discrete-time version of the transfer function (14.1) is obtained:

$$G(z^{-1}) = \frac{y(k)}{u(k)} = \frac{b_0 + b_1 z^{-1}}{1 + a_1 z^{-1} + a_2 z^{-2}} z^{-d}, \tag{14.2}$$

where: $a_1 = -1.7977 \times 10^{-1}$, $2_2 = 8.0520 \times 10^{-1}$, $b_8 = 7.7574 \times 10^{-3}$, $b_9 = 7.2169 \times 10^{-3}$ and delay $d = 8$. Simulation loop diagram is sketched in Fig. 14.1.

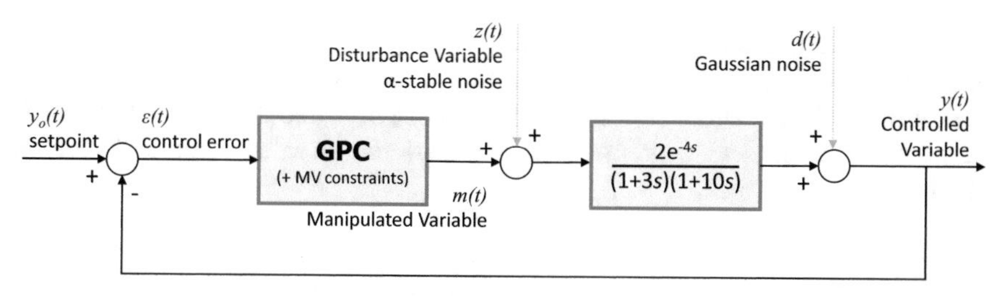

Fig. 14.1 Close loop GPC simulation environment

During GPC algorithm simulations the constraints imposed on the ranges of the manipulated variables are considered, $u^{\min} = -1$, $u^{\max} = 1$ together with the change rate limits: $\Delta u^{\min} = -0.05$, $\Delta u^{\max} = 0.05$. During the tuning further GPC parameters have been tuned, i.e.control horizon $N_{\mathrm{u}} = 3$, prediction horizons $N_1 = 9$, $N_2 = 25$ and the penalty term $\lambda = 0.5$. As the internal GPC model the perfect model (14.2) is used.

Loop disturbances are added into the loop simulation to reflect industrially common impacts and noises. Review of many industrial control loops has shown that majority of them was showing long-tail properties with the best function fitting by stable distributions [4]. Thus disturbance signal $z(t)$ generated by the α-stable distribution has been added before the process, while Gaussian noise $d(t)$ has been added to the process output. Four simulation scenarios has been designed to validate how disturbances and modeling mismatch may impact GPC control quality:

(SC_1) Real process gain differs from that of the GPC embedded model, $K = 2$.
Seven different model gains are used:
$K = \{0.4, 0.8, 1.2, 2.0, 2.8, 3.2, 3.6\}$.

(SC_2) GPC prediction horizon length changes from the optimal one $N_2 = 25$.
Seven horizon lengths are used:
$N_2 = \{10, 12, 15, 20, 25, 30, 35\}$.

(SC_3) Real process delay value differs from GPC embedded one, $T_0 = 8$.
Nine different process delays are used:
$T_0 = \{4, 5, 6, 7, 8, 9, 10, 11, 12\}$.

(SC_4) The real value of time constant, T_2, is varies from the optimal model $T_2 = 10$.
Seven different values are used:
$T_2 = \{0.5, 1, 5, 10, 15, 20, 40\}$.

Three different kinds of disturbances are taken into account for the simulation setup, i.e.:

(1) No disturbances,
(2) Three levels of normal noise (small, medium, large),
(3) Stability factors $\alpha = 1.5,\ 1.75,\ 2$ of α-stable probabilistic density function ($\alpha = 2$ means independent realization—normal distribution).

The optimal (called nominal) GPC control loop is simulated with different disturbance characteristics. For better visualization, two exemplary time trends generated during different simulation scenarios has been sketched in Figs. 14.2 and 14.3. Respective histograms are presented in Fig. 14.4 and rescaled range R/S plots in Fig. 14.5.

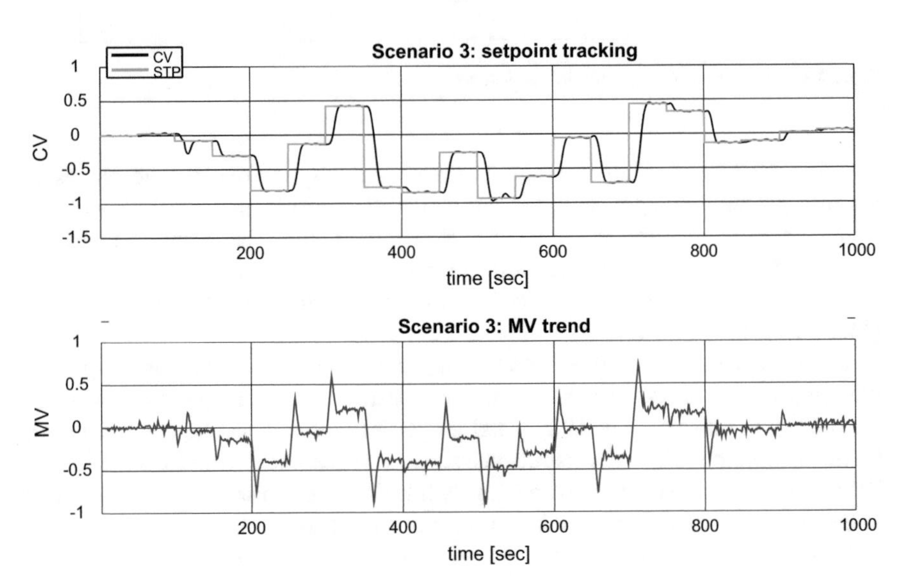

Fig. 14.2 Exemplary time trends for case with fat-tail disturbance and ideal parameters (Scen = 3)

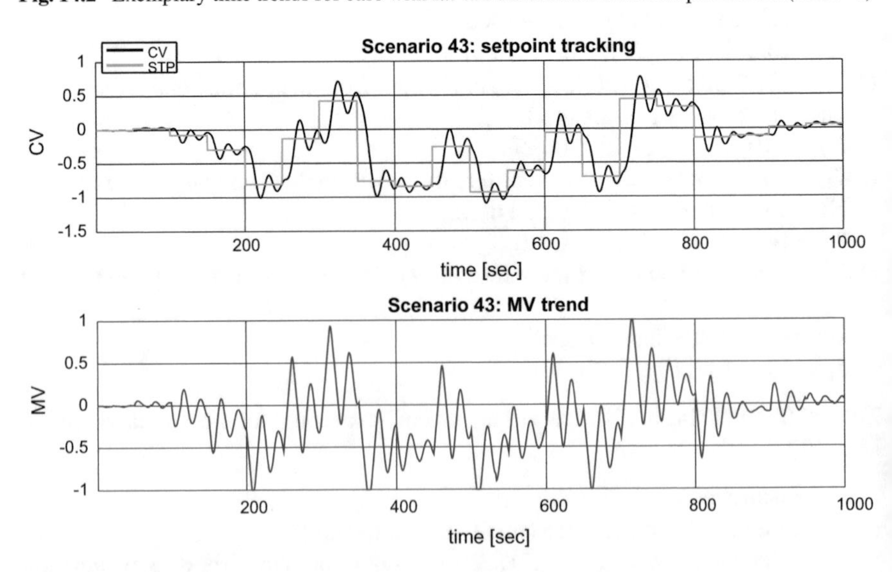

Fig. 14.3 Exemplary time trends for gain mismatch and Gaussian disturbance (Scen = 43)

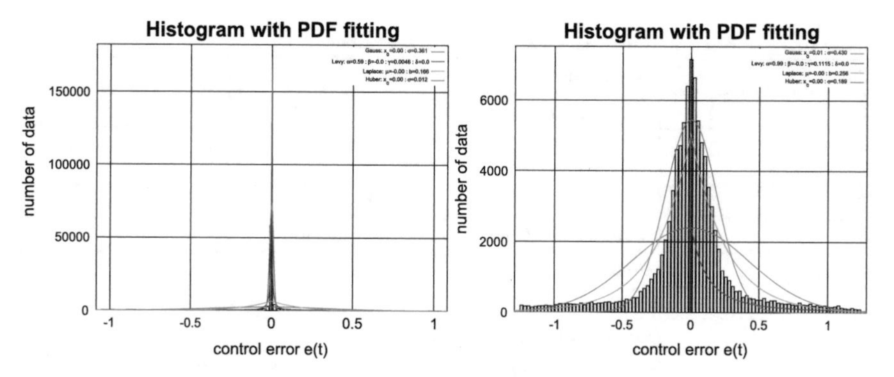

Fig. 14.4 Histograms for GPC control (Scen = 3 – left, Scen = 43 – right)

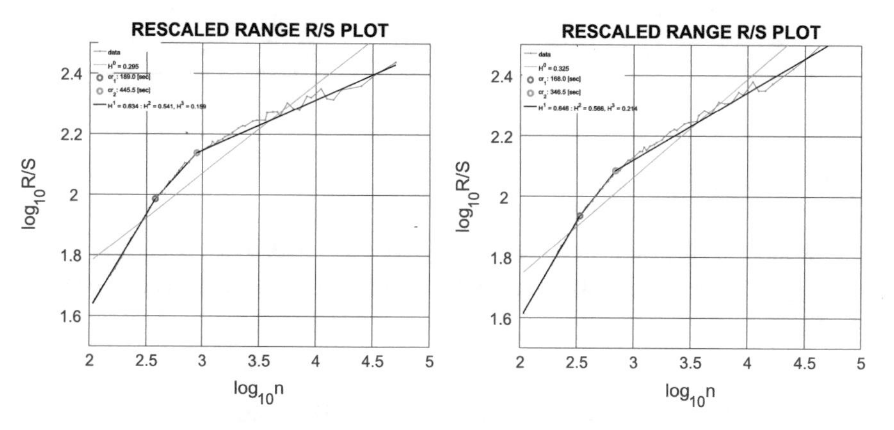

Fig. 14.5 RS plots for GPC control (Scen = 3 – left, Scen = 43 – right)

Several simulations have been run to address various hypotheses defined in form of the questions:

(H_0) Are the measures robust to disturbance properties (noise shadowing)? Can we evaluate loop quality despite disturbances?
(H_1) Does set-point have any impact on the results of loop quality assessment?
(H_2) Can we identify whether GPC gain is appropriate?
(H_3) Can we estimate if GPC horizon is set properly?
(H_4) Can we confirm whether GPC delay is appropriate?
(H_5) What can we say about GPC model dynamics (internal model time constant)?

The list of assessed control quality measures uses the previous results and observations and thus is limited to the list presented in Table 14.1.

Table 14.1 Selected CPA indexes used in the GPC assessment

No	Method domain	Index	Acronym
1	Time-domain methods	Mean Square Errors	MSE
2		Integral Absolute Error	IAE
3		Quadratic Error	QE
4	normal PDF (Gauss)	Standard deviation	σ
5	α-stable PDF factors	Stability	α
6		Scale	γ^L
7	Laplace PDF factors	Scale	b
8	Robust statistics	Standard deviation	σ^{rob}
9	Alternative	Multiple Hurst exponents	H^1, H^2
10		Differential entropy	H^{DE}
11		Rational entropy	H^{RE}

14.1.1 Simulations

The results of 162 simulations with different GPC parameters and loop disturbance profiles are organized systematically according to the predefined hypotheses $H_0 \dots H_5$.

14.1.1.1 Validation of the H_0 Hypothesis: Noise Shadowing

One of the main features of perfect loop quality assessment should be the ability to detect and identify controller tuning goodness despite any disturbances that might obscure investigation. From that perspective of controller the value of measure should be invariant. We are considering thirteen scenarios with different disturbances (see Table 14.2).

Analysis starts with comparison of undisturbed value with mean and variance of disturbed values for any measure. Six different simulations are considered: optimal GPC model with horizons equal to *12* and *25*, internal model with too small gain $K = 1.6$ for both horizons. Analogously, two simulations are run with too large gain $K = 2.4$ for both horizons.

The review of histograms for integral measures MSE (Fig. 14.6) and IAE (Fig. 14.7) shows clear consistency in the drawings. IAE detects wrong GPC tuning in relatively better way than MSE, especially for the worse controller model tuning. Very similar relationship is achieved for the QE index. The QE plot is not presented, as it does not bring about any additional information.

Disturbance effect on normal standard deviation (σ) is presented in Fig. 14.8. It seems that standard deviation is not a perfect choice. There is clear distinction

Table 14.2 Scenarios for disturbance robustness analysis

Experiment	$z(t)$	$d(t)$
$D0$	–	–
D	$\alpha = 2.00$, Amp = medium	–
$D2$	$\alpha = 1.50$, Amp = medium	–
$D4$	$\alpha = 1.75$, Amp = medium	–
$D5$	$\alpha = 2.00$, Amp = big	–
$D6$	$\alpha = 1.50$, Amp = big	–
$D7$	$\alpha = 1.75$, Amp = big	–
$D8$	–	$\alpha = 2.00$, Amp = medium
$D9$	–	$\alpha = 1.50$, Amp = medium
$D10$	–	$\alpha = 1.75$, Amp = medium
$D11$	–	$\alpha = 2.00$, Amp = big
$D12$	–	$\alpha = 1.50$, Amp = big
$D13$	–	$\alpha = 1.75$, Amp = big

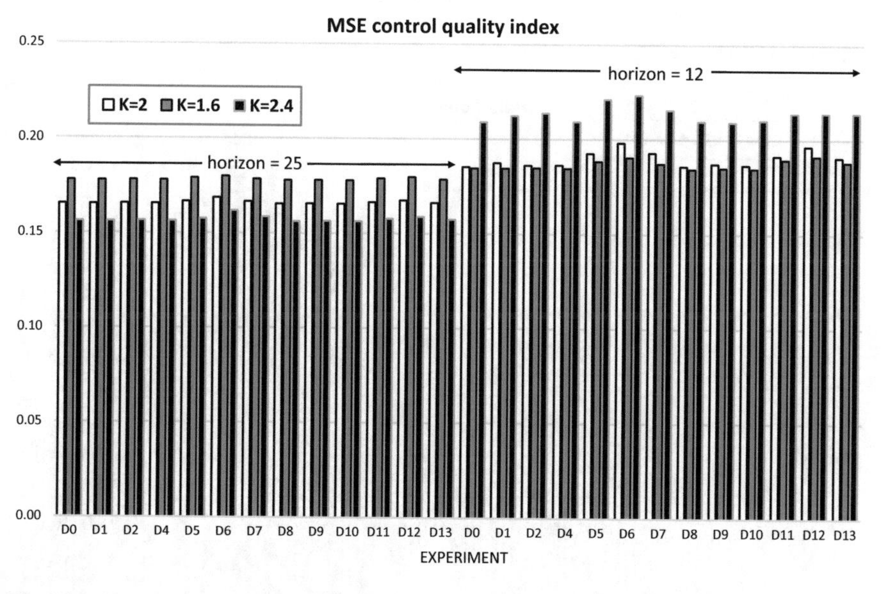

Fig. 14.6 Disturbance impact on MSE

between parameters and robust measure evaluation for better controller (larger horizon). However, we see significant variations impacted by disturbances for too short horizon. Next Fig. 14.9 shows the same diagram for robust distribution scaling. We see opposite situation. Detection is poorer for longer horizon (theoretically better GPC), than for short horizon case, when measure shows better performance and is not violated by disturbances.

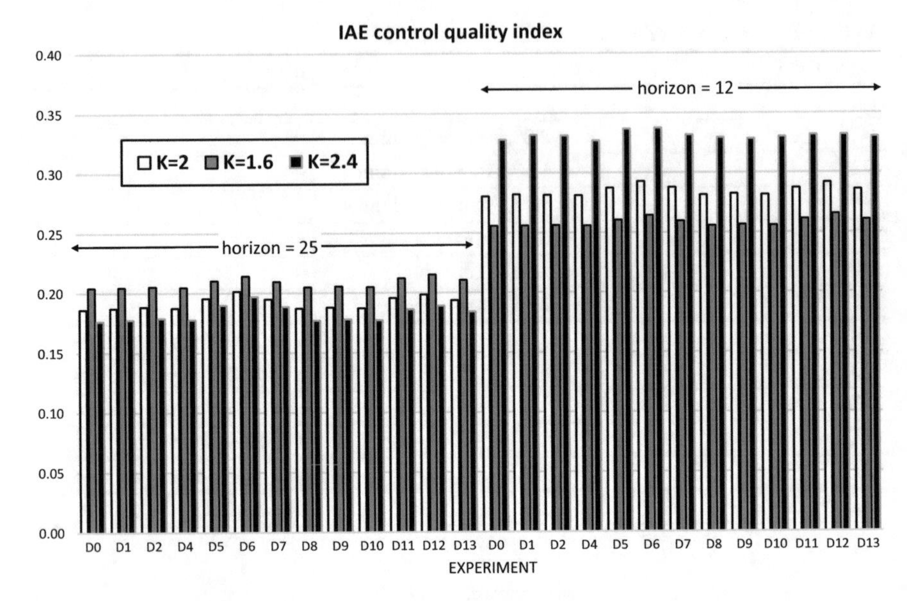

Fig. 14.7 Disturbance impact on IAE

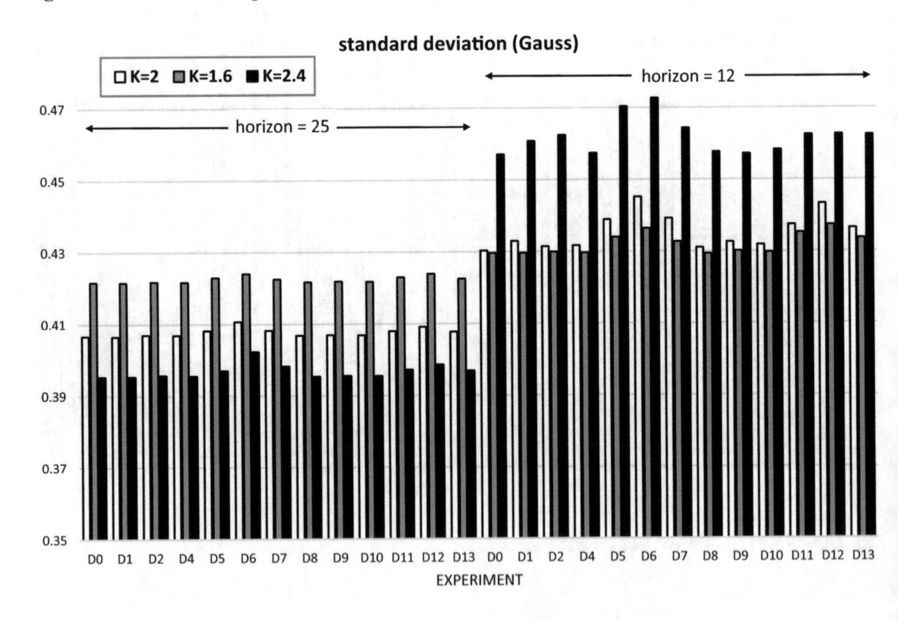

Fig. 14.8 Disturbance impact on σ index

Two other figures present disturbance impact on scaling factors γ (Fig. 14.10) for α-stable distribution and b for Laplace PDF (Fig. 14.11). Very clear separation is visible for the worse tuning of the controller despite disturbances. Conversely, GPC

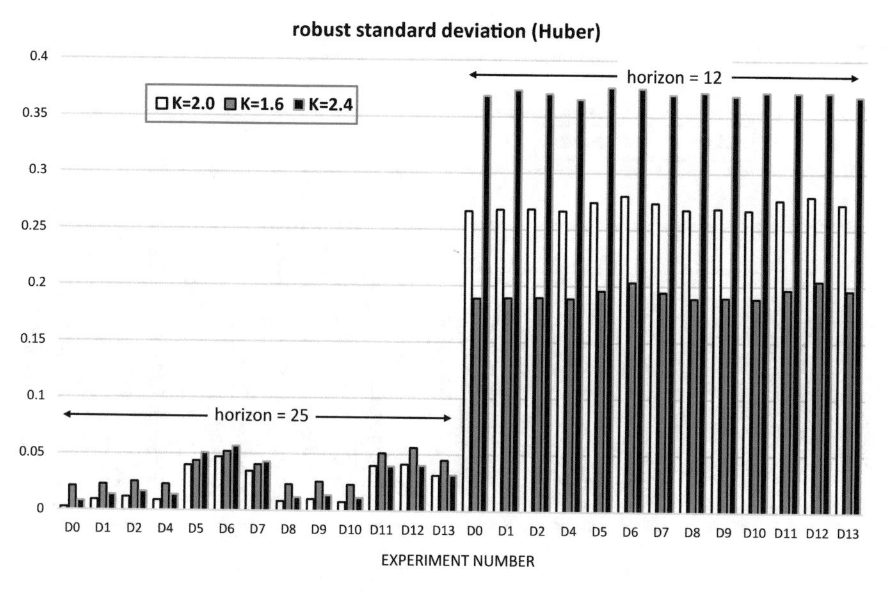

Fig. 14.9 Disturbance impact on σ^{rob} index

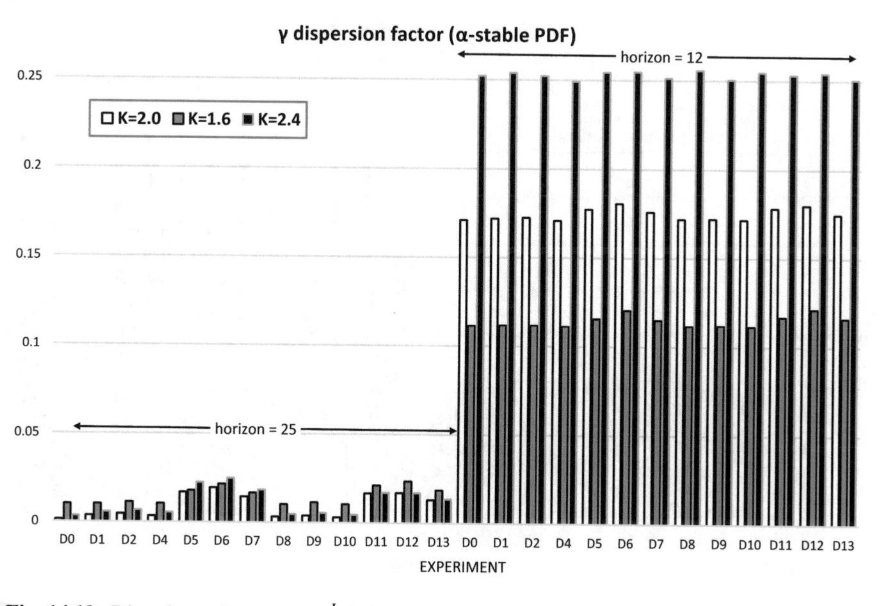

Fig. 14.10 Disturbance impact on γ^L index

regulation with optimal model is not well detected. Laplace scaling factor exhibits similar behavior as the IAE.

Stability factor α is sketched in Fig. 14.12. Very clear separation is visible with worse tuning. Unfortunately, GPC regulation with optimal model is not well detected.

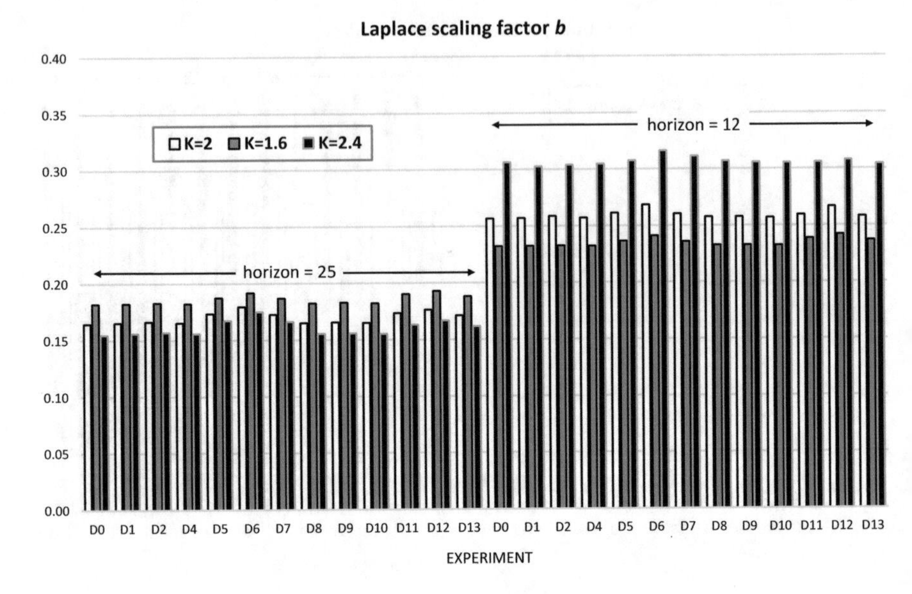

Fig. 14.11 Disturbance impact on b index

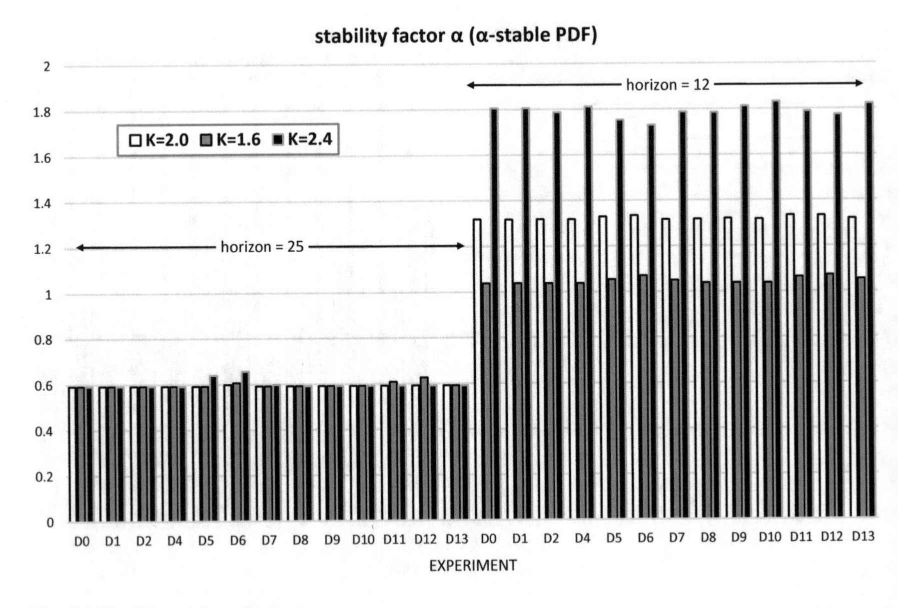

Fig. 14.12 Disturbance impact on α

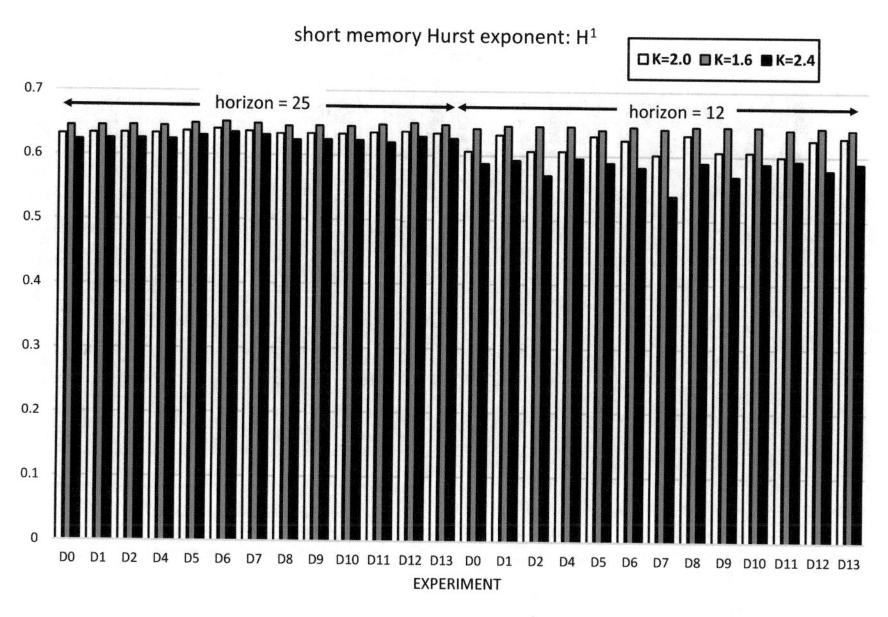

Fig. 14.13 Disturbance impact on H^1

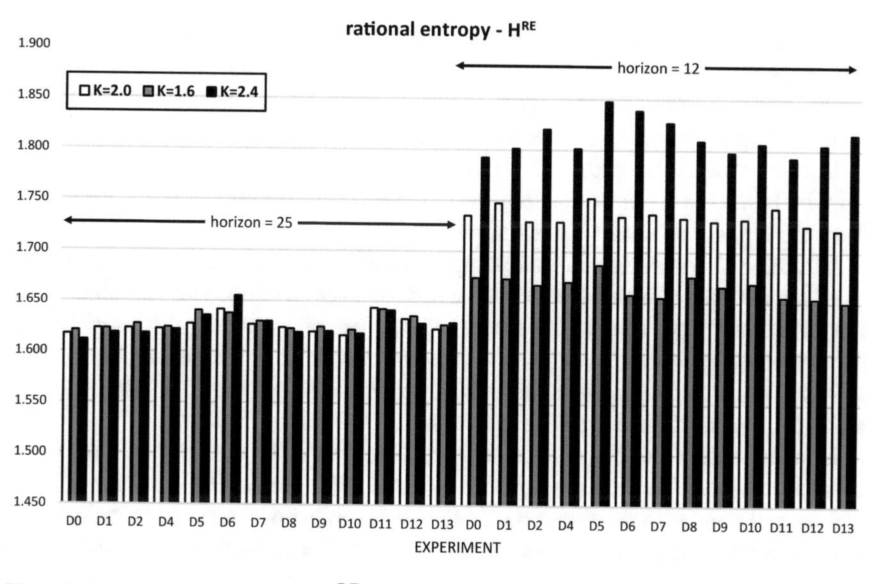

Fig. 14.14 Disturbance impact on H^{RE}

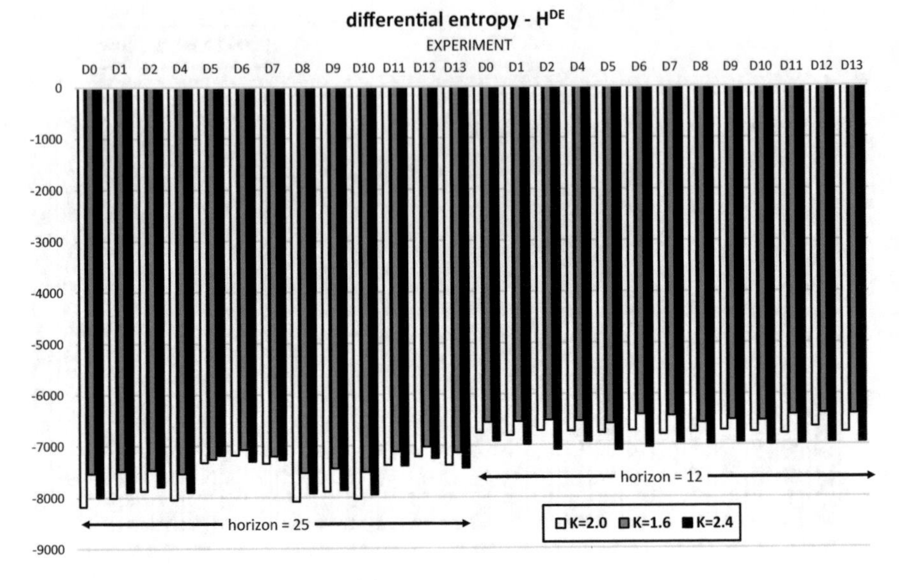

Fig. 14.15 Disturbance impact on H^{DE}

Short memory Hurst exponent H^1 is sketched is plot Fig. 14.13. It present consistent properties despite loop disturbances. One may notice from those plots behavior similar to the one from non-Gaussian statistics. More clear distinction between GPC model fitting for worse controller, rather than for the ideal one.

Similar analysis is sketched for both entropies: H^{RE} (Fig. 14.14) and H^{DE} (Fig. 14.15). It is very clear that they are completely not robust against disturbances. They are highly and randomly scattered.

The analysis presented in that paragraph evaluated potential robustness of the considered loop quality measures against disturbances embedded into the close loop. It seems that the parameters are mostly able to detect controller misfitting despite disturbances. On the other hand, in the group of fractal indexes it has been shown that R/S plot features distinctive crossover behavior. All single memory Hurst indexes are of no use and their estimation is not useful.

Simultaneously the equivalence between some of them is clearly visible, especially for IAE and Laplace scaling b and γ^L and σ^{rob}. The similar behavior of MSE and normal standard deviation σ and is also noticeable, however not so significantly. The following indexes will be taken into consideration in the further analysis:

- Integral indexes: IAE and MSE,
- Statistical parameters: Gauss standard deviation σ, robust standard deviation σ^{rob} and γ of α-stable distribution,
- Fractal Hurst exponent H^1.

14.1.1.2 Validation of the H_1 Hypothesis: Setpoint Shape Impact

During evaluation of above results there has been formulated hypothesis that results may be biased by the shape of setpoint. At first setpoint is in form of rectangular wave with varying amplitude. Second set of the same experiments is run to exclude that effect with setpoint filtered by first order inertia. It is to verify hypothesis that setpoint shape may affect results, especially the length of transient time in relation to steady state. Minimum, maximum, average and standard deviation of the measure error $\Delta\eta$ for each index are calculated. Results are shown in Table 14.3.

$$\Delta\eta = 100 \cdot \frac{\eta^{rect} - \eta^{filt}}{\eta^{rect}} \; [\%], \tag{14.3}$$

where η^{rect} denotes measure for rectangular setpoint and η^{filt} for the filtered one.

The comparison of setpoint effect reveals further issues. We have compare selected CPA measures. Observed error for Gaussian standard deviation has been several times larger, than the same one for fat-tail scalings. Even if we consider higher variability for γ, it would be safer to use non-Gaussian distribution. It is furthermore confirmed by the fact that stable distribution is better fitted to the control error histograms than the other ones. This is clear effect of dominating fat tail property. We also see that stability parameter has the largest standard deviation (variability). Comparison of two integral indexes favors IAE. It has twice smaller mean value with relatively similar standard deviation. The analysis indicates the reliability and the efficiency of the robust indexes.

14.1.1.3 Validation of the H_2 Hypothesis: Model Gain Effect

The simulation experiments to verify the hypothesis are organized as follows. We try to answer the question, if selected measures can detect proper selection of the GPC controller embedded model gain. Nine different gain values are tested: 0.4, 0.8, 1.2, 1.6, **2.0**, 2.4, 2.8, 3.2, 3.6. Three different disturbance scenarios are tested for each gain: no disturbances, or two disturbed ones with Gaussian and α-stable noise added

Table 14.3 Measures errors statistical properties

	MIN (%)	MAX (%)	MEAN (%)	STD. DEV
MSE	16.75	22.73	20.38	0.013
IAE	6.09	12.46	9.98	0.017
σ	8.76	12.10	10.77	0.007
σ^{rob}	−7.01	6.61	1.46	0.034
γ^L	−5.07	9.63	0.75	0.027
H^1	−8.06	5.33	0.39	0.020

before the process. Table 14.4 presents obtained values of the selected measures in the undisturbed simulations and Table 14.5 for the disturbed cases.

The relationship between the index value of the embedded model gain is graphically presented in the diagrams. Figure 14.16 shows the undisturbed case. The dotted blue line depicts index value equal to *0.5* to visualize the variability of the Hurst exponent. Further, the impact of the disturbance added before the process has been analyzed. Figure 14.17 presents the relations for the Gaussian disturbance properties and Fig. 14.18 reflects similar dependencies for the fat-tail noise.

The real value of model gain is $\mathbf{K} = \mathbf{2.0}$. Detection of optimal controller (the one that uses nominal model) does not depend on disturbances despite significant disturbance characteristics. The curves are in all three simulation cases almost the same. The detection slightly differs depending on the measure used. Robust indexes of γ^L (scaling factor of the α-stable distribution) and σ^{rob} (robust scale M-estimator) exactly point ot the optimal solution. Absolute error estimator (IAE) and Hurst expo-

Table 14.4 Obtained measures for the $\mathbf{H_2}$ hypothesis—undisturbed simulations

GAIN K	MSE	IAE	σ	σ^{rob}	γ^L	H^1
0.4	0.231	0.354	0.480	0.395	0.284	0.860
0.8	0.220	0.278	0.469	0.145	0.073	0.726
1.2	0.197	0.231	0.444	0.053	0.025	0.669
1.6	0.178	0.204	0.422	0.021	0.010	0.645
2	0.165	0.186	0.407	0.003	0.001	0.632
2.4	0.156	0.176	0.395	0.008	0.003	0.623
2.8	0.150	0.179	0.387	0.044	0.020	0.628
3.2	0.150	0.194	0.387	0.094	0.053	0.632
3.6	0.155	0.212	0.393	0.139	0.088	0.648

Table 14.5 Obtained measures for the $\mathbf{H_2}$ hypothesis—disturbed simulations

GAIN K	Scenarios: Gaussian disturbance						Scenarios: fat-tail disturbance					
	MSE	IAE	σ	σ^{rob}	γ^L	H^1	MSE	IAE	σ	σ^{rob}	γ^L	H^1
0.4	0.231	0.353	0.480	0.395	0.284	0.860	0.231	0.354	0.480	0.395	0.284	0.860
0.8	0.220	0.278	0.469	0.144	0.072	0.725	0.220	0.278	0.469	0.146	0.074	0.725
1.2	0.197	0.232	0.444	0.054	0.025	0.670	0.197	0.232	0.444	0.055	0.025	0.670
1.6	0.178	0.205	0.422	0.023	0.010	0.645	0.178	0.205	0.422	0.025	0.011	0.646
2	0.165	0.187	0.406	0.009	0.004	0.634	0.166	0.188	0.407	0.012	0.005	0.634
2.4	0.156	0.178	0.395	0.014	0.006	0.625	0.157	0.179	0.396	0.016	0.007	0.625
2.8	2.000	0.150	0.179	0.387	0.045	0.020	0.150	0.180	0.387	0.047	0.021	0.627
3.2	0.153	0.198	0.391	0.098	0.054	0.632	0.153	0.198	0.391	0.099	0.054	0.634
3.6	0.155	0.214	0.394	0.143	0.086	0.646	0.155	0.215	0.393	0.143	0.086	0.649

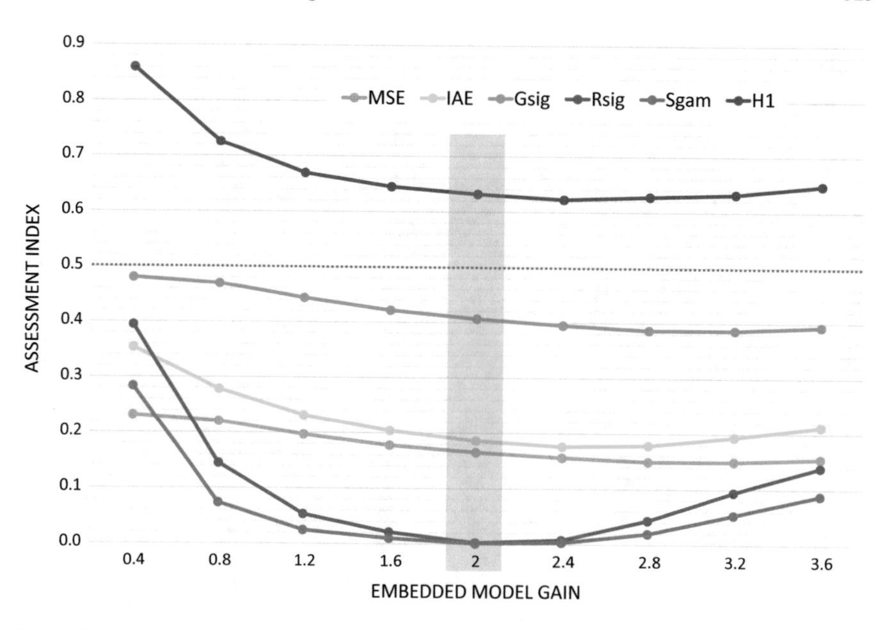

Fig. 14.16 Embedded model gain detection in the undisturbed case

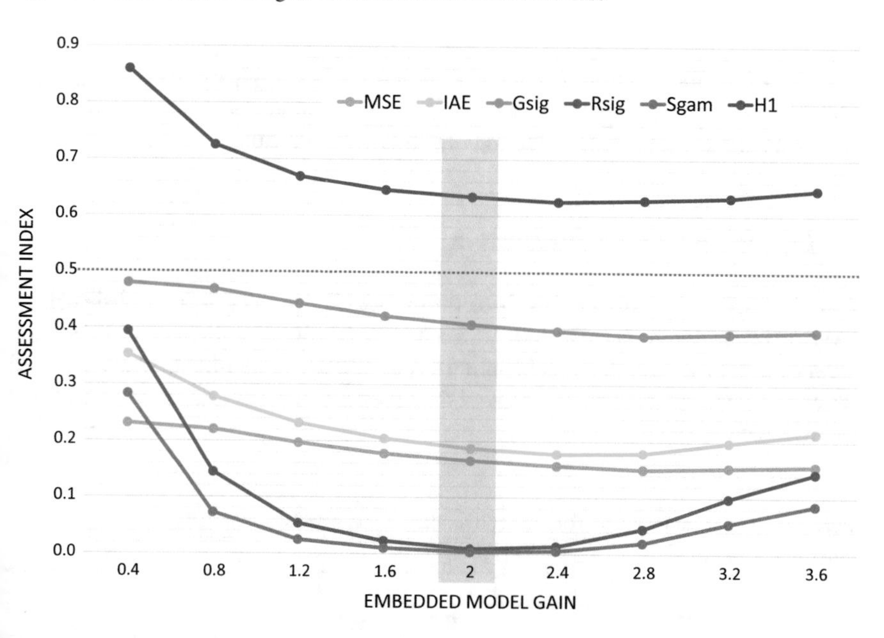

Fig. 14.17 Embedded model gain detection with the Gaussian disturbance

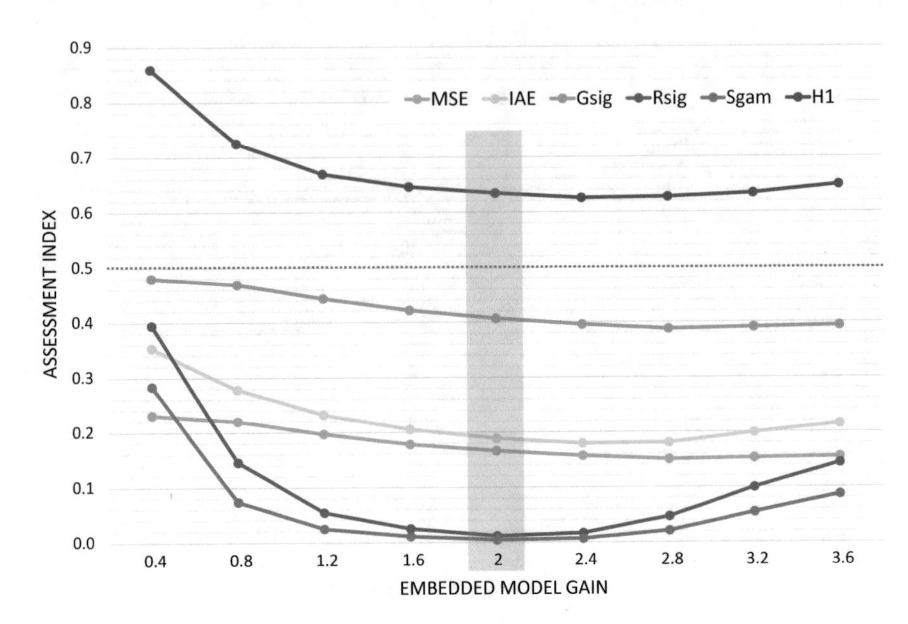

Fig. 14.18 Embedded model gain detection with the fat-tail undisturbance

nent H^1 points out the shifted value of $\mathbf{K} = \mathbf{2.4}$, while the other once are even more shifted towards much larger gain values. This observations shows up the reliability of the robust measures in the CPA task for the GPC controller.

14.1.1.4 H3—Impact of the Horizon

Next, the analysis is performed to check the proper GPC prediction horizon detection. Seven different horizons are tested: 10, 12, 15, **20**, 25, 30, 35 with three different disturbance properties designed and applied as previously. Table 14.6 presents obtained values of the selected measures in the undisturbed simulations and Table 14.7 for the disturbed cases.

Table 14.6 Obtained measures for the $\mathbf{H_3}$ hypothesis—undisturbed simulations

Horizon N_2	MSE	IAE	σ	σ^{rob}	γ^L	H^1
10	0.1918	0.3049	0.4379	0.3145	0.2038	0.6004
12	0.1852	0.2805	0.4303	0.2661	0.1710	0.6066
15	0.1689	0.2102	0.4110	0.0638	0.0303	0.6261
20	0.1651	0.1873	0.4063	0.0044	0.0021	0.6246
25	0.1654	0.1861	0.4066	0.0026	0.0013	0.6322
30	0.1661	0.1889	0.4076	0.0031	0.0016	0.6384
35	0.1669	0.1910	0.4085	0.0030	0.0016	0.6430

Table 14.7 Obtained measures for the $\mathbf{H_3}$ hypothesis—disturbed simulations

Horizon N_2	Scenarios: Gaussian disturbance						Scenarios: fat-tail disturbance					
	MSE	IAE	σ	σ^{rob}	γ^L	H^1	MSE	IAE	σ	σ^{rob}	γ^L	H^1
10	0.192	0.305	0.438	0.315	0.204	0.600	0.192	0.305	0.438	0.314	0.203	0.601
12	0.187	0.282	0.433	0.268	0.172	0.632	0.186	0.282	0.431	0.268	0.172	0.607
15	0.169	0.211	0.411	0.066	0.031	0.628	0.169	0.212	0.412	0.071	0.034	0.628
20	0.165	0.189	0.406	0.012	0.004	0.626	0.165	0.190	0.407	0.015	0.006	0.626
25	0.165	0.187	0.406	0.009	0.004	0.634	0.166	0.188	0.407	0.012	0.005	0.634
30	0.166	0.190	0.408	0.009	0.004	0.639	0.166	0.191	0.408	0.012	0.005	0.640
35	0.167	0.192	0.409	0.010	0.004	0.644	0.167	0.193	0.409	0.012	0.005	0.645

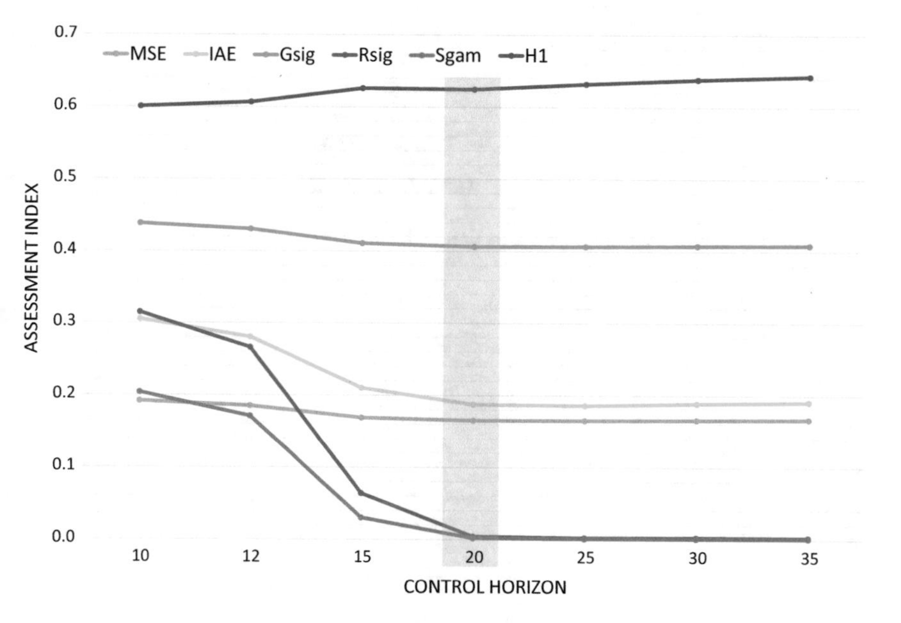

Fig. 14.19 Embedded control horizon detection in the undisturbed case

Theoretical speculations about predictive controller clearly show that the horizon should not be too short against process delay and dominant time constant (process dynamics). In contrary too large horizons will only result in larger calculation effort. This effect is clearly visible for all disturbance cases (Figs. 14.19, 14.20, 14.21). All the indexes except Hurst exponent diminish as the horizon increases. Once the horizon reaches its nominal selection $\mathbf{N_2 = 20}$, the indexes saturate and no further decrease is observed. It means that there is no reason to increase GPC horizon above that. We are unable to improve control performance behind this value. This effect is detected despite loop disturbances.

In contrary, Hurst exponent has opposite tendency, as it starts to from values closer to the $H^1 = 0.5$ and shows higher values as the horizon increases. The trend

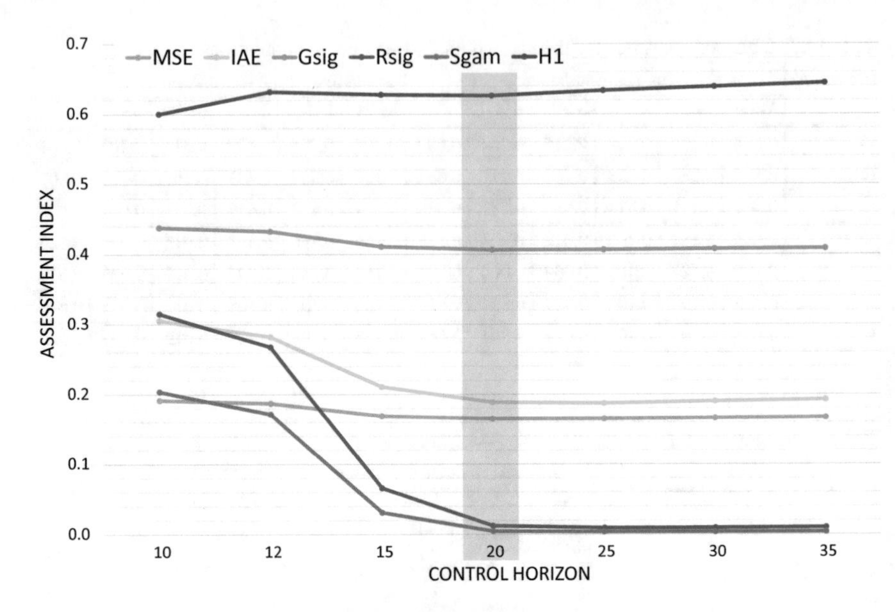

Fig. 14.20 Embedded control horizon detection with the Gaussian disturbance

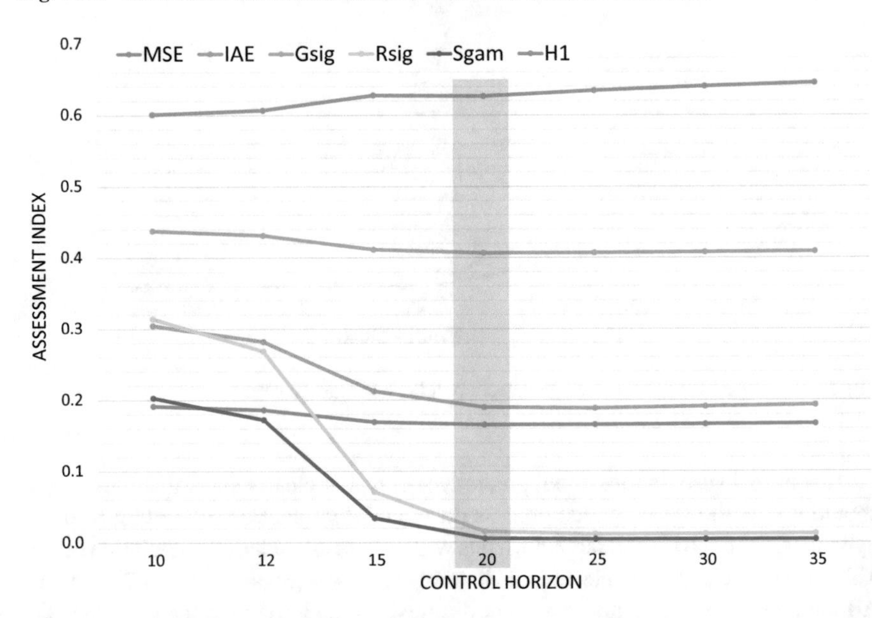

Fig. 14.21 Embedded control horizon detection with the fat-tail disturbance

is relatively small, however visible. This is negative observation from the perspective of the using Hurst exponent as the measure to assess GPC control quality with respect to the horizon length setting. Concluding, proper horizon value is identified in all cases, except the Hurst exponent.

14.1.1.5 H4: Impact of the Embedded Model Delay

Similar analysis may help in verification, whether the approach is able to assess correctness of the MPC embedded model delay. Various values of the delayed programmed inside of the MPC are tested: 4, 5, 6, 7, **8**, 9, 10, 11, 12. Similarly to the previous case, three different disturbance scenarios are taken into consideration. Table 14.8 presents the indexes calculate in the undisturbed simulations and Table 14.9 addresses the disturbed cases.

The relations between the index value of the embedded model gain is graphically presented in the diagrams. The undisturbed case results are presented in Fig. 14.22. The dotted blue line indicates value equal to *0.5* to better show the variability of the Hurst exponent against its optimal value. Next, the disturbance added before the process has been added. Figure 14.23 presents the dependencies for the Gaussian disturbance and Fig. 14.24 similar dependencies for the fat-tail noise.

Real value of the model delay has been set to $\mathbf{T_0} = 8.0$. It has been expected that plots should enable pointing out this value. Unfortunately, detection is not straightforward. It works only in one case, i.e. when there are no disturbances in the loop

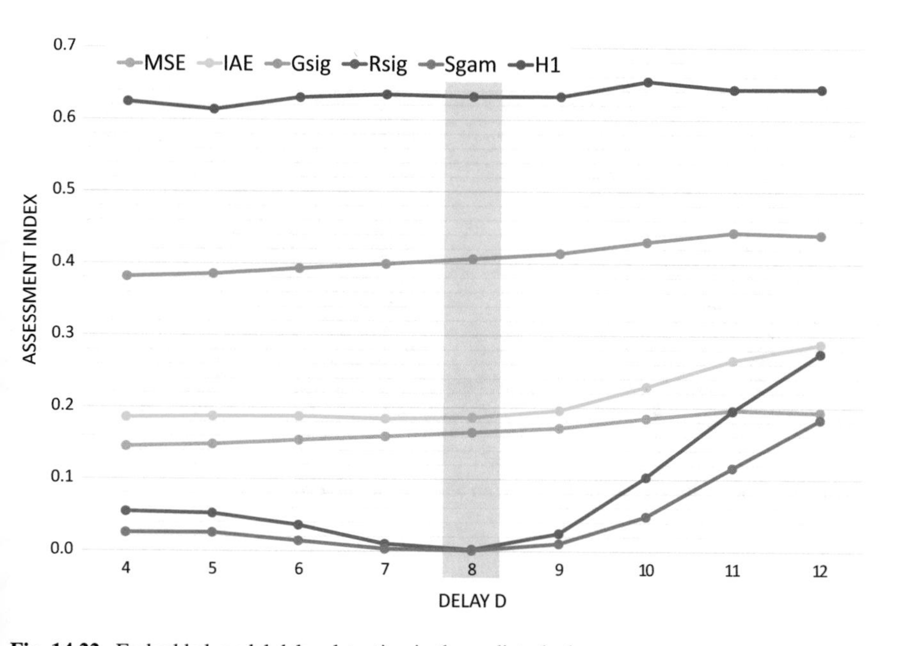

Fig. 14.22 Embedded model delay detection in the undisturbed case

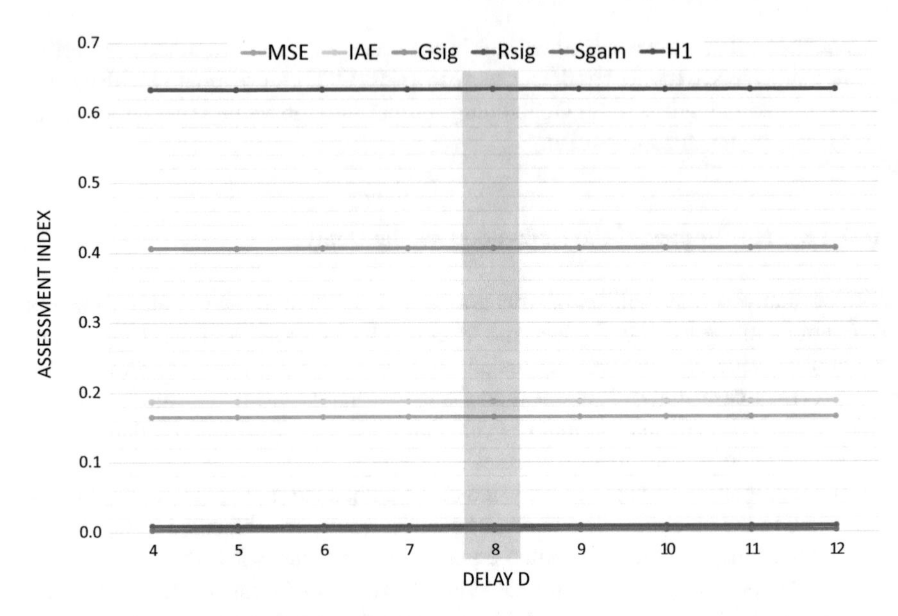

Fig. 14.23 Embedded model delay detection with the Gaussian disturbance

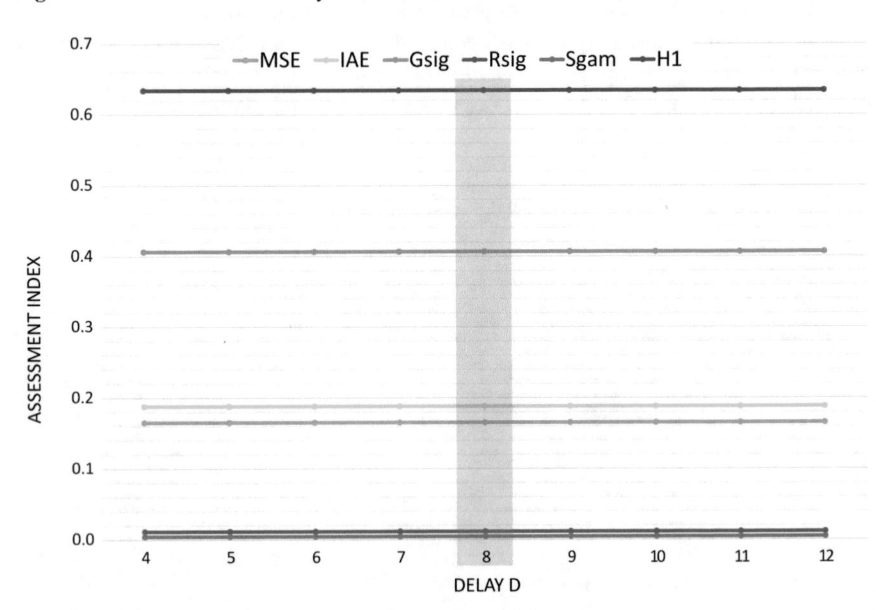

Fig. 14.24 Embedded model delay detection with the fat-tail disturbance

Table 14.8 Obtained measures for the $\mathbf{H_4}$ hypothesis—undisturbed simulations

Delay T_0	MSE	IAE	σ	σ^{rob}	γ^L	H^1
4	0.1456	0.1860	0.3816	0.0546	0.0256	0.6247
5	0.1485	0.1874	0.3854	0.0523	0.0256	0.6139
6	0.1545	0.1877	0.3931	0.0362	0.0144	0.6307
7	0.1596	0.1843	0.3995	0.0103	0.0035	0.6349
8	0.1654	0.1861	0.4066	0.0026	0.0013	0.6322
9	0.1716	0.1960	0.4143	0.0249	0.0108	0.6324
10	0.1851	0.2291	0.4303	0.1033	0.0486	0.6536
11	0.1967	0.2658	0.4435	0.1957	0.1162	0.6426
12	0.1937	0.2880	0.4401	0.2751	0.1836	0.6433

Table 14.9 Obtained measures for the $\mathbf{H_4}$ hypothesis—disturbed simulations

Delay T_0	Scenarios: Gaussian disturbance						Scenarios: fat-tail disturbance					
	MSE	IAE	σ	σ^{rob}	γ^L	H^1	MSE	IAE	σ	σ^{rob}	γ^L	H^1
4	0.165	0.187	0.407	0.009	0.004	0.634	0.165	0.188	0.407	0.012	0.005	0.634
5	0.165	0.187	0.406	0.009	0.004	0.633	0.165	0.188	0.407	0.012	0.005	0.634
6	0.165	0.188	0.407	0.009	0.004	0.634	0.165	0.188	0.407	0.012	0.005	0.634
7	0.165	0.187	0.407	0.009	0.004	0.634	0.165	0.188	0.407	0.012	0.005	0.634
8	0.165	0.187	0.407	0.009	0.004	0.634	0.165	0.188	0.407	0.012	0.005	0.634
9	0.165	0.187	0.407	0.009	0.004	0.634	0.165	0.188	0.407	0.012	0.005	0.634
10	0.165	0.187	0.407	0.009	0.004	0.633	0.165	0.188	0.407	0.012	0.005	0.634
11	0.166	0.187	0.407	0.009	0.004	0.634	0.165	0.188	0.407	0.012	0.005	0.634
12	0.165	0.187	0.407	0.009	0.004	0.633	0.165	0.188	0.407	0.012	0.005	0.634

(Fig. 14.22). It is especially clearly seen for the robust indexes of γ^L (scaling factor of the α-stable distribution) and σ^{rob} (robust scale M-estimator). The curves for undisturbed loop is in compliance with our expectation. First the index diminishes up to ideal value and than starts to rise in linear way. We may clearly point out optimal value of the delay and additionally see that increasing of the model delay constantly degrades control.

In contrary, similar indexes for disturbed loops are just flat for both types of the disturbance (Figs. 14.23 and 14.24). The measures are constant, independently on model delay. It seems that loop disturbances (whatever they are) screen (shadow) model delay effect on control performance.

Concluding, the proper model delay value is hardly detectable. Disturbances shadow any impact on control quality. All the curves flatten and show nothing.

14.1.1.6 H5: Impact of GPC Model Dynamics

The same methodology is finally adapted to verify if selected measures can detect proper GPC embedded model dynamics. Seven values for time constant T_2 are tested: 0.5, 1, 5, **10**, 15, 20, 40 with three different disturbances, as previously. Table 14.10 presents obtained values of the selected measures in the undisturbed simulations and Table 14.11 for the undisturbed cases (Fig. 14.25).

Real value of the model dynamics has been set to $\mathbf{T_2 = 10}$. We notice two types of the relations. We see clear ability to point out the right value of the model time constant for two robust indexes, i.e. γ^L (scaling factor of the α-stable distribution) and σ^{rob} (robust scale M-estimator). Both curves exhibit relatively similar shape. Strangely, the detection resolution is higher for both disturbed cases (Figs. 14.26 and 14.27). The relationship for the undisturbed case (Fig. 14.22) is not so clear, especially for the embedded model faster dynamics, i.e. $\mathbf{T_2 < 10}$. Too small values of dynamics does not significantly deteriorate control quality (in sense of the considered measure). In contrary, for high values of T_2 curve rapidly increases suggesting fast degradation of control quality. This behavior may be interpreted that underestimated dynamics is not so dangerous for GPC control in opposition to the opposite case.

Table 14.10 Obtained measures for the $\mathbf{H_5}$ hypothesis—undisturbed simulations

Dynamics T_2	MSE	IAE	σ	σ^{rob}	γ^L	H^1
0.5	0.1145	0.1297	0.3384	0.0000	0.0000	0.5761
1	0.1092	0.1243	0.3305	0.0001	0.0000	0.5747
5	0.1432	0.1619	0.3784	0.0009	0.0004	0.6030
10	0.1654	0.1861	0.4066	0.0026	0.0013	0.6322
15	0.1827	0.2062	0.4274	0.0083	0.0039	0.6509
20	0.1981	0.2260	0.4451	0.0226	0.0100	0.6688
40	0.2516	0.2984	0.5015	0.1374	0.0704	0.7381

Table 14.11 Obtained measures for the $\mathbf{H_5}$ hypothesis—disturbed simulations

Dynamics T_2	Scenarios: Gaussian disturbance						Scenarios: fat-tail disturbance					
	MSE	IAE	σ	σ^{rob}	γ^L	H^1	MSE	IAE	σ	σ^{rob}	γ^L	H^1
0.5	0.115	0.135	0.339	0.017	0.007	0.583	0.115	0.138	0.339	0.020	0.009	0.587
1	0.109	0.130	0.331	0.018	0.008	0.583	0.110	0.133	0.331	0.021	0.009	0.587
5	0.143	0.164	0.378	0.011	0.005	0.606	0.143	0.166	0.379	0.013	0.005	0.613
10	0.165	0.187	0.407	0.009	0.004	0.634	0.165	0.188	0.407	0.012	0.005	0.634
15	0.183	0.207	0.428	0.013	0.005	0.652	0.183	0.208	0.427	0.015	0.006	0.652
20	0.198	0.227	0.445	0.025	0.010	0.670	0.198	0.227	0.445	0.027	0.011	0.669
40	0.252	0.298	0.502	0.138	0.070	0.738	0.251	0.299	0.501	0.139	0.071	0.738

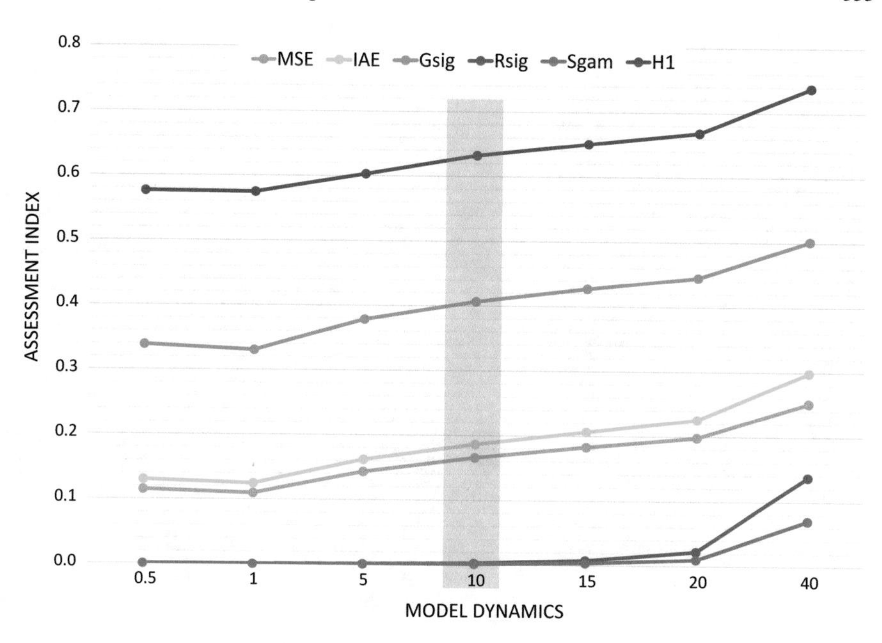

Fig. 14.25 Embedded model dynamics detection in the undisturbed case

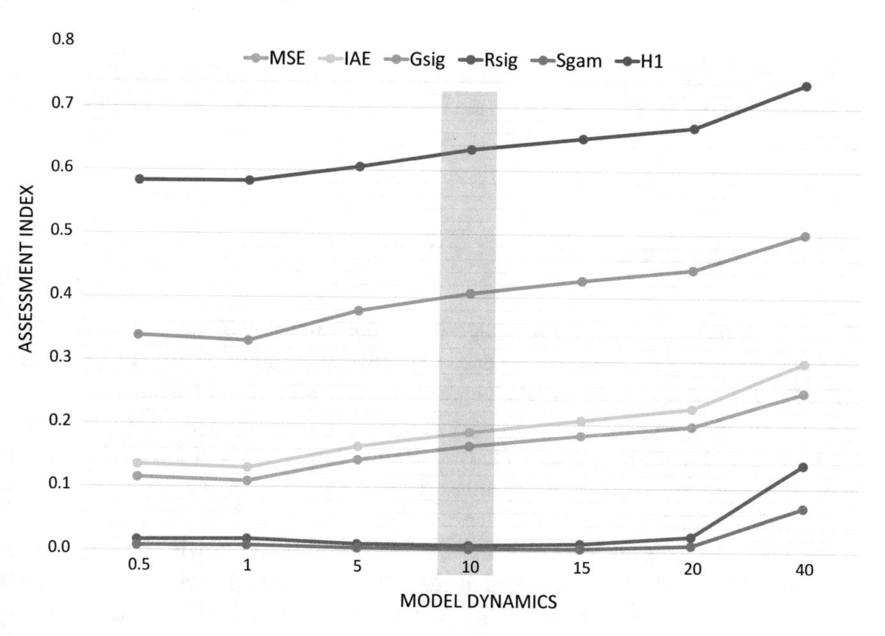

Fig. 14.26 Embedded model dynamics detection with the Gaussian disturbance

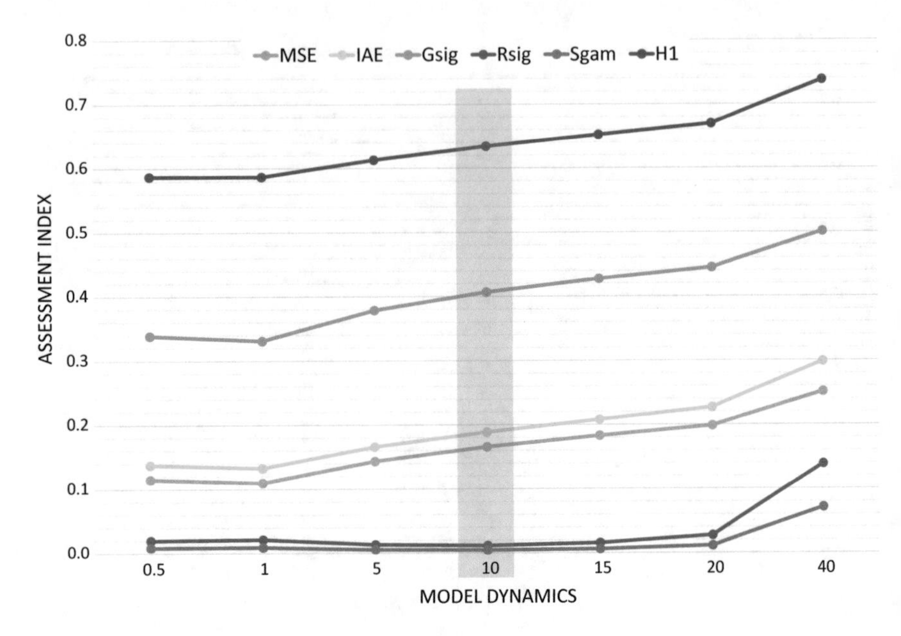

Fig. 14.27 Embedded model dynamics detection with the fat-tail disturbance

The assessment properties for the embedded model dynamics are opposite to the model delay impact case. More excited signals enable easier dynamics misfit exposure.

14.1.2 Observations

This chapter has addressed the investigation of the efficiency of the CPA analysis of the MPC control. The analysis has used extensive simulation results for SISO and MIMO processes. The used measures are evaluated using control error signal.

Different measures are presented and analyzed in investigations. Their selection has used the results of previous analyzes. They are listed in Table 14.1. Further analysis has shortened the list to six measures analysis selected of two robust indexes shown in Table 14.12.

Selected CPA indexes have been used to verify ability of the MPC tuning quality detection. The following parameters of the predictive control have been taken into consideration: model gain $\mathbf{K}$, delay $\mathbf{T_0}$, dynamics (time constant $\mathbf{T_2}$) and controller prediction horizon $\mathbf{N_2}$. The analysis has allowed to provide the following observations:

Table 14.12 Selected CPA indexes used in cascaded configuration

No	Method domain	textbfIndex	Acronym
1	Time-domain integrals	Mean Square Errors	MSE
2		Integral Absolute Error	IAE
3	normal PDF (Gauss)	standard deviation	σ
4	α-stable PDF	scale	γ^L
5	Robust statistics	standard deviation	σ^{rob}
6	Alternative	short memory Hurst exponent	H^1

(a) Some of the indexes like entropy measures fail as their assessment effectiveness is strongly influenced by loop disturbances. Similar observation may be find in the literature [6] in respect of the minimum variance measures as well.
(b) Verification of the statistical properties of control error has shown that though normal approach seems to be invariant against disturbances, it is biased by the character of setpoint signal. The same reason caused rejection of the stability parameter of the stable distribution function. The analysis has pointed out that two measures, i.e. scale factor of the α-stable distribution and robust standard deviation σ^{rob} seem to be the most effective. As control error histograms are significantly fat-tailed, the Gaussian approaches fail to work properly.
(c) Mean square error and absolute error indexes are compared. Pursuant to literature suggestions [8] it has been shown that absolute error index (IAE) is less invariant to the setpoint variations.
(d) The same robust measures, analogously to all previous analyzes, has proven its detection abilities. They are able to detect model gain and dynamics misfit together with GPC horizon selection. They only had problems with model delay. In that case any loop disturbance is screening detection.
(e) Detection ability is the most challenging for model delay misfit detection. It returns positive results only in ideal undisturbed situation. Introduction of disturbances effectively shadows the improper delay misfit and does not allow for reliable assessment.

The obtained results confirm previous observations in the literature [5, 6].

14.2 MIMO Distillation Column Example

The second MPC control performance assessment case deals with the predictive control of the linear MIMO process. In contrary to the previous example it uses the DMC predictive controller. The predictive controller is applied to the MIMO exemplary process of the the pilot scale binary distillation column developed by Wood and Berry [9]. It is exactly the same process that is addressed with PID control in Sect. 13.2 and described by the linearized model in continuous-time domain (10.27). The list of the controller settings that are used during the MPC design is similar to the one addressed earlier. It is longer than previously as the MIMO process is under consideration.

A mathematical model of the process being controlled is used for prediction evaluation and optimization of the control policy. Once the model is precise, the system exhibits very good control quality. Conversely, when there is a significant misfit between the process and the model, control quality deteriorates. Additionally, inappropriate selection of parameters of the MPC cost-function (10.6) may also impact negatively the overall system performance. Above stipulations lead towards the extended list of the potential MPC controller sources of the bad control quality:

(a) Too short horizon of model dynamics D, i.e the number of discrete-time step-response coefficients,
(b) Improper estimation of the model gain and/or delay,
(c) Too short prediction, N, and/or control, N_u, horizons,
(d) Improper selection of the tuning coefficients μ_m for the consecutive controlled variables ($m = 1, \ldots, n_y$),
(e) Improper selection of the tuning coefficients λ_n for the consecutive controlled variables ($m = 1, \ldots, n_y$).

In order to demonstrate the performance of the CPA measures the MIMO system consisting of the DMC controller and the TITO distillation column process has been simulated. The controlled variables, Y_1 and Y_2, are: the top product (distillate) composition and the bottom product composition, respectively. The manipulated variables, U_1 and U_2, are: the reflux flow rate and the vapour flow, respectively. The uncontrolled process input (the disturbance), D, is the flow rate of the input stream. All variables are deviations from a typical operating point.

The process is simulated in the control loop structure presented in Fig. 14.28. Selected setpoint trends for both controlled variables (CVs) are kept constant. The process is only driven by the variations of disturbances, which is a typical industrial scenario. The fat-tailed signal generated with the α-stable probabilistic density function generator is used as the raw material flow rate, i.e. the process disturbance variable (DV). Moreover, an additional fat-tailed signals are added to the manipulating variables (MVs) as the unmeasured disturbances. Furthermore, Gaussian noise (with insignificantly low variance) is added to the process outputs to represent possible measurement noise signals.

The nominal tuning parameters of the DMC algorithms are: $D = 100$, $N = 30$, $N_u = 5$, $\mu_1 = \mu_2 = 1$, $\lambda_1 = \lambda_2 = 1$. The sampling time of MPC is 1 minute. The following scenarios of improper MPC tuning and process-model mismatch are considered:

Scen$_1$ Too short control horizon: $N_u = 1$,
Scen$_2$ Incorrect weights: $\lambda_1 = 10$, $\lambda_2 = 1$ (the increments of the first MV are penalized 10 times more),

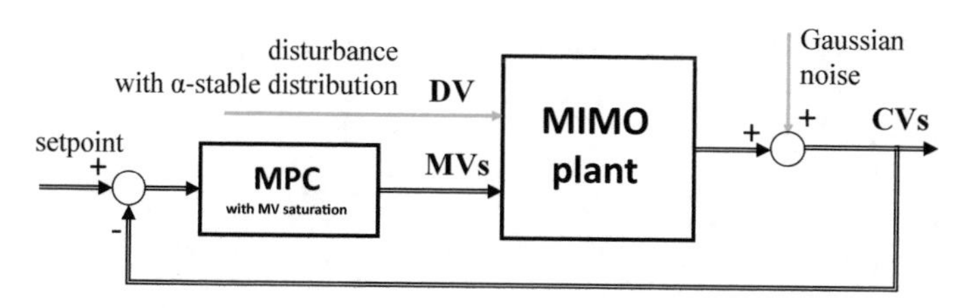

Fig. 14.28 MPC distillation column control environment (MVs, DVs, CVs)

$Scen_3$ Incorrect weights: $\lambda_1 = 1$ and $\lambda_2 = 10$.
$Scen_4$ Incorrect weights: $\lambda_1 = 10$ and $\lambda_2 = 10$.
$Scen_5$ Incorrect weights: $\mu_1 = 0.1$ and $\mu_2 = 1$ (predicted errors for Y_1 are taken into account in the MPC cost-function as 10 times less important than necessary),
$Scen_6$ Incorrect weights: $\mu_1 = 1$ and $\mu_2 = 0.1$,
$Scen_7$ Incorrect weights:: $\mu_1 = 0.1$ and $\mu_2 = 0.1$,
$Scen_8$ Too low (50%) model gain for Y_1,
$Scen_9$ Too low (50%) model gain for Y_2,
$Scen_{10}$ Too low (50%) model gain for both CVs,
$Scen_{11}$ Too large (150%) model gain for Y_1,
$Scen_{12}$ Too large (150%) model gain for Y_2,
$Scen_{13}$ Too large (150%) model gain for both CVs,
$Scen_{14}$ Increased by 3 sampling periods delay for Y_1,
$Scen_{15}$ Increased by 3 sampling periods delay for Y_2,
$Scen_{16}$ Increased by 3 sampling periods delays for both CVs.

Two different simulation cases (disturbance realizations) have been analyzed.

14.2.1 Simulations

All the simulation scenarios described in the previous section have been run. The control error signals have been collected and the selected CPA indexes (as in Table 14.12) have been evaluated. It must be noted that the Hurst exponent has been excluded from the list as some of its indications have been already indecisive in the MPC SISO case. The calculated indexes for the case 1 (the first disturbance realization) are presented in Table 14.13. Table 14.14 presents case 2, i.e. the second realization of disturbances. The first row (denoted by **Scen No = 0** and highlighted in green color) shows the measures evaluated for the nominal DMC settings.

The analysis of such large tabular data is difficult and ineffective. The multi-criteria radar plots has been prepared, analogously to the distillation column PID MIMO control example. The real values of the indexes have been scaled similarly to the the previous example (13.1).The difference is in the selection of the optimal index value. In the considered case the index value for the nominal parameters ($Scen_0$) are

Table 14.13 The results of the MPC controlled column simulation—case 1 (green color depicts nominal DMC settings)

Scen No	Output 1					Output 2				
	MSE	IAE	σ	σ^{rob}	γ^{L}	MSE	IAE	σ	σ^{rob}	γ^{L}
0	4.033	1.005	2.008	1.032	0.712	3.156	0.849	1.777	0.827	0.576
1	60.615	3.910	7.786	4.024	2.775	35.191	2.928	5.932	2.964	2.051
2	6.073	1.233	2.464	1.266	0.873	2.950	0.812	1.718	0.778	0.543
3	4.516	1.065	2.125	1.094	0.754	5.196	1.107	2.279	1.100	0.764
4	7.135	1.338	2.671	1.373	0.947	4.907	1.069	2.215	1.054	0.732
5	11.316	1.694	3.364	1.749	1.205	2.866	0.804	1.693	0.775	0.540
6	3.432	0.926	1.853	0.950	0.655	6.751	1.270	2.598	1.271	0.881
7	7.135	1.338	2.671	1.373	0.947	4.907	1.069	2.215	1.054	0.732
8	15.530	1.975	3.941	2.030	1.400	3.071	0.833	1.752	0.805	0.561
9	4.074	1.011	2.018	1.039	0.716	12.100	1.679	3.479	1.656	1.151
10	15.607	1.981	3.951	2.036	1.404	11.922	1.663	3.453	1.635	1.137
11	1.807	0.673	1.344	0.691	0.477	3.196	0.856	1.788	0.835	0.581
12	4.023	1.004	2.006	1.030	0.711	1.388	0.562	1.178	0.545	0.379
13	1.800	0.672	1.342	0.689	0.476	1.413	0.568	1.189	0.553	0.385
14	4.263	1.038	2.065	1.069	0.736	3.136	0.847	1.771	0.826	0.575
15	4.025	1.004	2.006	1.031	0.711	3.430	0.872	1.852	0.829	0.580
16	4.324	1.047	2.079	1.078	0.742	3.410	0.869	1.847	0.826	0.578

used (14.4). Such selection has used previously acquired knowledge about properties of the indexes and enables extended comparison.

$$Ind^{sc} = \frac{Ind^{nom}}{Ind} \tag{14.4}$$

Altogether four radar plots have been prepared. Each of the cases comprises of two separate plots, as the number of considered scenarios is large (16) and such a plot with so many curves would not be readable. Case 1 radar plots are presented in Figs. 14.29 and 14.30.

First group of the scenarios shown in Fig. 14.29 compares the incorrect settings of the prediction horizon ($Scen_1$) and modifications inside of the DMC penalty index ($Scen_2 \ldots Scen_7$). We may observe the results that confirm our intuitive expectations. Decreasing of the control horizon significantly deteriorates control quality, what is evidently detected. Four indexes may be clustered into two groups. $Scen_2 \ldots Scen_4$ addresses manipulations with the $\lambda_{1/2}$ tuning coefficient (different shades of green color in the radar plot), while $Scen_5 \ldots Scen_7$ show the results for the manipulations with the $\mu_{1/2}$ tuning coefficients (different shades of blue color in the radar plot).

We clearly see that when only one of the coefficients is biased then only one of the loops is affected (deteriorated) and the opposite one remains in good shape.

Table 14.14 The results of the MPC controlled column simulation—case 2 (green color depicts nominal DMC settings)

Scen No	Output 1					Output 2				
	MSE	IAE	σ	σ^{rob}	γ^L	MSE	IAE	σ	σ^{rob}	γ^L
0	2.218	0.952	1.489	1.037	0.718	1.738	0.809	1.319	0.836	0.583
1	33.318	3.698	5.772	4.035	2.796	19.362	2.779	4.400	2.983	2.071
2	3.339	1.168	1.827	1.271	0.881	1.625	0.774	1.275	0.788	0.550
3	2.483	1.008	1.576	1.098	0.761	2.860	1.053	1.691	1.110	0.772
4	3.923	1.266	1.981	1.379	0.956	2.702	1.018	1.644	1.065	0.741
5	6.219	1.602	2.494	1.753	1.215	1.579	0.766	1.257	0.784	0.547
6	1.887	0.877	1.374	0.954	0.661	3.716	1.207	1.928	1.281	0.890
7	3.923	1.266	1.981	1.379	0.956	2.702	1.018	1.644	1.065	0.741
8	8.538	1.869	2.922	2.037	1.412	1.692	0.794	1.301	0.814	0.568
9	2.240	0.957	1.497	1.043	0.723	6.662	1.599	2.581	1.672	1.164
10	8.580	1.874	2.929	2.043	1.416	6.565	1.583	2.562	1.652	1.150
11	0.994	0.637	0.997	0.694	0.481	1.760	0.816	1.327	0.844	0.588
12	2.212	0.950	1.487	1.035	0.717	0.764	0.535	0.874	0.551	0.384
13	0.990	0.636	0.995	0.692	0.480	0.778	0.541	0.882	0.559	0.390
14	2.345	0.981	1.531	1.073	0.742	1.727	0.807	1.314	0.835	0.582
15	2.213	0.951	1.488	1.035	0.717	1.889	0.831	1.375	0.840	0.587
16	2.379	0.989	1.542	1.080	0.747	1.878	0.829	1.371	0.836	0.585

Modifications in both $\lambda_{1/2}$ the $\mu_{1/2}$ tuning coefficients deteriorates the performance of both loops equally.

The second group of the scenarios sketched in Fig. 14.30 addresses simulations associated with the modifications inside of the embedded process models. $Scen_8 \dots Scen_{13}$ describes changes in the models gains and $Scen_{14} \dots Scen_{16}$ modifications in the delays. We see that setting of the too low model gain deteriorates the respective CV performance leaving the other CV control quality unchanged. In contrary overestimating of one model loop improves the operation withe the opposite CV and vice versa. This effect is overestimated with tme mean square error, wile all the pother indexes exhibit similarly. Effect of the delay changes in invisible with is compliant with previous observations for SISO case analyzed in Sect. 14.1.

Case 2 radar plots are sketched in Figs. 14.31 and 14.32. Although different disturbance realization is used in this case, the relative results are exactly the same.

14.2.2 Observations

The observations of the MIMO process controlled by the DMC controller are consistent with the outcome of the MPC SISO simulations and of the PID control of

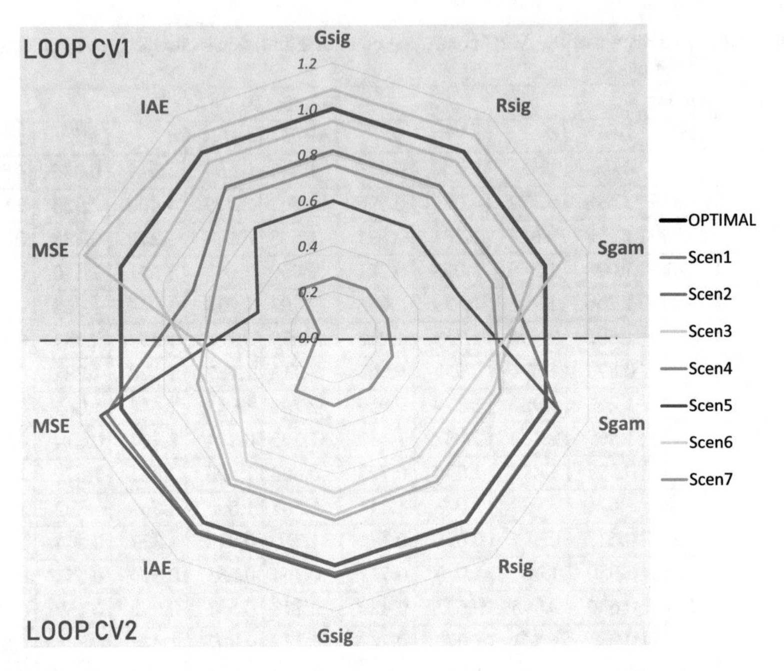

Fig. 14.29 MPC control assessment for distillation column: case 1, scenarios $Scen_1 \ldots Scen_7$

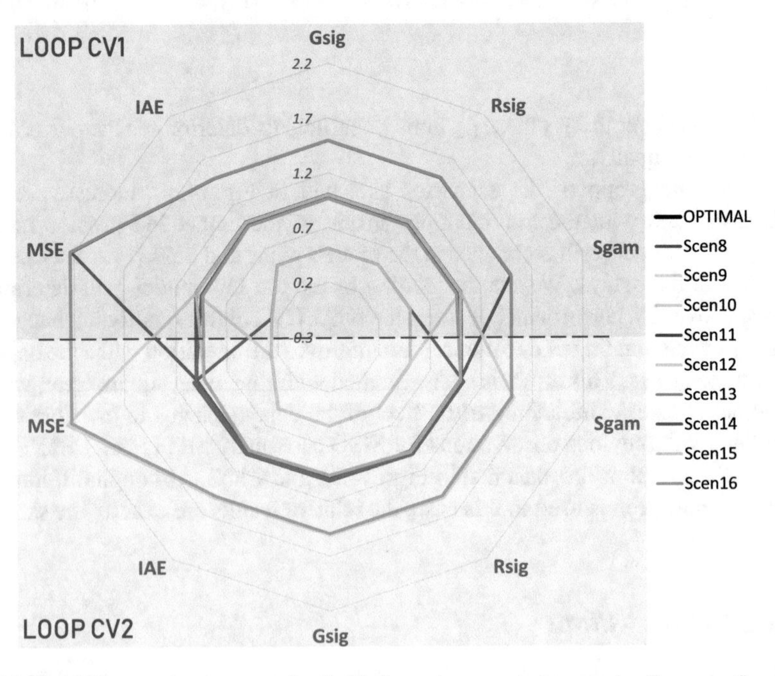

Fig. 14.30 MPC control assessment for distillation column: case 1, scenarios $Scen_8 \ldots Scen_{16}$

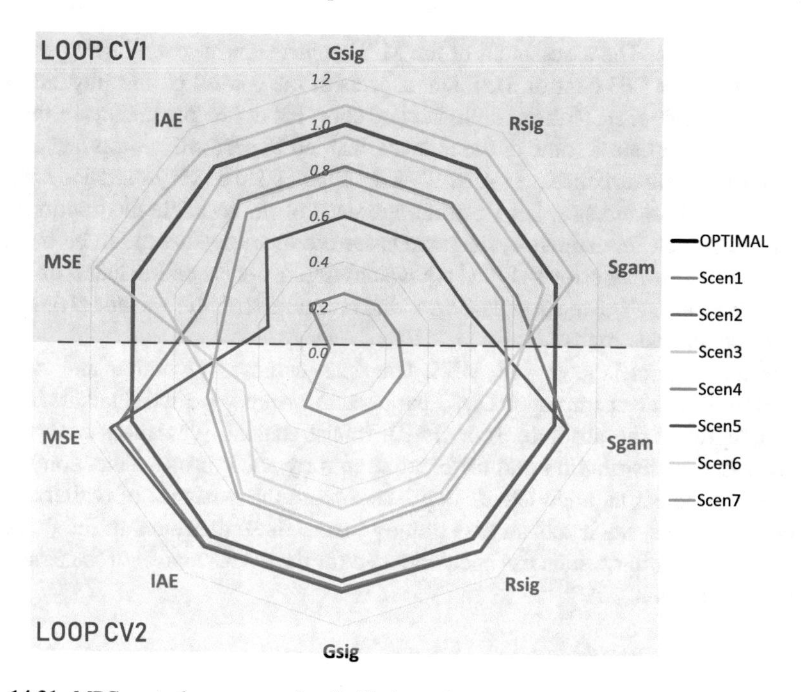

Fig. 14.31 MPC control assessment for distillation column: case 2, scenarios $Scen_1 \ldots Scen_7$

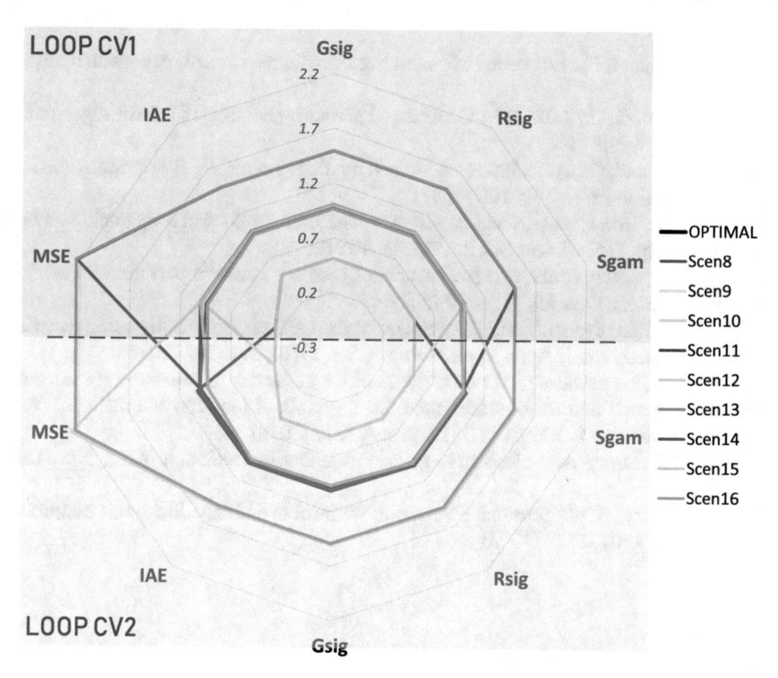

Fig. 14.32 MPC control assessment for distillation column: case 1, scenarios $Scen_8 \ldots Scen_{16}$

the same process. The assessment of the MPC control cause several new problems comparing to the PID control. Different aspects of the control philosophy has to be considered, comparing to the straightforward short list of few parameters of the PID algorithm. Nonetheless some of the indexes enabled to perform appropriate assessment in various disturbance scenarios. This list appeared to be very short and consists only of the robust indexes, i.e. γ^L (scaling factor of the α-stable distribution) and σ^{rob} (robust scale M-estimator). In some cases the measures list might be increase with integral absolute error IAE and equivalent Laplace scale coefficient b or by the Hurst exponent H. The mean square error and normal distribution standard deviation seem to be the least robust.

Further research shows that MSE, Gaussian standard deviation and rational entropy fail to detect improper DMC parameters or process-model misfit. In contrary, integral of the absolute error (IAE), robust standard deviation, scale factor of the α-stable distribution and differential entropy show proper detection facing strange, non-linear and non-Gaussian processes and the existence of outliers in the process variables. Such robustness strongly favors these measures in the practical applications. Similar results has been obtained for the DMC control of the nonlinear process [7] as well.

References

1. Clarke, W., Mohtadi, C.: Properties of generalized predictive control. Automatica **25**(5), 859–875 (1989)
2. Clarke, W., Mohtadi, C., Tuffs, P.S.: Generalized predictive control–I the basic algorithm. Automatica **23**(2), 137–148 (1987)
3. Clarke, W., Mohtadi, C., Tuffs, P.S.: Generalized predictive control—II extensions and interpretations. Automatica **23**(2), 149–160 (1987)
4. Domański, P.D.: Non-Gaussian and persistence measures for control loop quality assessment. Chaos Interdiscip. J. Nonlinear Sci. **26**(4), 043–105 (2016)
5. Domański, P.D., Ławryńczuk, M.: Assessment of predictive control performance using fractal measures. Nonlinear Dyn. **89**, 773–790 (2017a)
6. Domański, P.D., Ławryńczuk, M.: Assessment of the GPC control quality using non-Gaussian statistical measures. Int. J. Appl. Math. Comput. Sci. **27**(2), 291–307 (2017b)
7. Domański, P.D., Ławryńczuk, M.: Control quality assessment of nonlinear model predictive control using fractal and entropy measures. In: Preprints of the First International Nonlinear Dynamics Conference NODYCON 2019. Rome, Italy (2019)
8. Rousseeuw, P.J., Leroy, A.M.: Robust Regression and Outlier Detection. Wiley, New York, NY, USA (1987)
9. Wood, R.K., Berry, M.W.: Terminal composition control of a binary distillation column. Chem. Eng. Sci. **28**(9), 1707–1717 (1973)

Part III
Industrial Validation

Description of the industrial and practical issues concludes comprehensive presentation of the CPA approaches described in Part I and the simulation validation presented in Part II. This part summarizes the observations. The rationale for the research on the control quality assessment and and its further development, starting from the first works on benchmarking, has been always driven by the industrial demand. It is industry, which requires methods to assess their systems. There are several practical reasons, like the need for proactive knowledge about the elements requiring maintenance, finding out production bottlenecks caused by the inappropriate control, planning of the system improvement initiatives, migrations towards Industry 4.0 compliance or just natural engineering curiosity.

Once the research has been initiated by the industry, its outcome should come back to the industry. First of all, the industrial real process feedback is required to validated the scientific results. Any engineering research is not just an art for an art. Industrial acceptance expressed by the practitioners should be considered as the highest reward. Secondly, the industry just should gets the tools that would be able to improve control engineers daily work, make the process safer, more reliable and flexible, just open for the fast changing requirements of the modern digitally networked production.

The summarizing part of the book concentrates on the industrial issues of the control performance assessment. Two main aspects are addressed: the summary of the CPA reports describing outcomes of the control quality assessment in Chap. 15 and practical description of the Control Feasibility Study (Chap. 16) as the most popular industrial project utilizing CPA as a core functionality. It is intended that it might be used as the template for such an undertaking.

Chapter 15
CPA Industrial Applications

Achievement is talent plus preparation.

– Malcolm Gladwell

Nowadays we witness rapid industrial transition period towards Industry 4.0 production concept [57]. Although the initiative has started not long ago, only several years back (2011—Hanover fairs in Germany—Industrie 4.0), it has different faces and names, as "Industrie du futur" in France, "Made in China 2025" in China, "Industrial Internet of Things" (IIoT) in US, Fourth Industrial Revolution, etc. It has been driven by several factors, like the spread of the Internet, especially IoT (Internet of Things), integration of technical and business processes in the companies, digitalization and virtualization and "smart" production and products. It must be noted that control engineering plays one of the most important roles in that concept, as it enables production, integration and flexibility.

This industrial revolution brings about several new aspects that require practical perspective [10]. Horizontal and vertical integration together with all the Internet aspects (IoT, IoS, IoD) causes rapid increase of the amount of data available and thus the CPA task cannot be addressed without the consideration of the big data issues. Big data perspective also brings about other aspects of data availability, quality and safe storage. Communication and exchange of large amount of data, which is often fragile and valuable, with important company proprietary information require serious consideration about data safety and cyber security.

Cyber Physical Systems put much attention toward the human impact and interaction. There are two main situations important from the perspective of the Human-Machine interaction within the CPA task. As the assessed system is often impacted by human interventions the data analysis, measures evaluation and interpretation should take this factor into consideration. On the other hand the outcome of the CPA

P. D. Domański, *Control Performance Assessment: Theoretical Analyses and Industrial Practice*, Studies in Systems, Decision and Control 245,
https://doi.org/10.1007/978-3-030-23593-2_15

procedure should be easily visualized and presented to enable clear interpretation of results and easy decision making for further actions.

Flexible production and remodeling of the business models has even greater effect on the control systems. Existing task of the setpoint tracking and/or disturbance rejection will be soon extended with unceasing process reconfiguration. Once the process changes, the corresponding control system must follow. Thus, the control quality assessment tool/procedure has to follow and reconfigure itself as well. The complexity of the systems and their structural varying connections require further development of the adaptive large scale system general perspective approaches.

The close cooperation and true transparency between often competing companies is crucial to build new ecosystems and partnerships and to achieve the ultimate goal [39]. Professional staff is a key element in the whole concept, as they have a decisive role in the connected plant. New and continuous ways of learning with individual integration of the employee in the work process and new assistance functions and abilities should be further improved with wealth and well-being through adaptive workplace ergonomics. As the human expert is a very scarce and precious resource the emphasis on the CPA automated and autonomous operation with no or very limited operator interventions plays a crucial role as well.

15.1 Industrial Data

Existing control infrastructure enables to store an enormous, comparing to previous years, amounts of data. The concept of Big Data has just arisen. It has been expected that such a wide and cheap data availability should facilitate the CPA process. Unfortunately, the reports and industrial practice show that the reality is not so bright. There are still many limitations protecting the wider applicability [16], like for instance large mismatch between the used benchmarks and actually applied controllers, mostly due to the lack of consideration about process constraints, robustness and safety, excessively high implementation cost, data access and I&C infrastructure limitations and a difficulty to achieve the automated operation in long term. It seems that more data does not often mean better information.

The plant I&C infrastructure is often non-homogeneous. It may incorporate several different PLC/DCS and SCADA/MES systems in various versions and acquired from different vendors. Such a complex structure may cause serious problems with data access, as each of the systems often uses different data formats or protocols. Industrial practice often confirms that data collection from different systems and their synchronization should never be forgotten. Conducting the control study, the author has spent many hours bringing together obtained industrial data files before any real work could start.

The big data prospect of the CPA usage poses new considerations and issues to the industrial applications, like for instance the need for closer, more accurate and pragmatic approach to the monitoring; taking into consideration plant and loop specifics and issues associated with the data access, its storage and computational effort. New frameworks are expected, as there is a need to combine existing approaches with the use of the big data methods developed in the other information technology and data mining areas.

The raw data availability of good quality is a must. Any artificial or intended data manipulation, collection faults or compressions effects may highly bias the assessment or even make it impossible. Thus, the engineer performing a study should put much attention to the data collection process to ensure required data quality. The subjects of data conditioning, validation, preprocessing and security are carefully addressed in the following sections.

15.1.1 Data Conditioning, Validation and Preprocessing

Playing with data starts with the first visit to the site. We familiarize ourselves with plant systems, procedures and staff. Acquiring knowledge about the site and the personnel enables us to properly define requirements for data and/or the conditions for the plant experiments that might be required. One has to remember that each site and each project is different and cautiousness is required from the beginning.

Once the assessment procedure is initiated and defined, then the process can be well handled and we may avoid unexpected surprises. More difficult situation occurs in case of the on-line automatic performance assessment applications. In such an event, the definition of the good periods has to be defined and the proper period detection and isolation should be the first test before the competent assessment procedure.

Data acquirement in the industrial reality is not as straightforward as might be expected. The real process environment is not as pure and transparent, as in the scientific world or in the laboratory. First of all, the control system has to have the ability to collect, store and export process data. One has to be aware, that it is not always obvious. Even when the system is equipped with the SCADA or DCS historian package, all the required data has to be selected for historical acquisition. One should verify whether all data is properly configured for storing before the data acquisition starts. Once there are lacks of the required data, the data acquisition system should be configured accordingly. We must be aware that in some situations it might be just impossible.

The planning of the process experiments that should be run during the collection period might be also taken into consideration. Industrial processes often run around constant operating points, what disables or limits the assessment of the controls in the wider regimes, especially once the process exhibits nonlinear properties. The fact that the process operates around single operating point for almost the entire time period does not mean that it might not be changed, as for instance due to the modernization, some breakdown or outage/startup.

It frequently happens that once we obtain data, there are holes in the files or strange interpolations as in Fig. 15.1. The lack of all data strings is due to the data compression done by the historian software. One should be aware that such systems

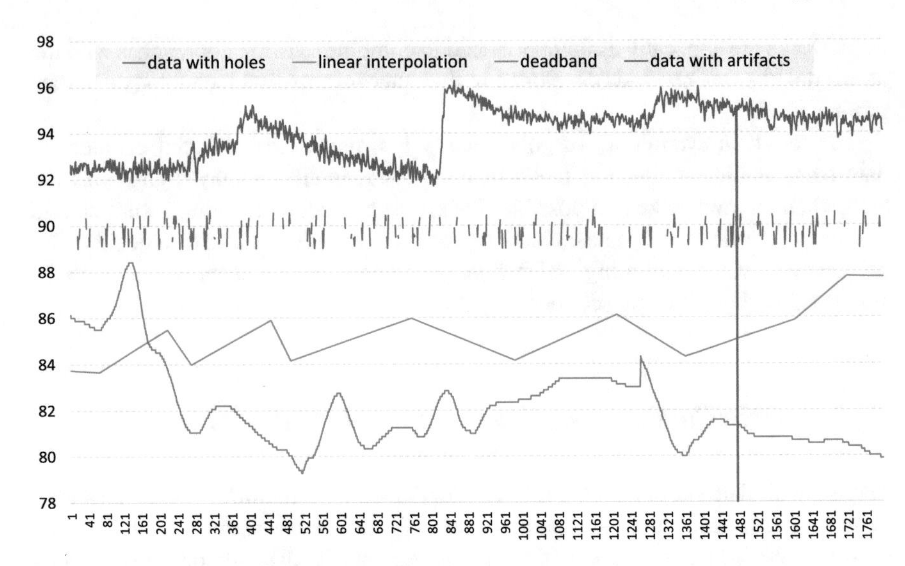

Fig. 15.1 Possible data acquisition artifacts (real data examples)

have other embedded tools that are dedicated to storage volume minimization. There are many different data compression mechanisms and algorithms applied. Data is filtered, saved with so called *deadband*, stored with long sampling intervals, linearly interpolated or contaminated with strange artifacts of unknown reasons.

Another aspect is associated with the storage sampling interval. From the perspective of the loop dynamic analysis we require short sampling periods. The storage interval should be significantly shorter than $\frac{1}{10}$ of the process dominant time constant. Each of these methods corrupts data adding strange effects that may significantly prevent proper further analysis. Having in mind all above features and concerns the assessment engineer should at first confirm data format and validate data.

Finally, stored data should have been exported to some readable external data format. Fortunately, majority of control systems is equipped with data interface, like for instance open OPC protocol, enabling storing of the historical data in some generally accessible format as for instance Excel, comma delimited or text file. Some systems cannot export big files so further preprocessing is often required to merge files or to standardize their format.

15.1.2 Cyber Security

Nowadays, the cyber security starts to play an important, even a crucial, role in the process industry. It is obvious that it also affects the control engineering as the control system, PLC or DCS, is a computer networked system. That aspect can be also visible in the recent reports [43, 63] adding new perspective to the research. The author selects five main performance areas that should be addressed:

- *manufacturing performance* of the continuous production process,
- *network performance* of the underlying network infrastructure,
- *computing resources performance* of all used devices: hardware and software,
- *protocol performance* of the underlying industrial communication protocols,
- and the *OPC Data Exchange performance* being the most popular link between various systems, SCADA/MES and other company IT systems.

The cyber security perspective places the standard CPA task into the *computing resources performance* area and opens new research areas and opportunities to enable comprehensive plant control performance assessment.

15.2 Human-Machine Interaction Within CPA

CPA analysis is performed with close human interaction. Control engineers support the CPA process and software with information. They interpret the results and later use them for the process improvement. On the other hand, the CPA tools are hardly delivering straightforward response with a single number. Each index has specific properties and the overall answer is often the result of the coordinative hybrid interpretation. Moreover, the assessed systems often consists of the dozens of the control loops. Such a complexity of different numbers and plots often requires some smart ways of data representation and interaction with the User.

The human interactions in the industrial production systems have been already well established and taken into considerations. The productivity improvement programs based on notions of: KAIZEN, Total Quality Management (TQM), 5S, Total Productivity Maintenance (TPM) take human factor into consideration successfully [52]. In control engineering the cyber-physical aspects are mostly considered from the perspective of the operating room and the process operator support [26]. Proposed framework has taken into account three steps of human interactions with the systems: sensing, processing, and physical/verbal responses.

Unfortunately, the CPA research almost does not take this aspect into consideration, with only a few reports. Mitchell et al. in the paper [50] have modified an English phrase *"A picture is worth a thousand words"* into the title of their paper *A picture worth a thousand control loops*. This phrase ideally gives a true view of the human-machine interaction associated with CPA. The authors have used the idea of the graphical analysis using the Tree Maps that has been proposed in [50] and further implemented in the commercial software solution. Tree Maps have been proposed in 1990s [61] and are primarily used to display data that is grouped and nested in a hierarchical (or tree-based) structure. In another report pattern recognition approach has been investigated and proposed in [23].

Table 15.1 Commercially available and university-based industrial CPA products

Company/website	Product name
ABB http://new.abb.com	Control Loop Monitoring Services Loop Optimizer Suite
Honeywell www.honeywellprocess.com	Control Performance Monitor powered by Matrikon Loop Scout
Emerson Process Management www2.emersonprocess.com	Process EnTech™Toolkit SmartProcess LoopMetrics
Schneider Electric www.schneider-electric.com	EcoStruxure Control Advisor Software
Metso ExperTune www.expertune.com	PlantTriage™ Process Control Performance Management
Yokogawa www.yokogawa.com	Loop Tuning (TuneVP)
Siemens www.siemens.com	Control Performance Analytics Control Performance Monitoring (CPM)
ControlSoft www.controlsoftinc.com	INTUNE+ Loop Optimization Software
OSIsoft www.osisoft.com	PI ControlMonitor
AspenTech www.aspentech.com	Aspen Watch Performance Monitor
Rockwell Automation http://locator.rockwellautomation.com	PlantESP
Control Arts Inc. www.controlartsinc.com	ControlMonitor
PAS www.pas.com	ControlWizard
SUPCON http://en.supcon.com	InPlant Total Solution for Process Automation
Control Station http://controlstation.com	PlantESP Control Loop Performance Monitoring
InTech www.intechww.com	Control Loop Performance Management
Migma www.migmasys.com	Process Performance Monitoring
I^2C^2 University of Auckland & AUT University www.i2c2.auckland.ac.nz	Control Performance Assessment
University of Alberta + NSERC www.sites.ualberta.ca/~control/	PATS Process Analysis Toolbox and Solutions

15.2.1 Control Assessment Software

Contrary to the research, the aspect of the human interactions has to be taken into account in the design of the CPA supporting software. Various companies are sharing their experience with the use of the Control Performance Assessment. A number of commercial control performance assessment software packages are available on the market (see Table 15.1). Many authors addressed in their research the review of the industrially available software packages [3, 29, 31, 36, 48].

15.3 Large Scale Root-Cause Analysis

All the research presented in the previous paragraphs has been addressing the control aspects for the single control element or single loop control algorithm. It could have been multivariate, however still it has been considered as a single entity. Industrial installations often constitute of dozens of control loops, which are directly or indirectly interconnected. The assessment of such large scale installations requires not only extensions of the existing methods, but also dedicated strategies and results interpretations.

In a large-scale industrial processes, installation units are interconnected. Any fault or problem can propagate from one unit through the whole plant along media or information flow paths. Therefore, the problem of the assessment, i.e. fault detection and isolation cannot be limited to a single unit, but has to be considered in the holistic perspective of the entire plant. Such an observation reveals new challenges. Any proposed approach has to take into account noise in the data, types of the faulty operation, common process interactions, time series characteristics and time frame and parameter selection.

First approaches to that subject have focused on the expert system approach [20], followed by the loop ranking and the selection of poor loops [14, 62] and an integrated framework of multivariable principal component analysis and the autoregressive moving average filter for building up the minimum variance performance bound of the MIMO feedback control system [8].

The research enables to develop and investigate several approaches, such as lag-based methods, adjacency matrix for the plant-wide oscillation propagation [33], conditional independence methods, like Bayesian methods [17, 24] and higher order statistics is one of the possible extensions of the classically defined CPA task.

Bayesian approach with the evaluation of various approaches and methods has been covered in a very comprehensive way in [17]. Both binary and continuous evidences have been considered. An interesting approach using decision fusion of different measures has been proposed in [35]. Such combined approach improves the assessment accuracy, especially in multi-loop control systems in the presence of loop interactions.

The research in this area is mostly addressed by the solutions called root cause diagnosis, fault propagation analysis or causality analysis [75]. Causality requires data analysis methods that enable to determine the embedded relationships between units and variables. A lot of the research in this area started for neuroscience applications. Process industry perspective has just followed recently.

The source of the faults can originate from the process itself or in the control system (sensors, actuators and control logic). Such anomalies mostly generate improper performance of the control loops, resulting in too sluggish or too aggressive operation. The attention of the observer of the process, i.e. the operator or the control engineer is most often attracted by resulting oscillations. The propagation of the oscillations throughout the large scale system has been addressed [41]. One may use one of the approaches, i.e. topology based models and data-driven causal analysis. Topology is reflected with the signed directed graphs, which can be generated automatically with Granger causality [77], cross correlation [44], frequency domain methods (Partial Directed Coherence) [41], hybrid approach based on the transfer entropy in conjunction with process connectivity information [40] or using existing process documentation. Obtained causality matrix is then analyzed using specific refinement and search methods. We may find out practical investigation of the above methods in the chemical industry [33, 40].

15.4 Reports on Industrial CPA Projects

Industrial applications often start from the simulations studies, which often use environment consisting of MATLAB as the evaluation and development tool for the CPA connected to the commercial full scale process simulation tools [1, 58, 59]. Such an approach makes the whole process safer and faster as the chance to negatively alter the plant is minimized.

Once the approach is tested it can be transferred in the on-site infrastructure in the reliable and safe way. As it has been stated before, the CPA story has started in the pulp and paper industry, like for instance paper-machine head-box [25], paper machine direction and cross direction control loops [46] or other pulp and paper related applications including business-like on-line approaches [21, 27, 34, 51].

Jelali in [30] has been considering the steel making industry with the application for the performance evaluation of control systems in a tandem cold rolling mill, i.e. a feedforward/feedback strip thickness controller and an internal model control of the strip flatness. Automatic gauge control systems in aluminum cold rolling mill has been analyzed in [76].

Further applications appeared in the chemical engineering companies like Eastman Chemical, DuPont or Grupa Azoty [11, 13, 22, 38], petrochemical plants, like BP [64], Shell [37, 65] and offshore oil production facilities [79]. An interesting comprehensive review of CPA applications for Eastman Chemical Company [54] shows a large-scale controller performance assessment system spanning over 14 000 PID loops worldwide. Loops can be sorted in order of performance to quickly identify the bottlenecks. The history is available to track improvement or degradation in performance for a single controller or an entire plant. Comprehensive summary of the CPA applications in these industries may be found in [29].

Other applications in process industry, like tank level control in Outokumpu pilot Tankcell for flotation cell project has been assessed in [28], grinding mill control performance in [73], pH control for pharmaceutical industries in [6], para-xylene production and poly-propylene splitter column processes in [15], kerosene and naphtha hydrotreating units is presented in [32] or delayed coking plant [5]. A flow control loop in a pharmaceutical plant has been taken into consideration in [66]. Further process industry projects have been reported in power industry, as for example in combined cycle power plant [7], steam temperature control assessment for pulverized coal fired unit [49], boiler control [42, 69–71] or fused magnesium furnace [74], while Haarsma and Nikolaou in [19] have presented a comprehensive review of the control quality assessment in food industry.

Mechanical engineering application of the control performance assessment for the aerospace industry has been addressed in [18]. The subject of the control of the multicopters has been analyzed in [12]. The authors use the concept of the degree of controllability (DoC) to define dedicated control performance index for multicopters subject to control constraints and off-nominal conditions. The proposed CPI is applied to a new switching control framework to guide the control decision of multicopter under off-nominal conditions. Nonlinear servomechanism control has been addressed in [55] with the use of the linearized NARX models.

Limited performance assessment studies of HVAC control may be found [2, 9, 60, 78]. A comprehensive review of control loop performance assessments in the context of building HVAC controls has been presented in [53]. The research over control performance applications in the semiconductor manufacturing has been connected with the Run-to-Run control systems and overlay monitoring [4, 56, 68].

Apart from the continuous production process the aspect of the batch processes has been addressed in some of the research [45, 67, 72].

An application in medicine to the control of atracurium-induced neuromuscular block [47] constitutes very interesting application in the non-engineering context.

References

1. Arjomandi, R.K.: Statistical methods for control loop performance assessment. In: 2011 International Conference on Communications, Computing and Control Applications (CCCA), pp. 1–6 (2011)
2. Bai, J., Li, Y., Chen, J.: Real-time performance assessment and adaptive control for a water chiller unit in an HVAC system. In: IOP Conference Series: Earth and Environmental Science, vol. 121, no. 005, p. 052 (2018)
3. Belli, P., Bonavita, N., Rea, R.: An integrated environment for industrial control performance assessment, diagnosis and improvement. In: Proceedings of the International Congress on Methodologies for Emerging Technologies in Automation (ANIPLA) (2006)

4. Bode, C.A., Ko, B.S., Edgar, T.F.: Run-to-run control and performance monitoring of overlay in semiconductor manufacturing. Control Eng. Pract. **12**(7), 893–900 (2004). pC-B02-Process Control IFAC 2002
5. Botelho, V.R., Trierweiler, J.O., Farenzena, M., Longhi, L.G.S., Zanin, A.C., Teixeira, H.C.G., Duraiski, R.G.: Model assessment of mpcs with control ranges: an industrial application in a delayed coking unit. Control Eng. Pract. **84**, 261–273 (2019)
6. Campbell, I., Uduehi, D., Ordys, A., van der Molen, G.: pH process control system benchmarking. In: Proceedings of the 2001 American Control Conference, vol. 6, pp. 4332–4335 (2001)
7. di Capaci, R.B., Scali, C., Pestonesi, D., Bartaloni, E.: Advanced diagnosis of control loops: experimentation on pilot plant and validation on industrial scale. IFAC Proc. **46**(32), 589–594 (2013). 10th IFAC International Symposium on Dynamics and Control of Process Systems
8. Chen, J., Wang, W.Y.: Performance assessment of multivariable control systems using PCA control charts. In: 2009 4th IEEE Conference on Industrial Electronics and Applications, pp. 936–941 (2009)
9. Clarke, J.A., Cockroft, J., Conner, S., Hand, J.W., Kelly, N.J., Moore, R., O'Brien, T., Strachan, P.: Simulation-assisted control in building energy management systems. Energy Build. **34**(9), 933–940 (2002)
10. Crnjac, M., Veza, I., Banduka, N.: From concept to the introduction of industry 4.0. Int. J. Ind. Eng. Manag. **8**, 21–30 (2017)
11. Domański, P.D., Golonka, S., Jankowski, R., Kalbarczyk, P., Moszowski, B.: Control rehabilitation impact on production efficiency of ammonia synthesis installation. Ind. Eng. Chem. Res. **55**(39), 10366–10376 (2016)
12. Du, G.X., Quan, Q., Xi, Z., Liu, Y., Cai, K.Y.: A control performance index for multicopters under off-nominal conditions, arXiv:1705.08775v2, Cornell University Library (2017)
13. Dziuba, K., Góra, R., Domański, P.D., Ławryńczuk, M.: Multicriteria control quality assessment for ammonia production process (in Polish). In: Zalewska, A. (ed.) 1st Scientific and Technical Conference Innovations in the Chemical Industry, Warszawa, Poland, pp. 80–90 (2018)
14. Farenzena, M., Trierweiler, J.O.: LoopRank: a novel tool to evaluate loop connectivity. IFAC Proc. Vol. **42**(11), 964–969 (2009). 7th IFAC Symposium on Advanced Control of Chemical Processes
15. Gao, J., Patwardhan, R., Akamatsu, K., Hashimoto, Y., Emoto, G., Shah, S.L., Huang, B.: Performance evaluation of two industrial mpc controllers. Control. Eng. Pract. **11**(12), 1371–1387 (2003). 2002 IFAC World Congress
16. Gao, X., Yang, F., Shang, C., Huang, D.: A review of control loop monitoring and diagnosis: prospects of controller maintenance in big data era. Chin. J. Chem. Eng. **24**(8), 952–962 (2016)
17. Gonzalez, R., Qi, F., Huang, B.: Process Control System Fault Diagnosis: A Bayesian Approach. Wiley, Chichester, UK (2016)
18. Gupta, P., Guenther, K., Hodgkinson, J., Jacklin, S., Richard, M., Schumann, J., Soares, F.: Performance monitoring and assessment of neuro-adaptive controllers for aerospace applications using a Bayesian approach. In: Proceedings of the AIAA Guidance, Navigation, and Control Conference and Exhibit, AIAA 2005, San Francisco, CA, USA, p. 6451 (2005)
19. Haarsma, G.J., Nikolaou, M.: Multivariate controller performance monitoring: Lessons from an application to a snack food process. In: AIChE Annual Meeting (1999)
20. Harris, T.J., Seppala, C.T., Jofriet, P.J., Surgenor, B.W.: Plant-wide feedback control performance assessment using an expert-system framework. Control Eng. Pract. **4**(9), 1297–1303 (1996)
21. Hölttä, V.: Plant performance evaluation in complex industrial applications. Ph.D. thesis, Helsinki University of Technology, Control Engineering, Espoo, Finland (2009)
22. Hoo, K.A., Piovoso, M.J., Schnelle, P.D., Rowan, D.A.: Process and controller performance monitoring: overview with industrial applications. Int. J. Adapt. Control Signal Process. **17**(7–9), 635–662 (2003)

23. Howard, R., Cooper, D.: A novel pattern-based approach for diagnostic controller performance monitoring. Control Eng. Pract. **18**(3), 279–288 (2010)
24. Huang, B.: Bayesian methods for control loop monitoring and diagnosis. J. Process Control **18**(9), 829–838 (2008). Selected Papers From Two Joint Conferences: 8th International Symposium on Dynamics and Control of Process Systems and the 10th Conference Applications in Biotechnology
25. Huang, B., Shah, S.L., Kwok, K.E., Zurcher, J.: Performance assessment of multivariate control loops on a paper-machine headbox. Can. J. Chem. Eng. **75**(1), 134–142 (1997)
26. Ikuma, L.H., Koffskey, C., Harvey, C.M.: A human factors-based assessment framework for evaluating performance in control room interface design. IIE Trans. Occup. Ergon. Hum. Factors **2**(3–4), 194–206 (2014)
27. Ingimundarson, A., Hägglund, T.: Closed-loop performance monitoring using loop tuning. J. Process Control **15**(2), 127–133 (2005)
28. Jämsä-Jounela, S.L., Poikonen, R., Halmevaara, K.: Evaluation of level control performance. In: 15th Triennial IFAC World Congress (2002)
29. Jelali, M.: An overview of control performance assessment technology and industrial applications. Control. Eng. Pract. **14**(5), 441–466 (2006)
30. Jelali, M.: Performance assessment of control systems in rolling mills—application to strip thickness and flatness control. J. Process Control **17**(10), 805–816 (2007)
31. Jelali, M.: Control Performance Management in Industrial Automation: Assessment, Diagnosis and Improvement of Control Loop Performance. Springer, London (2013)
32. Jiang, H., Shah, S.L., Huang, B., Wilson, B., Patwardhan, R., Szeto, F.: Performance assessment and model validation of two industrial MPC controllers. IFAC Proc. Vol. **41**(2), 8387–8394 (2008)
33. Jiang, H., Patwardhan, R., Shah, S.L.: Root cause diagnosis of plant-wide oscillations using the concept of adjacency matrix. J. Process Control **19**(8), 1347–1354 (2009). Special Section on Hybrid Systems: Modeling, Simulation and Optimization
34. Jofriet, P., Bialkowski, W.: The key to on-line monitoring of process variability and control loop performance. In: Proceedings Control Systems '96, p. 187 (1996)
35. Khamseh, S.A., Sedigh, A.K., Moshiri, B., Fatehi, A.: Control performance assessment based on sensor fusion techniques. Control. Eng. Pract. **49**, 14–28 (2016)
36. Knierim-Dietz, N., Hanel, L., Lehner, J.: Definition and verification of the control loop performance for different power plant types. Technical report, Institute of Combustion and Power Plant Technology, University of Stuttgart (2012)
37. Kozub, D.J.: Controller performance monitoring and diagnosis. Industrial perspective. IFAC Proc. Vol. **35**(1), 405–410 (2002). 15th IFAC World Congress
38. Kozub, D.J., Garcia, C.: Monitoring and diagnosis of automated controllers in the chemical process industries. In: Proceedings of the AIChE (1993)
39. Krause, H.M.: Turning ideas into reality: the journey of one leading manufacturer into the industrial IoT. In: Manufacture 2017, Tallin, Estonia (2017)
40. Landman, R., Jämsä-Jounela, S.L.: Hybrid approach to casual analysis on a complex industrial system based on transfer entropy in conjunction with process connectivity information. Control Eng. Pract. **53**(Complete), 14–23 (2016)
41. Landman, R., Kortela, J., Sun, Q., Jämsä-Jounela, S.L.: Fault propagation analysis of oscillations in control loops using data-driven causality and plant connectivity. Comput. Chem. Eng. **71**, 446–456 (2014)
42. Li, S., Wang, Y.: Performance assessment of a boiler combustion process control system based on a data-driven approach. Processes **6**, 200 (2018)
43. Li, X., Zhou, C., Tian, Y., Xiong, N., Qin, Y.: Asset-based dynamic impact assessment of cyberattacks for risk analysis in industrial control systems. IEEE Trans. Ind. Inform. **14**(2), 608–618 (2018)
44. Lindner, B., Auret, L., Bauer, M.: Investigating the impact of perturbations in chemical processes on data-based causality analysis. Part 1: defining desired performance of causality analysis techniques. IFAC-PapersOnLine **50**(1), 3269–3274 (2017). 20th IFAC World Congress

45. Liu, K., Chen, Y.Q., Domański, P.D., Zhang, X.: A novel method for control performance assessment with fractional order signal processing and its application to semiconductor manufacturing. Algorithms **11**(7), 90 (2018)
46. Mäkelä, M., Manninen, V., Heiliö, M., Myller, T.: Performance assessment of automatic quality control in mill operations. In: Proceedings of Control Systems (2006)
47. Mason, D.G., Edwards, N.D., Linkens, D.A., Reilly, C.S.: Performance assessment of a fuzzy controller for atracurium-induced neuromuscular block. Br. J. Anaesth. **7**(3), 396–400 (1996)
48. McNabb, C., Ruel, M.: Best practices for managing control loop performance—roadmap to success. In: Conference Proceedings from PAPTAC Annual Meeting (2009)
49. Meng, Q.W., Zhong, Z.F., Liu, J.Z.: A practical approach of online control performance monitoring. Chemom. Intell. Lab. Syst. **142**, 107–116 (2015)
50. Mitchell, W., Shook, D., Shah, S.L.: A picture worth a thousand control loops: an innovative way of visualizing controller performance data. In: Invited Plenary Presentation, Control Systems (2004)
51. Nissinen, A., Nuyan, S.: Performance improvements and sustainability through a remote monitoring, analysis, and warning system—benchmarking performance for long-term stability. In: Tappi PaperCon 2012 Conference (2012)
52. Oborski, P.: Social-technical aspects in modern manufacturing. Int. J. Adv. Manuf. Technol. **22**(11), 848–854 (2003)
53. O'Neill, Z., Li, Y., Williams, K.: HVAC control loop performance assessment: a critical review (1587-RP). Sci. Technol. Built Environ. **23**(4), 619–636 (2017)
54. Paulonis, M.A., Cox, J.W.: A practical approach for large-scale controller performance assessment, diagnosis and improvement. J. Process Control **13**(2), 155–168 (2003)
55. Pillay, N., Govender, P., Maharaj, O.: Controller performance assessment of servomechanisms for nonlinear process control systems. In: Proceedings of the World Congress on Engineering and Computer Science (WCECS), vol. II (2014)
56. Prabhu, A.V.: Performance monitoring of run-to-run control systems used in semiconductor manufacturing. Ph.D. thesis, University of Texas at Austin, US (2008)
57. Rojko, A.: Industry 4.0 concept: Background and overview. Int. J. Interact. Mob. Technol. **11**(5) (2017)
58. Salahshoor, K., Arjomandi, R.K.: Comparative evaluation of control loop performance assessment schemes in an industrial chemical process plant. In: 2010 Chinese Control and Decision Conference, pp. 2603–2608 (2010)
59. Salahshoor, K., Karimi, I., Fazel, E.N., Beitari, H.: Practical design and implementation of a control loop performance assessment package in an industrial plant. In: Proceedings of the 30th Chinese Control Conference, pp. 5888–5893 (2011)
60. Seem, J.E., House, J.M.: Integrated control and fault detection of air-handling units. HVAC R Res. **15**(1), 25–55 (2009)
61. Shneiderman, B.: Tree visualization with treemaps: a 2-D space-filling approach. ACM Trans. Graph. **11**(1), 92–99 (1992)
62. Streicher, S.J., Wilken, S.E., Sandrock, C.: Eigenvector analysis for the ranking of control loop importance. In: Klemeš, J.J., Varbanov, P.S., Liew, P.Y. (eds.) 24th European Symposium on Computer Aided Process Engineering, Computer Aided Chemical Engineering, vol. 33, pp. 835–840. Elsevier (2014)
63. Tang, C.Y.: Key performance indicators for process control system cybersecurity performance analysis. Technical Report NISTIR 8188, National Institute of Standards and Technology (NIST), U.S. Department of Commerce (2018)
64. Thornhill, N.F., Sadowski, R., Davis, J.R., Fedenczuk, P., Knight, M.J., Prichard, P., Rothenberg, D.: Practical experiences in refinery control loop performance assessment. In: UKACC International Conference on Control '96 (Conf. Publ. No. 427), vol. 1, pp. 175–180(1996)
65. Thornhill, N.F., Oettinger, M., Fedenczuk, P.: Refinery-wide control loop performance assessment. J. Process Control **9**(2), 109–124 (1999)
66. Veronesi, M., Visioli, A.: An industrial application of a performance assessment and retuning technique for PI controllers. ISA Trans. **49**(2), 244–248 (2010)

67. Wang, J., Wang, Y.: Control performance assessment for ILC-controlled batch processes based on MPC benchmark. In: 2018 IEEE 7th Data Driven Control and Learning Systems Conference, pp. 129–134 (2018)
68. Wang, J., He, Q.P., Edgar, T.F.: Control performance assessment and diagnosis for semiconductor processes. In: Proceedings of the 2010 American Control Conference, pp. 7004–7009 (2010)
69. Wang, J., Su, J., Zhao, Y., Pang, X., Li, J., Bi, Z.: Performance assessment of primary frequency control responses for thermal power generation units using system identification techniques. Int. J. Electr. Power Energy Syst. **100**, 81–90 (2018a)
70. Wang, J., Pang, X., Gao, S., Zhao, Y., Cui, S.: Assessment of automatic generation control performance of power generation units based on amplitude changes. Int. J. Electr. Power Energy Syst. **108**, 19–30 (2019)
71. Wang, Y., Gao, Y., Su, W., Zheng, W., Jiang, X., Huang, Q.: The control system assessment based on a class of disturbance characteristics. In: 2018 37th Chinese Control Conference, pp. 8596–8600 (2018b)
72. Wang, Y., Zhang, H., Wei, S., Zhou, D., Huang, B.: Control performance assessment for ILC-controlled batch processes in a 2-D system framework. IEEE Trans. Syst. Man Cybern. Syst. **48**(9), 1493–1504 (2018c)
73. Wei, D.: Development of performance functions for economic performance assessment of process control systems. Ph.D. thesis, Faculty of Engineering, Built Environment and Information Technology, University of Pretoria (2010)
74. Wu, Z., Ran, Z., Xu, Q., Wang, W.: Dynamic performance monitoring of current control system for fused magnesium furnace driven by big data. In: 2018 IEEE International Conference on Consumer Electronics-Taiwan, pp. 1–2 (2018)
75. Yang, F., Xiao, D.: Progress in root cause and fault propagation analysis of large-scale industrial processes. J. Control Sci. Eng. **478**, 373 (2012)
76. Yuan, H.: Process analysis and performance assessment for sheet forming processes. Ph.D. thesis, Queen's University, Kingston, Ontario, Canada (2015)
77. Yuan, T., Qin, S.J.: Root cause diagnosis of plant-wide oscillations using granger causality. J. Process Control **24**(2), 450–459 (2014). aDCHEM 2012 Special Issue
78. Zhao, F., Fan, J., Mijanovic, S.: Pi auto-tuning and performance assessment in HVAC systems. In: 2013 American Control Conference, pp. 1783–1788 (2013)
79. Zheng, Y., Qin, S.J., Barham, M.: Control performance monitoring of excessive oscillations of an offshore production facility. IFAC Proc. **45**(8), 25–32 (2012). 1st IFAC Workshop on Automatic Control in Offshore Oil and Gas Production

Chapter 16
Control Feasibility Study

I did everything, everything I wanted to.

– Ian Curtis

16.1 Control Feasibility Study

Practical implementation of the process improvement project is always (or at least should be) a big challenge or even a dream for any researcher working in that field. It is the biggest fun and excitement to see that the fruit of the long term research and scientific interest works. Or even more, that it brings profits and makes the end user, called in commercial world a client, satisfied [3].

The study goal is to improve installation performance with knowledge and best practice methodology. Problem formulation, focused on an ultimate target, enables appropriate project harmonization and scheduling allowing selection of optimal tools and methodologies. It has one additional benefit. It does not point out any predefined and assumed technology to be used. The tools are selected according to the targets.

Process improvement almost always is called *optimization*. Optimization is a technology calculating the best possible utilization of the resources (people, time, processes, equipment, materials, money, etc.) needed to achieve a desired result. It might be minimizing process and environmental costs or maximizing throughput or profits. Quite often the goal comprises of more than one criteria and an engineer faces the multi-criteria, often constrained, problem. Optimization technology improves decision making by providing business with responsive, accurate, real-time suggestions and solutions. Thus, we may call the control system improvement as the control system optimization.

P. D. Domański, *Control Performance Assessment: Theoretical Analyses and Industrial Practice*, Studies in Systems, Decision and Control 245,
https://doi.org/10.1007/978-3-030-23593-2_16

Actually, it frequently appears that an improvement can be reached after extensive road and the final result migrates away from the initial assumptions. The biggest challenge encountered in commercial world is that there are a lot of misunderstandings what optimization really is. There are many interpretations and every vendor wants to claim to have optimization.

Improvement project is often synonymous with cost reduction, lower environmental carbon footprint, higher productivity or profitability. Optimization should speed up the development and simultaneously empower the developers to create "smarter" control philosophies, which improve business even further. These benefits help to generate the highest Return On Investment (ROI) from the improvement project. Payback time is usually measured in months—sometimes in days!

Control improvement project can benefit the business and its profitability in many ways, as for instance through:

- the optimized controls enabling more accurate installation operation,
- the automated solutions operating in the autonomous mode with the limited human interventions,
- the improved business flexibility, responsiveness to changing circumstances, and ability to test "*what-if*" scenarios,
- the higher installation throughput,
- operation according to the flexible goals and constraints,
- meeting of the technology, environmental, business or any other limitations.

It occurs that the biggest problem is to convince the user to the proposed solution. Sometimes it is easy, in other situations extremely difficult. If we are not able to get common understanding and self confidence, then it is not worth to go on with the project. The work may not be accepted nor used without close cooperation with the customer. Such implementation is often discarded and cannot be considered as the reference case study anyhow.

Convincing the customer has to be supported with the tangible arguments. Required numbers can be delivered by the CPA study. We have to make the user to understand, what is our idea and what is its advantage over other available solutions. We have to convince him that we have the knowledge and abilities, and our proposal will work. Simultaneously, we also should convince ourselves that we are able to meet all the goals an requirements and that the results bring measurable profit. The feasibility study (CPA study) is a common mean of such an approach.

Webster's New Collegiate Dictionary defines the adjective *feasible* as (1) capable of being done or carried out; (2) capable of being used or dealt with successfully: suitable; (3) reasonable, likely–synonym: see possible. This definition really captures the essence of the feasibility study. Consideration needs to be given to whether we are capable of completing the project and carrying out all of the steps necessary to successfully develop a project in today's competitive environment.

The analysis of a problem has to determine, whether it can be solved effectively. The operational, financial and technological aspects need to be addressed in the study. The study provides information that helps to determine the potential return on investment. It gathers benchmarking data, allowing the potential user to gouge the impact of planned modifications. The results determine whether the solution is worth to be implemented. The feasibility study is often done as an initial phase of a comprehensive plant optimization program. It provides economic and technical justifications for the planned initiative. This low risk and multi-staged approach delivers key benefit estimation and cost justification limiting initial investment and the risks.

Bottom-top project formulation allows orchestration of the tasks and monitoring of the achieved milestones. Initial project planning plays crucial role for the whole process. Goals are confronted with technology, site instrumentation, control philosophy, I&C infrastructure, available algorithms, economical expectations and limitations. Analytical part covers four main areas: instrumentation, control philosophy, control system and calculation procedures to evaluate KPIs and efficiency baseline.

CPA study activities consist of the two main phases. First, comprehensive installation review is performed. This stage is done by project stakeholders expert team: technology owner, control system provider and research organization supporting the parties with scientific expertise. These activities are performed on-site and include collection of the historical data stored in the plant historian database, review of the available documentation (P&ID drawings, SAMA control logics, construction and system layouts, etc.) and interviews with key staff to review information and fully understand the system including inputs from various plant teams: operations, engineering, planning and scheduling, control and instrumentation, maintenance, management and finance. Project team reviews and analyzes obtained information. The team holds several meetings with plant personnel, especially members of the operation team (engineers, shift supervisors, operators) to gather comments on daily operation and common issues.

Second part is performed in the office. The collected information is sorted, data are analyzed and appropriate Key Performance Indicators are proposed. The team identifies opportunities through engineering analysis in terms of the degree of freedom available for optimization. Problem complexity is assessed and confronted with the identified degrees of freedom. Historical process data are analyzed to identify the level of change, to which the system is subject. In some situations a simulation model is constructed to assess the identified opportunities, ensuring an appropriate rigor to handle the complexity associated with the opportunities.

Control loops are analyzed, the measures are calculated and their performance is evaluated with the preselected approach. Loops dynamic quality is assessed according to the accepted best practices:

(a) Review of the loop mode of operation, i.e. AUTO versus MANUAL.
(b) Calculation of the simple statistical measures (min, max, mean, median, variance) for selected typical period and graphical representation of the time trends for setpoint, controlled variable and manipulated variable.
(c) If the loop works in AUTO, control error may be calculated and analyzed. Its histogram can be plotted and fitted with curves describing different Probabilistic Distribution Functions (PDFs).
(d) Selected measures are evaluated depending on the loop properties, operational regimes, available data and project requirements.

(e) Static characteristics of actuators and loops, i.e. CV versus MV are calculated to identify eventual nonlinearities or strange clusters of points.
(f) If required, additional data is collected from plant historian to clarify inconsistencies or to investigate discovered potential.

Control loops quality assessment should be accompanied with control instrumentation review, i.e. inspection of plant actuators and sensors. The status of all actuators (valves, dampers, pumps, fans, etc.) must be verified. Static characteristics can be calculated using historical data. Operation and calibration of sensors is investigated to confirm proper performance of the measurements. Control system related issues, like operators screen panels and information visualization, data archivization, trending and alarming system might be carefully checked. Finally, all the observations should be confronted with plant maintenance and technology teams.

There is one more, very important thing that often is missed. The study should define very clearly and in details how the profits are calculated. The results should explain representative economic benefits expected from the optimization. The profit improvement possible over the full range of change in operating modes should be estimated and translated into the economic KPIs. Typically these cases might be based on operating data collected over previous years, along with future operating scenarios. The cases can use historical, current or future predicted data. When compared with project cost estimates, this provides a solid economic basis to make a decision about the improvement project. This estimation should be included in the resulting project offer that follows the study. Such detailed and clear definition gives safety that the final results will not be controversial.

The observations are summarized, the improvement opportunities are calculated, the conclusions are formulated accompanied by the definition and presentation of the resulting improvement project. Identified issues and problems enable to formulate recommendations and implementation roadmap disclosing possible bottlenecks and risks backed with the risk management plan. All the CPA study project findings and definitions are included in a comprehensive CPA study report. The document describes all performed activities, results, synthesis of the proposed rehabilitation schedule and risk management plan.

There should be noted one, very important issue at that point. To assure future success of the project, the study must be performed in close collaboration with the user's site personnel. This condition must be met. The user/customer must feel committed the study project. The success should be the team success. On completion, the study documentation is delivered to the customer, presented and discussed. They should help to make reasonable and optimal investment decision allowing to remove identified bottlenecks.

Above description gives an overview of the CPA feasibility study project. This description is general and presents several important aspects that should be taken into account during project execution. It is obvious that each project is unlike, as the processes differ and the demands vary. Four aspects of the CPA feasibility study project are brought closer to facilitate potential project realization: study project plan, loop performance assessment procedure, estimation procedure for the economic benefits of control improvements and CPA feasibility study report template.

16.1.1 Loop Performance Assessment Procedure

Automatic evaluation of indexes without reflection about loop environment is misleading [5]. Visual inspection of data in form of time series trends, histograms, basic statics, cross-correlations and auto correlation functions, static X-Y relations supported with curve estimation is strongly recommended. More advanced relations, like rescaled range R/S plots are welcome as they may reveal the persistence properties of data or suggest that the fractal analysis is practically impossible. Although such observation may be discouraging, they are reasonable. Thus, the following procedure is proposed:

(1) Review the time series of the variables. Verify if the raw data are proper, i.e. without historian data compression effects, unknown artifacts and that they originate from normal (desired) operating regimes.
(2) Evaluate time of the AUTO mode operation for each loop.
(3) Calculate basic statistics for data, like: minimum, maximum, mean, median, standard deviation, skewness, kurtosis, etc.
(4) If the step response is available or might be evaluated in the reliable way, the step response measures, mostly the settling time and the overshoot should be calculated.
(5) If required, calculate auto correlation function and cross correlation function between variables.
(6) Prepare static characteristics drawings (X-Y plots) for the analysis of the loop eventual nonlinearities.
(7) Select loop variable for further index evaluations. Control error signal is suggested.
(8) Calculate the integral indexes. MSE should be used with caution, although IAE seems to be a reliable suggestion.
(9) In case the model is known or easy to be estimate consider the use of the model-based estimates like for instance minimum variance benchmark index.
(10) Draw control error histogram to evaluate statistical properties of the selected variables. Check the normality, the existence of fat tails and the shape of histogram.
(11) Fit the underlying distributions, select the best fitting probabilistic density function and estimate its coefficients.

   a. In Gaussian case normal standard deviation should be used as the loop quality measure.
   b. Once fat tails are detected, factors of Cauchy or α-stable distribution should be used. Scaling γ is the suggested choice, as it is the most robust against disturbances.
   c. Calculate robust estimators of the standard deviations σ^{rob} and consider it as the possible index candidate.
   d. Otherwise, select factors for any other best fitted custom probabilistic density function.

(12) In case of the existence of fat tails, when data is non-stationary or when the self-similarity is suspected, perform the persistence analysis using rescaled range R/S plots and estimate according Hurst exponents. Check for multipersistency and evaluate respective multiple Hurst exponents and crossover points.
(13) Validate the indexes robustness against being biased by disturbance shadowing.
(14) Translate the numbers into verbal observations, like for instance:

   a. frequent MAN mode operation is the evidence of the operators lack of confidence to the loop design and/or tuning or the equipment breakdown,
   b. steady state error can be observes with the step response or in position factor of the underlying distribution,
   c. sluggishness or aggressiveness can be assessed using step response indexes or Hurst exponent,
   d. asymmetric operation can be estimated using statistical analysis (skewness factors, histogram, time trends),
   e. loop nonlinearity (process or actuator) can be decided upon static X-Y plots or loop asymmetric behavior,
   f. violation of the constraints may cause the asymmetric variable properties,
   g. signal persistence may be the evidence of fat tails, which may be caused by the uncoupled disturbances or the embedded disturbance properties,
   h. multifractal properties may be the evidence of frequent human interventions (manual mode, operators biasing) into the control process.

(15) Each of the above observations may suggest some improvement actions, as for instance:

   a. low shift operators confidence → review/change of the control philosophy or malfunctioning equipment,
   b. steady state error → adding of the integral action inside of the feedback loop or verification of the operating points against limiting saturations,
   c. sluggishness or aggressiveness → loop tuning,
   d. actuator nonlinearity → controller output linearization with the `f(x)` SAMA block,
   e. process nonlinearity → gain scheduling,
   f. impacting measurable disturbances → feedforward decoupling,
   g. impacting unmeasurable disturbances → filtering,
   h. oscillations → significantly aggressive tuning, relay operation, inappropriate operating point or actuator malfunctioning.

Above procedure is evidently nor universal, neither comprehensive, however should be helpful as the starting point. Process uniqueness demands flexibility and creativity from an assessment engineer. Comprehensive control performance assessment is an engineering art that cannot be fully dehumanized and any software packages should be considered as the decision-making support.

16.1.2 Predicting Economic Benefits of Control Improvements

There arises a need for methodologies to compare control rehabilitation cost against expected economic benefits. Such decisions are mostly based on the financial basis. Estimation techniques allowing calculation of the benefits resulting from the control system improvement have been proposed by [2, 9]. The cost element of the decision is simple as it may be easily derived from past projects or obtained from control system vendor. The benefit part is evaluated specifically for each case. The algorithm is based on the mitigation of process variability, leading towards quantitative results [1]. Frequently, one may assume upper or lower limitation for the variable. Reduction of its variability through better control enables to shift it closer to the constraint and thus to generate benefit. As the variable is explicitly linked with the performance, the benefits may be calculated. The method assumes that the shape of the variable histogram is Gaussian and standard deviation is used as the variability measure. Apart from that, other methods were proposed, like probabilistic optimization approach [10].

The task to predict possible improvements associated with upgrade of a control system exists in literature for a long time [8]. From the early days it was mostly associated with the implementation of advanced control. There are three well established approaches called: *the same limit*, *the same percentage* and *the final percentage* rules [2]. All of them are based on the evaluation of the normal distribution for selected variable keeping information about economic benefits and its modifications. The method assumes Gaussian properties of the process behavior.

Lately, there have appeared non-Gaussian extensions of the algorithm. The modification of the algorithm towards the Laplace and Cauchy probabilistic distribution functions has been proposed in [4], while more general version of the approach using α-stable distribution is presented in [6]. In some process time series are biased and thus some detrending is required [7]. This aspect has been discussed in details in Chap. 9.

Practically, the estimation procedure is as follows:

(1) Evaluate histogram of the selected variable or the performance index.
(2) Fit appropriate distribution to the obtained histogram.
(3) Apply the right method depending on the selected PDF to evaluate the safe shift of the distribution function.
(4) Finally the improvement can be calculated (9.2).

16.1.3 Study Report Template

It has been shown that the study is often similar to the "*micro simulated*" version of the full scale project. The study document should address the following aspects:

(1) definition of the project goals and limitations—review with plant personnel,
(2) selection of the according performance measures and their connection with control system—KPIs formulation,
(3) review of the existing I&C infrastructure, i.e. site DCS/PLC/SCADA and interfaces,
(4) separation of the control loops and logics under project consideration,
(5) analysis of the historical trends relevant to the study subject,
(6) identification and understanding of the currently implemented control strategies and interlocking,
(7) review of the plant instrumentation (sensors and actuators) estimating available degrees of freedom for optimization,
(8) finding out known control challenges and bottlenecks,
(9) design and realization of some dedicated unit tests (load change tests with different parameters, control configuration and at different load levels),
(10) collection of the historical data,
(11) benchmark of current controls performance and constraints against other cases and examples,
(12) review of the collected verbal plant information gathered during meetings and converstations,
(13) understanding of the existing control philosophy,
(14) loop performance assessment procedure,
(15) observations summary and the development of the improvement variants,
(16) description of the proposed solution together with the improvement opportunities,
(17) development of the improvement roadmap (covering primary and secondary goals) and associated business aspects,
18) project implementation offer with the schedule and cooperation plan,
19) capital and operating cost estimate versus revenue estimates—ROI analysis,
20) report executive summary.

Finally, it has to be noted that the study should be prepared using a comprehensible language. The observations have to be clearly distinguished, while all the conclusions, especially the financial ones, must be well justified. The study should start with one page executive summary bringing together main observations and study outcome. The document often forms technical material and annex for business negotiations and contracting. The text document is frequently accompanied with the slide show, that is presented in person, explained and discussed.

References

1. Ali, M.K.: Assessing economic benefits of advanced control. In: 5th Asian Control Conference, Process Control in the Chemical Industries, Chemical Engineering Department, pp. 146–159. King Saud University, Riyadh, Kingdom of Saudi Arabia (2002)
2. Bauer, M., Craig, I.K., Tolsma, E., de Beer, H.: A profit index for assessing the benefits of process control. Ind. Eng. Chem. Res. **46**(17), 5614–5623 (2007)
3. Domański, P.D.: Optimization projects in industry—much ado about nothing. In: Proceedings of the VIII National Conference on Evolutionary Algorithms and Global Optimization KAEiOG 2005, Korbielów, Poland (2005)
4. Domański, P.D.: Non-Gaussian assessment of the benefits from improved control. Preprints of the IFAC World Congress 2017, pp. 5092–5097, Toulouse, France (2017)
5. Domański, P.D.: Statistical measures for proportional-integral-derivative control quality: simulations and industrial data. Proc. Inst. Mech. Eng. Part I J. Syst. Control Eng. **232**(4), 428–441 (2018)
6. Domański, P.D., Marusak, P.M.: Estimation of control improvement benefit with α-stable distribution. In: Kacprzyk, J., Mitkowski, W., Oprzedkiewicz, K., Skruch, P. (eds) Trends in Advanced Intelligent Control, Optimization and Automation, vol. 577, pp. 128–137. Springer International Publishing AG (2017)
7. Domański, P.D., Golonka, S., Marusak, P.M., Moszowski, B.: Robust and asymmetric assessment of the benefits from improved control industrial validation. In: 10th IFAC Symposium on Advanced Control of Chemical Processes ADCHEM 2018, IFAC-PapersOnLine, vol. 51(18), pp. 815–820 (2018)
8. Tolfo, F.: A methodology to assess the economic returns of advanced control projects. In: American Control Conference 1983, pp. 1141–1146. IEEE (1983)
9. Wei, D., Craig, I.: Development of performance functions for economic performance assessment of process control systems. In: 9th IEEE AFRICON, pp. 1–6 (2009)
10. Zhao, C., Xu, Q., Zhang, D., An, A.: Economic performance assessment of process control: a probability optimization approach. In: International Symposium on Advanced Control of Industrial Processes, pp. 585–590 (2011)

Printed in the United States
By Bookmasters